Degradation of Plastics

Edited by

Inamuddin[1], Rizwana Mobin[2], Mohd Imran Ahamed[3] and Rajender Boddula[4]

[1]Department of Applied Chemistry, Zakir Husain College of Engineering and Technology, Faculty of Engineering and Technology, Aligarh Muslim University, Aligarh-202002, India

[2]Department of Industrial Chemistry, Govt. College for Women, Cluster University, Srinagar, Jammu and Kashmir-190006, India

[3]Department of Chemistry, Faculty of Science, Aligarh Muslim University, Aligarh 202 002, India

[4]CAS Key Laboratory of Nanosystem and Hierarchical Fabrication, National Center for Nanoscience and Technology, Beijing 100190, PR China

Published by **Materials Research Forum LLC**
Millersville, PA 17551, USA

Published as part of the book series
Materials Research Foundations
Volume 99 (2021)
ISSN 2471-8890 (Print)
ISSN 2471-8904 (Online)

Print ISBN 978-1-64490-132-8
eBook ISBN 978-1-64490-133-5

Distributed worldwide by

Materials Research Forum LLC
105 Springdale Lane
Millersville, PA 17551
USA
https://www.mrforum.com

Manufactured in the United States of America
10 9 8 7 6 5 4 3 2 1

Table of Contents

Preface

Synthetic plastic production has progressed rapidly in the rising field of global trade owing to its simple processability, low-cost, and natural resistance. Synthetic plastics such as polyethylene, polypropylene, polystyrene, polyvinyl chloride, polyurethane, and polyethylene terephthalate are typical plastics we encounter in our daily life. Despite the very fact that plastics have been utilized for a hundred years, the start of industrial-scale production dates back to 1950 and reached production of at least 350 to 400 million tons each year and will be almost quadruple by 2050.

However, poor recycling and low circular use of plastics leads its accumulation in the ecosphere including oceans, landfills, and lakes in the form of accumulated plastic wastes leading to serious environmental pollution and social issue. Thus, skyrocketing concerns in plastic wastes degradation and solutions are required to remove plastic waste in the ecosphere. This book describes a comprehensive overview of plastics degradation. The topics include, plastic degradation methodologies, mechanistic actions, factors effecting, various aspects, recycling, and potential implications are covered.

The book is an excellent resource for faculty, students, and industrialist working in polymer science and technology, chemistry, physics, and environmental science.

Key features:

- Discusses cutting-edge technology for plastics degradation
- Overview of potential solutions to plastic degradation challenges
- Written by internationally renowned scientists
- Delivers current research progress and development trend of plastic degradation technologies

Summary

Chapter 1 discusses the harmful effects of plastic that arise due to the mismanagement of plastic waste. Different plastic degradation methods and factors affecting them have been discussed in detail. Plastic waste management techniques have also been reported to reduce the environmental effects of plastic waste.

Chapter 2 details the synthesis of biodegradable polymers from various renewable raw materials. It also gives brief information on the classification of bioplastic, its advantages over conventional polymer, and methods involved in the production of bio-based polymers using various renewable resources such as biomass, waste, microorganism, etc.

Chapter 3 highlights the potential challenges of recycling and its probable elucidations. It compares the recycling efficiency of degradable plastics with conventional plastics. It also emphasizes how the world is committed to reuse bioplastics to promote an eco-friendly environment. It divulges that prioritized reduction of plastic consumption and turning towards recycling is the only safe route to meet the demands of the growing population worldwide.

Chapter 4 explores different catabolic enzymes responsible for the degradation of plastic polymers. Most of these enzymes are naturally produced by well-known bacteria and fungi. The chapter further discusses the mode of action on enzyme-substrate interaction, hydrolysis of bonds and conducive environment for maximum degradation.

Chapter 5 details the current scenario of plastic biodegradation, along with the recent advancements, opportunities, and future challenges. It also outlines the factors affecting the biodegradation process and the characterization techniques being employed to assess degradation extent. The overall work focuses on thrust areas to be improved concerning environmental sustainability.

Chapter 6 summarizes the recovery of biodegradable bioplastics polyhydroxyalkanoates (PHAs) from different activated sludge processes by using wastewater as a renewable feedstock. A brief idea about PHA structure, synthesis pathways, types of wastewater and activated sludge used with reactor parameters and environmental factors affecting PHA productions are also described.

Chapter 7 highlights the environmental problem associated with ever-increasing plastic waste. Due to the non-biodegradable nature of plastic, its disposal has become a significant issue. Single-use plastics are the highest contributor to plastic waste which is also least recycled. Furthermore, the potential of different photo-catalysts is described for the environment-friendly disposal of plastics.

Chapter 8 delineates the dominance of degradable plastics in the global market. It narrates the journey of degradable plastics in overtaking the international market of plastics. It guesstimates the worldwide growth of bioplastic in the business market. It also creates a genuine opinion of consumer acceptance towards these products making them a favorable item in the circular economy.

Chapter 9 details various types of plastics, bioplastics and their impacts on the environment. Biodegradation strategies to manage plastic pollution are discussed. Bioplastics, their sustainable sources, and their uses are explained comprehensively. In short comparison between the characteristic features of plastics and bioplastics is the main focus of this chapter.

Chapter 10 describes numerous applications of degradable plastics across a wide range of industries (e.g. packaging, medical, agricultural, automobile, aerospace, consumer product application). Furthermore, the chapter discusses the growth of the degradable plastic industry and its prospect of limiting the adverse environmental effects of conventional plastics.

Chapter 11 discusses the production of polyhydroxyalkanoates by cyanobacteria as an attractive alternative to replace conventional chemical plastics. Initially, an overview of biodegradable plastics is addressed, as well as its production by cyanobacteria. Furthermore, factors that influence and production processes are discussed. Finally, commercial applications, economic aspects, and prospects are presented.

Chapter 12 provides compiled information about various polymers used in plastic preparation. Environmental concerns of plastic with comprehensive details about the degradation of various types of plastics by different methods have also been discussed. State-of-the-art on degradable plastics market with an emphasis on principle design for recyclable plastics and biodegradable plastics from renewable raw materials have also been included.

Degradation of Plastics
Materials Research Foundations 99 (2021) 1-36

Materials Research Forum LLC
https://doi.org/10.21741/9781644901335-1

Chapter 1

Introduction, Past and Present Scenarios of Plastic Degradation

Neema Pandey[1], Bhashkar Singh Bohra[1], Chetna Tewari[1], S.P.S. Mehta[2], Nanda Gopal Sahoo[1]*

[1]Prof. Rajendra Singh Nanoscience and Nanotechnology Centre, Department of Chemistry, DSB Campus, Kumaun University, Nainital, India

[2]Department of Chemistry, DSB Campus, Kumaun University, Nainital, India

*ngsahoo@yahoo.co.in

Abstract

The fascinating properties of plastic make its use widely possible in every field for the ease of human life. On the other hand, these properties make plastic non-biodegradable in nature. Hence the increasing accumulation of plastic in the environment is the arising concern for environmentalists and human society. This growing concern has motivated researchers and technologist to promote research activity for finding new degradation methods for plastics, synthesis of new plastic with biodegradable nature or find alternatives to plastics. Considering these facts, this book chapter briefly discusses the management of post-consumer plastic products and all the possible methods for plastic degradation such as photo and thermo-oxidative, catalytic, mechano-chemical and chemical along with the factors affecting these degradation methods.

Keywords

Plastics, Degradation, Biodegradation, Thermo-Oxidative, Photo-Oxidative, Catalytic Degradation

Contents

1. Introduction

The resilience behavior of plastic against degradation and its indiscriminate use by humans leads to the accumulation of plastic in the environment, which arises as a serious concern for global ecology and biodiversity [1-2]. The word plastic is taken from the Greek word plastikos which means ability to deform into different shapes without breaking [3]. Plastics are synthetic, organic long chain, high molecular mass polymeric molecules usually synthesized from petrochemicals (oil, coal and natural gas) and often contain N, O, Si, Cl etc. [4-6]. Hence those polymers which have plasticity property (in a class of moldable solid polymers) are known as plastics [7].

Due to the ease of fabrication, low cost, flexibility, toughness, resistivity towards corrosion, microorganisms and water, good barrier properties, durability and high stability, plastics are used in various products in different fields with varying scale from paper tie up pin to spacecraft [4, 8-11]. Hence plastic materials are considered as a group of materials which is resistant to many environment influences. With these properties, synthetic plastics have prevailed/ substitute the traditional/ natural materials such as malleable metals, wood, glass, ceramic, leather, cotton etc. and became an unavoidable part of human lives. These features make plastic a very popular material, used massively in every field but on the other side these features make it an environmental threat [9, 12-19].

The proliferation of synthetic plastic in different fields is due to its multitude applications. The global production of plastic was 381 million tons in 2015 of which 42% (146 million tons) was used in packaging industry like pharmaceuticals to consumer items, grocery to vegetables products, cosmetic to chemicals and nearly 19% (65 million tons) in building and construction field like piping, vinyl siding etc. The commonly used plastics for packaging application are polyethylene (low density polyethylene (LDPE), high density polyethylene (HDPE), medium density polyethylene (MDPE), linear low density polyethylene (LLDPE), polypropylene (PP), polystyrene (PS), polyvinyl chloride (PVC), polyurethane (PUR), poly (ethylene terephthalate) (PET), poly (butylene terephthalate) (PBT), nylons. The plastics used in packing have a very short lifetime than the other sectors [20- 22].

According to British plastic federation report (2009) nearly 25,000 companies in India are involved in plastic production and produced nearly 5.27 million tons of plastics (in 2009) of which packaging and agriculture sector consume maximum amount of plastic.

Materials Research Forum LLC
https://doi.org/10.21741/9781644901335-1

The plastic industry development rate in India is one of highest in the world, with the increasing plastic consumption rate of 16% per annum as compared with other countries such as 10% per annum in China and nearly 2.5 % per annum in United Kingdom (UK). According to the report of Statista Research Department (2016), the per capita consumption (in kilogram) of plastic materials in different region of the world is as follow: NAFTA i.e. North America Free Trade Agreement countries: Canada, the United States and Mexico (139 kg per capita), Western Europe (136 kg per capita), Japan (108 kg per capita), Central Europe and Commonwealth of Independent States (CIS) (48 kg per capita), Asia (excluding Japan) (36 kg per capita), Middle East and Africa (16 kg per capita), World total (45 kg per capita). Plastic consumption by an Indian in a year is 11 kg (global average is 28 kg per year) while an American can consumes nearly 109 kg per year [23,19,24]. As stated by central pollution control board in 2015, nearly 6.92% plastic waste of the total municipal solid waste was generated in India. Approximately 50% of plastic is produced for use and throw purpose but their accumulation in the environment is for several hundred years. The World Bank analysis reveals that the world generates maximum 242 million tons of plastic waste-nearly 12% of all municipal solid waste (MSW), in which 9% plastic waste has been recycled, nearly 12% incinerated and the remaining plastic waste (79%) dumped in the land or natural environment and nearly 8 million tons of plastic ends up in the ocean [25, 26]. Hence plastic waste reaches from the niche of the ocean to Mount Everest. Plastic waste in the oceans originates from both land bases (80%), marine bases (20%). The plastic debris ingestion by the animals leads to many problems such as persistence of plastic in the stomach, gastrointestinal blockage, decreased secretion of gastric enzymes, fertility problems [1]. Most of the rivers located in Asia are highly polluted and deliver 86% of plastic in the oceans. The Chinese river "Yangtze" contains maximum amount of plastic debris (1,469,481 tons) and is highly polluted in the world [20]. Eriksen et al. [27] estimated that there was approximately 269000 tons of plastic in the surface water across the world. The commonly found plastic waste is plastic bottles, bottles caps, food wrappers, straw and stirrer.

Plastic, the magical material has a flipside, it is not degrade naturally. Usually the commodity plastics (~99%) originates from petrochemicals (i.e. synthetic plastic) and is non- biodegradable in nature or degrades in a very slow rate [28]. The main obstacle for plastic degradation is the high molecular weight of the plastic. The microorganism cannot grasp the higher molecular weight compounds [29]. The continuous improvement in plastic properties with the help of additives such as plasticizers, stabilizers, cross linking agents etc. according to our requirement, makes its degradation very difficult [30]. The alarming situation originating from the plastic is due to its non-biodegradable nature, short life and inappropriate dumping such as throwing it in water streams, fields, from running car windows, filled garbage containers etc. which is a subject of great concern such as environmental, economic and waste management problems [31].

The increasing accumulation of plastic waste in our environment has innumerable effects like soil, air and water pollution which disturbs every habitat in the world. The plastic waste causes floods by clogging drainage system, reduces fertility power of soil, and causes respiratory problems when burnt, reduces the lifespans of animals, and contaminate the water bodies by releasing harmful chemicals when dumped in rivers or oceans [32]. Another problematic thing originating from plastic waste is "the generation of microplastic" which is developed by the degradation of plastic by any of the physical factors such as sunlight, temperature, air, water, microorganism degradation etc. These microplastics are invisible to our eyes but their presence spoils the food chains and ecosystem [33]. The disposal methods such as landfilling, incineration and recycling are not sufficient to overcome this problem. Now it is the time to make people aware of its harmful effects and at the same time, takes some useful scientific steps towards this menace so that the upcoming generations can get a better planet to live on [1, 34].

This growing concern has led to use of degradable polymers, promoted research activity for new degradation methods for plastics and synthesis of new plastic with biodegradable nature or find alternatives to the plastics. Nowadays some products are available in the market which are based on biodegradable plastics such as poly(lactic acid) (PLA), poly(ε-caprolactone) (PCL), poly(butylene succinate) (PBS), or poly(butylene succinate-co-butylene adipate) (PBSA) [35,36]. Degradation or the up gradation of the plastic waste by means of various methods such as biodegradation, photolytic, catalytic, thermally etc. is the better solution to minimize this problem. Some synthetic polymers absorb UV radiation and degrade by photolytic, photo and thermo oxidative reactions. Several polymers are susceptible to the microorganism and degrade in a natural way. In this book chapter a brief discussion about the problems originating from the plastic waste, their uses and lifespan in the ambient environment, present approach to deal with the plastic waste and various degradation and up-cycling approaches is presented [29, 30, 37].

2. History of plastic

Mesoamericans are the first who used processed natural rubber for bands, balls etc. in 1600 BC [38]. The first man made polymer is Parkesine (Nitrocellulose) or celluloid, invented by Alexander Parkesin in 1856 [21]. It is prepared from cellulose biomolecules. Cellulose is treated with nitric acid and the resulting product is dissolved in alcohol to get harden and elastic material which is easily molded on heating [39]. In 1907, the first synthetic polymer was formed by Belgian Chemist Leo Baekeland by using phenol and formaldehyde i.e. Bakelite. In 1909 Baekeland coined the term "plastic" for the new group of material. After First World War, polymer technology developed effectively and produced various polymers making life very easy. The mass production of these polymers started during 1940-1950s [21, 22].

3. Types of plastic and their uses

Polymers are known for their diverse application in different fields. Plastics are used in food industry, health care products, building materials, furniture and make living comfortable by making things affordable and cheaper without compromising their strength. Polymers can be divided into different groups based on the different parameters such as on the basis of structure of polymers (linear, Branch chain and cross-linked polymers), mode of polymerization (addition and condensation polymers), thermal behavior (thermoplastic and thermosetting) and source of origin (synthetic and natural polymers) [40, 41]. However, the easier way to classify is the source of origin.

3.1 Natural polymers

These polymers are made by renewable sources such as biomass obtained from tress, microorganisms and animals and are found in abundance in nature. Cellulose is highly abundant polymer formed by the components of the cell wall. Chitin (N-acetylglucosamine) is prepared by the exoskeleton of shrimp and insects, starch is present in potato [42]. Generally, plastics derived from biomass are biodegradable in nature and their use does not harm ecosystem.

3.2 Synthetic polymers

These are man-made polymers and are prepared by non- renewable sources (petrochemicals) and are non-biodegradable and their accumulation in the environment is harmful to ecosystem. Some examples of these polymers are polyethylene, nylon fibre, polyvinylchloride, and polypropylene, polyethylene terephthalate etc. Synthetic polymers are of two types thermoplastic and thermosetting. Thermoplastic polymers are held together by weak interaction such as van der Waal's forces, hydrogen bonding etc. There is an absence of cross links in these polymers. These polymers are moulded repeatedly on heating without breaking i.e. it can be reheated, remould repeatedly while thermosetting polymers can be shaped only once and cannot be remoulded again on heating. Thermosetting polymers develop three dimensional molecular networks on heating. These polymers have strong crosslinking between polymer chain [2, 21, 43]. Hence thermosetting polymers have higher strength than thermoplastic polymers. Some common examples of synthetic polymers are polyethylene, polypropylene, polyethylene terephthalate etc. [44, 46]. Fig.1 indicates the percentage of different plastics demand distribution on the basis of resin type in global market on 2018 year and Table 1 shows applications of different polymers [43].

Table 1 *Shows a list of different plastics with their properties and application.*

Types of Plastic	Properties	Applications	References
Polyethylene terephthalate (PET)	Highly flexible, colorless, resistance to moisture, solvent higher strength, toughness, barrier to gas	Used for Carbonated soft drink bottles, sleeping bag, textiles fiber, Containers, fleece clothing, peanut butter and jam bars mouthwash bottles	[22,29]
High-densityPolyethylene (HDPE)	Weather proof, low temerature toughness, chemical resistance, permeability to gas,	Mainly used for Packaging of food, fruits, vegetables and drainage pipes, motor oil bottles, shampoo bottles	[22, 46]
Polyvinyl Chlorides (PVC)	Easy processibility, chemical resistance, flexible	Mainly used for electronic pipes, plumbing pipes and shoe soles, automobiles seat covers, juice bottles, medical equipment	[22, 34]
Low- density polyethylene (LDPE)	Quite flexible and tough, ease of processing, moisture and chemical resistance	Container lids, shopping bags, squeezable bottles, shipping envelopes, carpet, furniture, clothing, floor tile, dry cleaning bags, and frozen food w66or bread bags.	[46,35]
Polypropylene (PP)	Semi-rigid. Translucent. Good chemical resistance. Tough. Good heat resistance.	Drinking straws, medicine bottles, furniture, car batteries, bottles cap, plumbing pipes and carpet	[22,29]
Polystyrene (PS)	Brittle, gamma radiation resistance, rigid, good electrical properties , poor chemical and UV resistance	Food containers laboratory ware, plates, cutlery, disposable tea cups, CD and cassette boxes and certain electronic uses	[22, 35]
Others (often Polycarbonate/ Acrylonitrile Butadiene styrene)	Stiff, rigid, brittle, chemical and stains resistant, good electrical insulators, non-remouldable.	used for making the lens of glasses, rear lights of cars, street lighting, Electronic casing, baby milk bottles, protective headgear,	[22, 29]

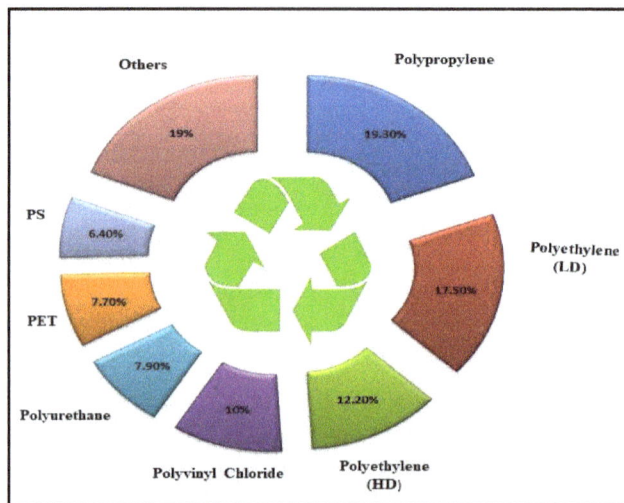

Fig.1 *Synthetic polymers demand distribution globally on the basis of resin type in 2018.*

4.   Hazards of plastic

The non- biodegradable nature and toxic components makes plastic a threat to humans, animals and environment. Plastic creates pollution in environment from the manufacturing process to its usage and finally the improper disposal. Plastic releases harmful chemicals in environment such as vinyl chloride, dioxins, benzene, plasticizers, phthalates, bisphenol, formaldehyde and these harmful chemicals reach humans through air, soil, and water. The plastic manufacturing industries release toxic chemicals into the environment such as carbon dioxide, dioxin, and hydrogen cyanide. Methane (greenhouse gas) released during bio-degradation process of plastic causes global warming [47]. Some of the major compounds released into environment include vinyl chloride (in PVC), dioxins (in PVC), benzene (in polystyrene), phthalates and other plasticizers (in PVC and others), formaldehyde, and bisphenol-A, or BPA (in polycarbonate). Many of these are persistent organic pollutants (POPs)-some of the most damaging toxins on the planet, owing to a combination of their persistence in the environment and their high levels of toxicity; however, their unmitigated release into the environment affects all terrestrial and aquatic life with which they come into contact [2, 22].

Materials Research Forum LLC

https://doi.org/10.21741/9781644901335-1

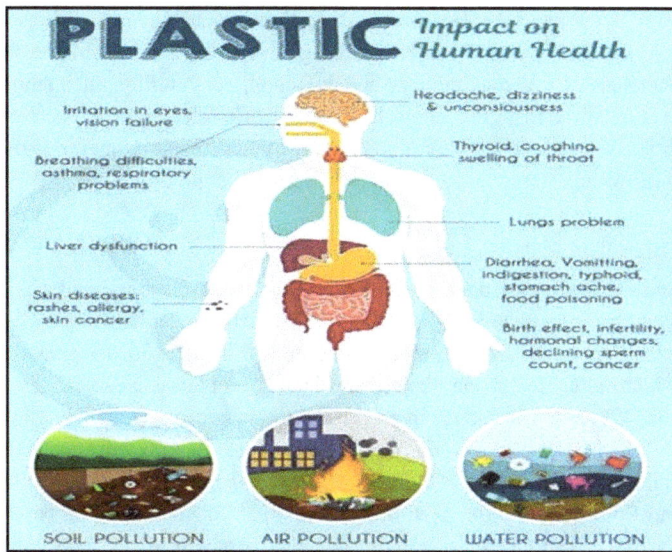

Fig.2 *Harmful effects of plastic on human health. Adapted with permission from Proshad et. al. [50].*

Polyethylene is used in shopping bags; toys, bottles etc. and are carcinogenic in nature. Phthalates present in emulsion ink, toys, footwear cause hormonal disturbance, are carcinogenic, reduce sperm count, infertility, and reduce immunity. Polyvinyl alcohol is used in packaging, vinyl siding, utility items and in cosmetics, cause birth, genetic diorders, skin disease, vision problems, deafness, digestion problem, liver related problems and are carcinogenic. Polyvinyl alcohol produces dioxin (potent synthetic chemical) cause cancer, reduce immunity and reproductivity. Bisphenol A is monomeric building block for polycarbonate plastic. During polymerization of polycarbonate, certain percentage of bisphenol A remains unbound. These unbound bisphenol are released from the containers to the food/drinks at elevated temperature. The presences of these chemicals in our body disturb natural hormonal system, causes cancer, reduces immune system, and trigger obesity and diabetes [48, 49]. Fig. 2 [50] shows the harmful effects of plastic on human health.

Plastic waste is also disposed into the oceans. In oceans, the plastic is broken down into small fragments (in micron level) due to the weathering effect. This plastic debris makes physical harm to marine mammals through ingestion or entanglement such as sea turtle, sea bird, fish and other mammals often ingesting plastic debris mistakenly as food. Such ingestion can lead to internal injuries and infection, malnutrition in mammals when plastic debris is collected in the stomach. Plastic debris releases toxic chemicals which cause death and reproductive failure in mammals. Entanglement in marine system causes suffocation drowning, starvation, vulnerability to predators [47,50].

5.   Degradation of plastic

The long term existence of plastic in nature is due to its chemical and physical inert behavior. But plastics are naturally degraded in the environment at a very slow rate. The term "degradation" means any physical and chemical changes in the properties of the material which alter its shape, color, mechanical, optical, electrical and thermal characteristics [21, 29]. The change in properties is also known as "aging". The degree of degradation of plastic depends on the properties of the polymer as well as the environmental factor to which it is exposed. The environmental factors which help in the degradation of plastics are light, heat, humidity, pH, air, chemicals (acid, base, salt) and biological activity. The molecular weight, chemical structure of the plastic and mixed additives are the factors which effect degradation of plastic [51]. Complete degradation of polymers occurs when it gets converted to water, methane, CO_2, ammonia, hydrogen, sulfide [52]. Salem et. al. [53] studied the biotic and abiotic degradation rate of pro-oxidant filled three types of polyethylene (PE) thin films with 30 and 70 μm thickness. These films are presumed to be oxo- biodegradable in 12 months and they contain nearly 11 wt% $CaCO_3$. The results of weathering tests revealed that UV radiation triggered the aging mechanism and 50% weight loss occurred after weathering exposer which triggered the fragmentation of plastic films while 83% weight loss occurs after 12 months of soil burial. The degradation of the polymers results in deterioration of functionality of polymers, crazing, erosion, cracking, bond scission, phase transformation, development of new structural unit or functional groups moiety reduces molecular weight of polymers which makes it favorable for biodegradation (degradation by living organism) [54, 55]. The plastic degradation occurs through abiotic and biotic processes [22]. Fig. 3 indicates the degradation route of polymers.

Materials Research Forum LLC
https://doi.org/10.21741/9781644901335-1

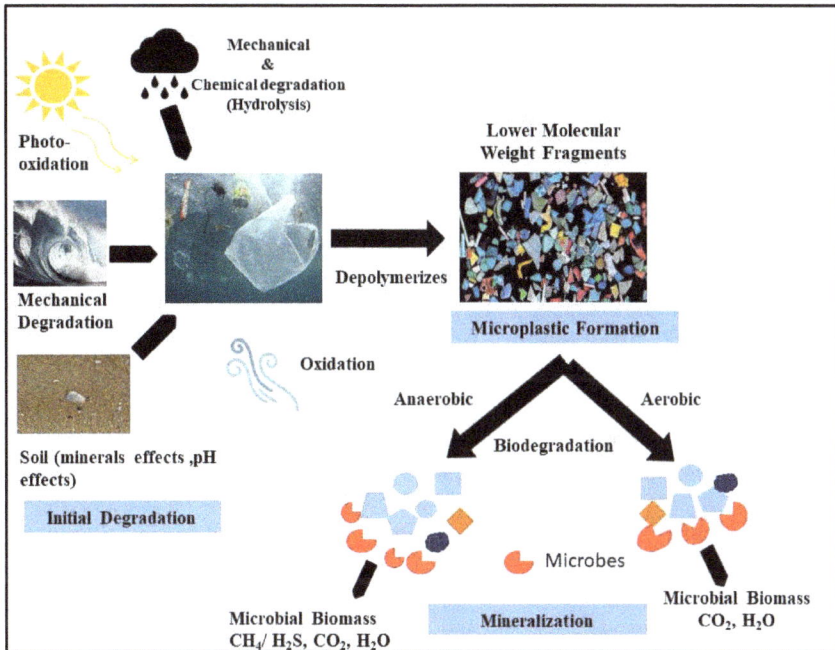

Fig.3 *Plastic degradation via various means.*

5.1   Abiotic degradation

Abiotic degradation takes place in the intra-molecular level of plastic by physical and chemical process which results discoloration, reduction in molecular weight, size, tensile strength and makes it visually disappear i.e. total loss of its functionality [56, 57]. The physical and chemical aging of plastic converts this inert material into small fragments of lower molecular weight, makes it susceptible for biological degradation. These smaller fragments easily cross the cellular membrane and biodegrade by cellular enzymes in microbial cells i.e. easily metabolized by the microorganism. Hence, abiotic degradation of plastic is a very important step for its biodegradation. Abiotic degradation is initiated by thermal, chemical, hydrolytical and UV-radiation [1, 2]. The beginning of natural degradation of plastic is photodegradation which induces thermo-oxidative degradation. Most of the plastics absorb high energy UV- radiation which provides high activation power to activate their electrons and leads to oxidation, cleavage and further degradation. This causes embrittlement and disintegration of the plastic. Thermal degradation occurs due to overheating of plastic and causes molecular scission. The chemical aging of plastic is a result of chemical process on plastic. The change in colour and crazing of the surface are the first visual changes of plastic degradation [21, 22]. López et. al. [58] studied the effects of accelerated weathering (AW), natural weathering (NW) and thermal oxidation (TO)[abiotic degradation] on different plastics (high density polyethylene (HDPE), oxodegradable high density polyethylene (HDPE- oxo), compostable plastic, Ecovio ®; metalized polypropylene (PP) and oxodegradable metalized polypropylene (PP-oxo). The plastic films exposed to AW and NW showed general loss in mechanical properties and no substantial loss was found on TO. Oxo-plastics show higher degradation rate than their counterparts.

5.2   Biotic degradation

The degradation of plastics through the living organism such as fungi, bacteria is called biotic degradation. This biological process converts organic materials into water, CO_2, methane, ammonia, hydrogen, energy and new biomass with the help of microorganism. The conversion of materials into small organic or inorganic fragments with the combined effect of abiotic and biotic effects is known as mineralization. Abiotic degradation of plastic converts it into a very small molecular weight fragments which are easily metabolized by microorganism [29-31]. Biotic degradation is of two types: aerobic and anaerobic biodegradation.

5.2.1 Aerobic biodegradation

In aerobic degradation or respiration process, microorganism requires oxygen as an electron acceptor for the deformation of the materials into the smaller products: carbon dioxide (CO_2) and water. Plastics are aerobically degraded in open environment [59, 60].

5.2.2 Anaerobic biodegradation

In anaerobic biodegradation process, microorganisms degrade plastic into carbon dioxide, water and methane in the absence of air (oxygen) [61]. Anaerobic degradation takes place in sediments and landfills. The mineralization of plastic needs several different organisms, some are involved in fragmentation of polymers into monomers, some organisms consume these monomers and excrete smaller compounds, and others use these compounds and excrete biomass and finally few use waste biomass [29]. There are two pathways for anaerobic biodegradation: anaerobic respiration and fermentation. Anaerobic bacteria use nitrate, sulphate, Fe, Mn and CO_2 as an electron acceptor during anaerobic respiration. The organic materials serve as both electron donor as well as electron accepter and form smaller molecules in anaerobic fermentation process [62, 63].

Materials Research Forum LLC

https://doi.org/10.21741/9781644901335-1

6.   Types of degradation method

Plastic degradation techniques are classified according to their causing factors. These methods are photo, thermo-oxidative, mechano-chemical, catalytic, chemical and biological degradation [30, 64] (Fig.4).

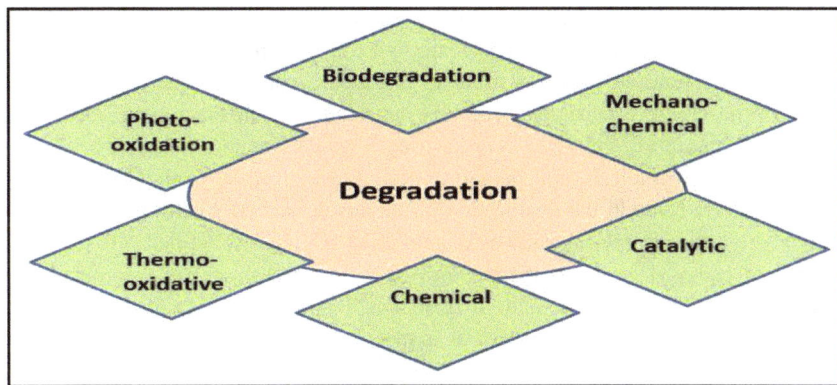

Fig.4 *Different types of degradation method.*

6.1   Photodegradation

Plastic degradation carried out in presence of light is known as photodegradation. An appropriate amount of energy is absorbed by the photo responsive functional groups present inside the polymer chain. This absorbed light provides activation energy to lead chain breaking into small fragments [65]. Photodegradation is of two types; UV and oxidative degradation. In ultraviolet (UV) degradation, UV light is used to disintegrate polymer materials while in the oxidative method polymers degrade by the combined effect of light and air [21, 30]. UV and infrared portion of light have capabilities to degrade plastic in the presence of air. The exposure to ultraviolet radiation (UV) creates alteration in the properties of plastic by the addition of oxygen molecules into polymer chain, hence this process is known as photo-oxidative degradation. The photo-oxidative degradation is considered as an initial step to degrade plastic into small fragments at ambient condition [66-68]. This degradation results in breaking of polymer chains, generates free radicals, reduces molecular weight and lowering the mechanical strength, variation in optical properties and yellowing the surface of plastic [69-77]. As one or more oxygen molecules enter into the polymer chain, it becomes brittle and easy to degrade into smaller fragments i.e. it is easily metabolized by microorganisms into carbon dioxide and biomass. However, this process takes a significant time for complete mineralization of plastic, nearly 50 years or more. The photo-oxidative degradation of plastic becomes much more difficult and protracted in subaquatic levels due to lower temperature and oxygen availability and slower rate of hydrolysis of many plastics in ocean [78-82]. Most of the plastics absorb UV radiations (290- 400 nm) and result in bond cleavage in polymer chains. The bond cleavage takes place in the soft segment of plastic, where it generates ester, formate, aldehyde, propyl groups [83]. The wavelength of UV radiation causing plastic destruction is varied according to the nature of bond present in the plastic e.g. polyethylene (300 nm) and 370 nm for polypropylene. The presence of photosensitizers, catalyst residues and metallic impurities are also helpful for initiation of free radicals by homolysis of polymer chain. For example TiO_2 (photosensitizer) helps in the degradation of polyamide and polyolefin at 480 nm and forms highly reactive species such as atomic oxygen, $\cdot OH$, $\cdot OOH$ and $\cdot O_2$ [30]. The presence of carbonyl groups (chromophores) in polymer chains helps to produce free radicals via Norrish type I and II reaction. Photodegradation of polyethylene, polypropylene and polystyrene brings loss in mechanical strength, molecular weight due to extensive bond breaking in polymer chains [71-74]. Chiantore et al. [84] performed photo-oxidative degradation of acrylic and methacrylic polymers by using artificial solar irradiation. FT-IR results show that acrylate units are more reactive towards oxidation than methacrylate.

6.2   Thermo-oxidative degradation

Thermal degradation of polymers results in molecular deterioration due to overheating. The long chains of polymer break into small fragments at high temperature and these fragments react with one another to make change in the properties of plastic. Thermal degradation of plastic results in change in the molecular weight and reduces some physical properties such as ductility and embrittlement, fading of colour, cracking [85]. The products formed on the thermal degradation of the plastic depend on the additives, temperature, time of exposure and environment factors and it is a complex mixture of compounds such as CO, ammonia, furans, dioxin, ketones, hydrogen cyanide, nitriles, aliphatic amines which is toxic and irritating [86]. The photochemical and thermal degradation are similar under normal condition. But these methods are differentiating by some processing steps. The first difference is the initiation step leading to the auto-oxidation and second is that thermal degradation occurs in bulk of the material whereas photodegradation occurs at the surface of the material [87]. The depolymerisation reaction in thermal degradation is not initiated at the terminal end of the macromolecules whereas it takes place in a weak link (a peroxide or ether link) of polymer chain. The polymers synthesized by addition polymerization degrade thermally at very high

Materials Research Forum LLC
https://doi.org/10.21741/9781644901335-1

temperatures. These are some techniques for the analysis of thermal degradation of polymer for example thermogravimetric analysis (TGA), differential scanning calorimetric (DSC). The impurities present in the polyolefins makes it very susceptible for thermal degradation at high temperature [88]. Polyester degradation results in different types of products such as formaldehyde, acetaldehyde, formic acid, acetic acid, CO_2 and H_2O and some other compounds in small quantity [89].

6.3 Mechano-chemical degradation

It involves the degradation of plastic by mechanical forces and ultrasonic irradiation [90]. Breakdown of molecular chains under shear or mechanical force is often aided by a chemical reaction and is known as mechano-chemical degradation. High speed stirring, regrinding, agitation or ball milling, injection moulding, extrusion are some methods applied for the machno-chemical degradation of plastic. The basic fundamental is to apply shear force on polymer that breaks polymer chain [91]. Polymer breakdown during mechanical stress is aided by chemical reagents, hence, known as machno-chemical degradation. Under the influence of mechanical stress, the main backbone of polymer chain (where amorphous region is connected with crystalline region) gets broken into free radicals. These free radicals form peroxy radicals in the presence of oxygen: for example of mastication of rubber [92]. The mastication of rubber in open environment shows polymer chains breaking and origination of plasticity under shear while in N_2 atmosphere it does not show any change in plasticity and molecular weight. On the other hand, in the presence of O_2 it shows fast degradation of rubber. This is because under the influence of mechanical stress, polymer produces free radicals; these radicals react with oxygen (radical scavenger) and lead to the permanent chain breaking whereas nitrogen is not a radical scavenger hence radicals are recombined and show no change in molecular weight [93]. The polymethlymethacrylate (PMMA) decomposes into macro free radicals in the presence of nitroxide (chain terminating agent) which is used in free radical polymerization reaction [94].

Materials Research Forum LLC
https://doi.org/10.21741/9781644901335-1

6.4   Catalytic degradation

This kind of degradation technique is basically used for the up-cycling of plastic waste. The presence of catalyst helps to transform the plastic waste into highly valued added products such as oil, gases and nanomaterials as well as it gives the selective formation of a particular product; reduce the decomposition temperature of plastic. Some examples of catalysts used in polymer degradation are zeolite and non-zeolite catalysts, transition metal catalysts (Cr, Ni, Mo, Co, Fe) on the support (Al_2O_3, SiO_2), zeolite and zirconium hydride [95- 98]. Jia et al. [99] converts the polyethylene into liquid fuel and wax by using tandem catalytic cross alkane metathesis method under mild condition. This method is performed in batch autoclaved instrument.

Tiwari and co-workers [100] converted waste plastic except polyvinylchloride and polyethylene terephthalate into fuel range hydrocarbons i.e. petrol, diesel and kerosene etc. by using zeolite catalyst. The effect of polymer catalyst ratio on this conversion is also studied. Ratio (4:1) shows good result for yielding liquid product [100]. Zhang and co-workers [101] produced carbon nanotubes from the pyrolysis of plastic (high density polyethylene) with stainless steel mesh loaded with nickel catalyst.

6.5   Chemical degradation

Polymer disintegration by chemical reaction is a constitutive process. In this process polymers are depolymerized into monomers or oligomers or small chemical compounds. Polymers are made from organic and inorganic moieties which easily break from the polymer chain by using reagents such as water, alcohol, amine, organic and inorganic solvent [102]. Hence chemical degradation is classified into different groups.

6.5.1 Solovolysis or hydrolysis

This method involves depolymerisation of polymers with the help of different solvents. In hydrolysis, water is added to the polymer in the presence of acid or base or neutral environment. For example hydrolysis of polyurethane produced polyol, diamine, and CO_2 under the presence of alkali metal hydroxide catalyst at elevated temperature (250- 450) °C and pressure (15 atm to 50 atm). At higher temperature, recovery yield of polyol is reduced while at lower steam temperature, the produced polyol forms unstable polyurethane foam [103, 104]. The other solvents such as alcohol, amine, glycol and acids are also used for the degradation of plastic or recycling of plastic. But hydrolysis process is very slow as compared to the methanolysis and glycolysis because water is a weak nucleophile and this process also needs high temperature and pressure [105].

6.5.2 Ozonolysis

The degradation of plastic in the presence of ozone causes oxidation of plastic in normal condition. The small concentration of ozone in the air makes significant destruction of polymer in a very short period while other oxidation process at a very slow rate. Polymer exposure to the ozone produces active radicals with change in molecular weight, mechanical and electric properties: saturated polymers forms active oxygen compounds; aliphatic ester, ketones, lactones and aromatic carbonyl associated with styrene produce a variety of carbonyl, ethers, terminal vinyl groups, hydroxyl and unsaturated carbonyl, products based on the polymer type [30]. Ozonolysis of polyvinyl alcohol occurs through the oxidation of alcohol group with the formation of ketone group. This group is further oxidized and leads to the chain scission. FT-IR study reveals the formation of oligomers with numerous ketone group and oligomers with carboxylic end groups [106].

6.6 Biodegradation

Biodegradation method involves the biochemical transformation of materials into small molecular weight fragments that are easily mineralized by microorganism [30]. The biodegradation is also defined as the property of a material to degrade into its constituent molecules by natural process (microbial degradation) and the resulting constituents are non-toxic in nature and these are further used in carbon, nitrogen cycle. Biodegradation rate is influenced by different factors: physical and chemical properties of polymer, types of organism and environmental factors [21]. The large polymeric molecules are not easily biodegraded because they cannot pass through the cellular membrane of microorganism hence polymers must be depolymerized first into small monomers to be metabolized by microorganisms. The initial breakdown can be achieved by the weathering effect such as heat, air, temperature, water etc. [29, 31]. The environmental degradation of synthetic polymers was initially started by the abiotic hydrolysis. Biodegradation of polymers proceeds via four different mechanisms: solubilization, charge formation followed by dissolution, hydrolysis and enzyme catalysed degradation [2, 35]. Hydration of polymers depends on the hydrophilic nature of polymer and results due to the disruption of secondary and tertiary structure stabilized by weak van der Waals forces. Hydration of polymers makes it water soluble and results in bond cleavage by chemical or enzyme catalysed reaction [41, 51]. Ionization, protonation or hydrolysis of the pendent groups (anhydride or ester groups) of polymer chain makes water insoluble polymer soluble; for example polylactic acid become water soluble at high pH. Poly(methacrylate) and poly(methylmethacrylate) are water insoluble polymers and these polymers become water soluble on the hydrolysis of pendent groups (ester) and further ionization of carboxylic groups [53, 56].

Natural polymers undergo degradation by hydrolysis whereas synthetic polymers are water insoluble. They tend to be more crystalline and this property accounts for their water-insolubility [30]. For hydrolysis to occur, polymer must have hydrolytically favorable bonds i.e. reasonably hydrophilic for the access of water. Biodegradable polymers including esters and ester derivative polymers are favorable for this purpose [21, 64]. Microorganisms and enzymes play a very important role for biodegradation process of polymer. Enzymes act as catalysts and a particular enzyme catalyze only a particular reaction. Those, polymers which do not contain any hydrolysable groups such as polyethylene, polypropylene and polystyrene are the most stable polymers and do not degrade easily [21, 30, 53]. Laccase enzyme helps in oxidation process of polyethylene and its degradation process is analysed by gel permeation chromatography. For this process, putative laccase enzyme was isolated from actinomycete R. rubber. This enzyme reduces the average molecular weight (20%) and average molecular number (15%) of polyethylene. Laccase usually found in lignin biodegrading fungi, where they function as a catalyst for the oxidation of aromatic compound. However, laccase also shows activity on nonaromatic substrate [21,30]. Lipase enzyme catalyzed degradation of poly (vinylacetate) was studied by Chattopadhyay and Madras [107]. They observed that the ester bond is broken from the side chain of the polymer to produce oligomers with alcohol and acid [107].

Microbial degradation of plastic is achieved by the action of microorganisms such as bacteria, fungi, algae etc. [29]. Microorganisms are best for the contamination destruction because they contain enzymes which help them to easily metabolize it [21]. The conversion of plastic (contamination) into biomass by microorganisms occurs because they utilize contamination as food for their growth and reproduction. Organic contamination fulfills two purpose of microorganisms: first they provide carbon to organisms which is a basic need or basic building block for the development of new cell and second one is that they provide electron to the organism to gain energy [21, 30]. Organisms gain energy by catalyzing exothermic chemical reaction i.e. breaking of chemical bonds, transferring electron from the contamination. Such reactions are known as oxidation-reduction reaction (National Research Council, 1993). The destruction of organic substance in the presence of air is known as aerobic destruction. In this process oxygen molecules are used to oxidize carbon chain and produce CO_2 and remaining carbon is used for cell manufacturing of organisms while oxygen molecules get reduced and produce water. Many organisms cause degradation of plastic in the absence of oxygen molecules are known as anaerobic degradation. In this process nitrate, carbon dioxide, sulphate and some metals such as iron and manganese are used in the place of oxygen and accepting electron from the organic contamination and produce nitrogen gas,

Materials Research Forum LLC
https://doi.org/10.21741/9781644901335-1

methane, hydrogen sulphide, reduced metals according to the nature of electron acceptor [2, 21, 53]

Polyolefins are non-biodegradable and very strong polymers. Biodegradation of polyethylene occurs by two mechanisms: hydro and oxo- biodegradation. These two mechanisms successfully degrade modified polyethylene (small addition of starch in polyethylene) in the presence of amylase enzyme. The selection of proper test procedure for biodegradation experiment depends on the type of plastic and environment condition. Soil burial and composting methods are used frequently for the determination of degradability behavior of plastic [2, 30, 36]. Biodegradation of plastic can be measure or analysed by weight loss, change in tensile strength, dimensions or physical and chemical properties, carbon dioxide production etc. In soil burial method, degradation test can be performed under natural or laboratory conditions. Testing specimens with definite weight and dimensions is buried in soil in specific depth for a definite time intervals. After a particular time interval, sample is taken out from soil, washed with distilled water and then dried. This dried specimen is analyzed to measure degradation rate. In composting method, dried plastic sample is buried in a definite amount of mature compost in the presence of air and moisture. The nature of compost also affects the degradation rate [30, 21]. Vaverkova and co-workers evaluated the degradation of plastic bags of high density polyethylene (HDPE) with TDPA additive, polypropylene with prooxidant and compostable (starch, polycaprolactone) polymer in composting condition. 12 weeks analysis reveals that the plastic with additives are not degraded (no change in color and physical properties) while other sample are decomposed [108].

7.   Factors affecting plastic degradation

Several factors affecting the degradation of plastic (Fig. 5) are as follows:

7.1   Chemical and physical properties of plastic

The biodegradation of plastic is influenced by several factors which include chemical structure, molecular weight, functional groups (hydrophilic and hydrophobic in nature), crystalline and amorphous nature, number and types of bonds, plasticizer or additives added to the plastic [21]. The previous study [60, 78, 109, 110] reveals that plastic with lower molecular weight is more favorable for bio-degradation than the higher molecular weight such as polyolefin with lower molecular weight was degraded by some microorganism much faster than the higher molecular weight polyolefin [30]. The polymer containing only long carbon chains makes it non-susceptible for bio-degradation. The introduction of hetero atom and unsaturation makes it susceptible for the thermal and oxidative degradation, and biodegradation. The presence of carbonyl group and metal- metal bond accelerates the photodegradation of plastic [30, 31]. The amorphous polymers degrade faster than crystalline one due to the permeability of oxygen molecules [21, 22]. Hydrophilic and flexible polymers are degrading much faster than branched polymers. The microbial degradation of the polymers according to the functionality is as follows: ester > etheric >amide > urethane [111]. The additives such as pigments, fillers, catalyst and secondary products remaining during the processing affect the rate of degradation. Metal additives in polyolefins act as pro-oxidant and enhance the thermo-oxidative degradation such as manganese (Mn). Pro-oxidant accelerates the reaction of chain scission and produces small fragments with these moieties –COOH, -OH, C=O [112]. The addition of photo stabilizer (benzotriazoles, hindered amine, phenyl esters etc.) in the polymer reduces photodegradation [29-31].

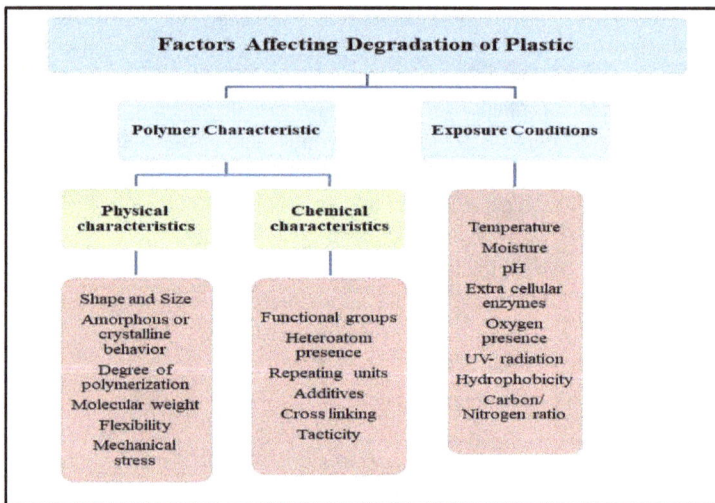

Fig.5 *Factors influencing the biodegradation of polymers.*

7.2 Environment factors

The biodegradation of plastic is also affected by environmental condition such as moisture, temperature, pH, oxygen, and suitable population of microorganisms. The hot wet weather is more favourable for all types of degradation rather than hot dry weather [29, 30]. The increase in moisture (above 70%) and temperature increase the rate of hydrolytic degradation, photodegradation and microbial activity [21, 31]. Under high humidity conditions, the water-soluble photo-stabilizers leak out from the plastic and accelerate the photodegradation. Higher temperature and UV radiation enhance thermal

Materials Research Foundations **99** (2021) 1-36   https://doi.org/10.21741/9781644901335-1

degradation [53]. The thermal stability of polymers (polyethylene, polyethylene terephthalate, polycarbonate polypropylene) is reduced in the presence of air (O_2) compared to the inert environment. Molecular oxygen chemically reacts with the plastic and forms reactive species to enhance the rate of degradation [30].

8. The plastic waste management

8.1 Past scenario

Plastic waste attracts attention of environmentalists due to their durability and non-biodegradability nature and their unavoidable presence. The management rules for plastic waste are not regulated properly and also peoples are not aware of the problems originating from the plastic leakage into the environment. The progressive growth in population, enhanced the consumption rate of plastic which simultaneously increased plastic waste, that promoted the researchers to develop new techniques for plastic disposal /recycling / incineration rather than dumping in the land. Improper dumping of plastic such as throwing everywhere in the land, rivers, drainage system and burning of the plastic in open environment is a serious concern and also creates environmental problems as it impacts natural habitat, wildlife, marine environment and also humans [113]. From the generation of plastic materials (prior to 1980s), nearly all plastic waste was discarded that is it was simply dumped in the land, after 1980s, incineration process was adapted for plastic waste disposal and then recycling process was adapted in 1990s. In 2015, globally 55 % of plastic waste was discarded, 25% was incinerated and 20 % was recycled [20]. In this chapter various methods of plastic disposal and their limitations are briefly discussed.

8.1.1 Landfills

This method is widely used for the disposal of plastic. The drawbacks associated with this method are: landfills occupy a vast area of land which may be utilized for any other productivity means; the occupied space is unavailable for a long time due to the slow degradation rate of plastic, the chemical components and energy exhibited in plastic debris are lost due to this disposal method and the pollutants (phthalates and bisphenol A) are released to the ground water, soil and it also emits toxic gases such as dioxin. Due to the unavailability of O_2 in the land fill area, the plastic degrade anaerobically only [1,114, 115]. Despite these disadvantages the landfill is needed.

Degradation of Plastics   Materials Research Forum LLC
Materials Research Foundations **99** (2021) 1-36   https://doi.org/10.21741/9781644901335-1

8.1.2 Incineration or waste to energy conversion

Incineration method of plastic disposal involves heating of plastic nearly to 1000°C for 1-2 hour and evolved heat is converted into thermal energy or electricity [111]. The PlasticEurope analysis represents that only 16% of plastic waste is combusted with energy recovery while 4% of plastic waste is combusted without energy recovery [111]. This disposal technique reduces plastic waste and generates energy but evolves some air borne mixture of toxic gases such as dioxins ($C_4H_8O_2$) and furans (C_4H_4O) and most commonly the greenhouse gas CO and some other compounds are also emitted according to the type of polymer such as hydrogen cyanide, hydrogen chloride, phosgene, phosphine, phenol, formaldehydes etc. [1, 111]. Hence plastic should be burn in an appropriate chamber not in open environment. With these drawbacks of landfills and incineration method, new method developed for plastic disposal that is recycling [113].

8.1.3 Recycling

Many post-consumer plastic can be used further after reprocessing. This reprocessing gives second life to these materials. There are two types of recycling process first one is mechanical recycling in which plastic is converted to new product without changing the basic structure of materials. Mechanical recycling involves grinding, washing, separating, drying, granulating, and compounding to prepare new product. Injection moulding or extrusion is used for the process of re-melting or reprocessing of products [116]. This recycling requires the separate collection of the plastic which is the main problematic issue during the process. This process helps to regenerate material with small expenditure on the material sorting and cleaning [113]. Another method is chemical recycling in which the plastic is treated with water or any other chemicals at high temperature or pressure, originate lower molecular weight fragments or monomers for the preparation of new materials [1]. There are different depolymerisation routes for the plastic recycling depending on the chemical agent such as solovolysis, ozonolysis, hydrolysis, glycolysis and methanolysis etc. [117]. PET and Nylon 6 polymer is easily depolymerized by the solovolysis using water, alcohol and ether [113]. The advantage of this process is that it is cost effective, very efficient and does not need the separate collection of plastic [113].

8.2 Present approach and prevention for plastic waste management

The above mentioned methods are frequently used today but there are some new techniques also developed for the disposal of plastic. As is known that prevention is better than cure: the new research is focused on this phrase and develops new materials as an alternative for plastic. Here we mention some new disposal techniques and prevention steps.

8.2.1 Organic disposal

This method involves the biochemical transformation of the synthetic and natural plastic into stable small organic fragments by the function of aerobic and anaerobic microorganism in different temperature range. The biological or organic recycling method is divided into two types: composting and anaerobic digestion [113]. Digestion by anaerobic microorganism in the absence of oxygen while composting method is a human derived process performed under a specific set of circumstances in the presence of oxygen and moisture [106, 108, 118]. This method is generally favourable for biodegradable polymers. The presence of bio-component in new generation of plastic shows positive effect on the rate of biodegradation of the plastic and produced biomass with no harmful gases and materials released [2, 30, 117]. Previous study analysis [119,120] reveals that the C/N ratio is a very important factor for the composting process and it should be nearly to 30/1 [113]. Presence of this ratio in biomass prepared for composting method effects the metabolic activity of microorganisms and thus makes composting effective. Kim and Rhee [121] analysis shows that the inadequate availability of key energy component i.e. carbon or key structure components i.e. nitrogen for microorganism may reduce their metabolic activity [119]. The best environment for this method is high temperature, humidity, pH variation, air, light and groups of microorganisms [21, 30, 113]. However, water is a main element for the environmental degradation of plastic. Water effect depends on the physical and chemical properties of plastic. According to the study of Adamcová and co-workers [122], the presence of water molecules initiates the hydrolysis of a bond between polymer and filler and forms active reactive groups such as hydroxyl radical or free radicals. On the other hand, presence of water also helps in the development of microorganisms and proceeds enzymatic reactions causing microbiological degradation.

8.2.2 Reduce the use of plastic

Plastic use can be reduced by choosing alternative or by reducing the use of plastic. For the prevention of plastic eco-friendly approach must be followed: reduce reuse and recycle. Plastic waste can be reduced by dropping plastic use in our life or source reduction occurs by changing the design, manufacturing process. Some steps for the sustainable use of plastic: use cloth bags instead of use and throw carry bags, avoid plastic water bottles and foods in plastic packaging can be introduced. Store food materials in steel or glass container rather than plastic box, don't drink and eat food in use and throw glass and plates. Each single step to the sustainable use of plastic makes our earth and environment safe for all living organisms [123].

8.2.3 Biodegradable plastic

Biodegradable plastics are decomposed by living organism. It is an alternative to the non-biodegradable synthetic plastic [29,30]. These plastics are nearly equivalent to the synthetic petrochemical based plastics in all aspects with an additional quality of being decomposable naturally into safe bio- products in a short period as compared to the conventional plastic [123]. The manufacturing technology of biodegradable plastic are new and very expensive at present. These limitations (high production cost and decreased durability) make it difficult to enter into mainstream yet. Hence the further development of the new technology with energy efficiency and more cost effectiveness is required [113]. The promotion of biodegradable plastic is a progressive and promising prospect and helps to reduce the pressure on the consumption of fossil fuels. The main advantage of these polymers over to conventional polymer is that it is totally manufactured by renewable biomass such as vegetables, tree leaf, food waste, animal shells and by microorganism rather than fossil fuel which are very cheap. There are two main approaches for the development of biodegradable polymer: the production of the synthetic polymer with enhanced degradability nature without reducing the material properties such as addition of functional moieties or bio-component which helps in the deterioration of polymer chain via post polymerization treatments [30, 55, 113]. The main purpose behind this is to enhance the microbial attack on the polymer chain. The degradation rate for conventional plastic will become much better when producing block copolymer with hydrolysable polymeric molecules such as starch, ethylene glycol, lactic acid, caprolactone [124]. New synthetic polyester based biodegradable polymers are frequently used in the packaging industry, paper coating and garbage bags. The polyester blended with polylactic acid and starch- based thermoplastic are degraded into carbon dioxide and water in environment when exposed to microorganism and in compost disposal. Polylactic acid is a synthetic polymer which is completely mineralized and frequently used in medical field since 1990 [113]. Unfortunately, these polymers show less durable nature than synthetic polymer in some cases. The second approach is based on the development of plastic entirely from the biological molecules such as starch, proteins, cellulose acetate, lactic acid, caprolactone and polysaccharides [113, 124]. The mechanical properties of biologically derived polymers have been increased by the addition of plasticizers, nanomaterials or by controlling the production condition.

Example of bioplastic or sustainable polymer is poly(hydroxyalkanoates) (PHAs), which is synthesized by many bacteria such as Pseudomonas, Bacillus, Ralstonia, Aeromonas, Rhodobacter, and Azotobacter as well as certain Archaea, such as Haloferaxsulfurifontis are used in packaging, biomedical and coating application [125].

There are some obstacles which prevent the use of biodegradable polymers: the cost of biodegradable polymers is higher than petrochemical based polymers. Another drawback is limited availability of monomers, their processability, cost, performance limitation and rate of degradation. Bio-plastic such as starch and cellulose are not favorable for mechanical recycling as well as for incineration because their calorific value is very low. Next obstacle is the absence of suitable infrastructure for the development of biodegradable polymers such as higher cost of manufacturing technology of biodegradable polymers. Hence further research is required for making the biodegradable polymers manufacturing cost more effective.

8.2.4   Up conversion of plastic into value added product

Waste conversion into wealth is the basis of upcycling technique. In upcycling process, plastic is converted into a better quality and higher value products. Upcycling helps in plastic waste management by convert plastic into value added products such as chemicals for monomer feedstock, fuels, carbon nanomaterials, etc. due to the presence of carbon content in major amount [113]. For examples polyethylene contains 85.6%, polystyrene (92.2%) polypropylene 85.6 %. carbon content [126]. Hence it can be used as a carbon source for carbon based value added products for examples activated carbon, carbon nanotubes, graphene, carbon fibers etc. Pandey et. al. [127] reported an invincible method for the development of graphene nanosheets in bulk amount from plastic waste by two step pyrolysis process: first one at 400°C and second is at 750°C in presence of catalyst (nanoclay). The obtained black residue is graphene nanosheets with confirmation from Raman spectroscopy data. The methodology flow chart of graphene nanosheets formation is described in (Fig. 6) [127]. The non-wetting nature of plastic makes its use in road construction. Water penetration across the bitumen film makes pothole in road however, plastic coating over the aggregates prevents the penetration of water due to the non-wetting nature of plastic. Polymers also have higher softening temperature which reduces the stripping of aggregates during summer season [113]. Raoand co-workers [128]and team prepared oil from waste plastic. His team took 1 ton of plastic waste after washing and drying processing. This plastic mixture was pyrolized in the presence of catalyst (aluminum silicates nearly 2.5% of the plastic mass) at (603- 723) K. Three products are obtained: obtained pyrolysis oil (60- 70%), gas (15-20%) and carbon black (20-30%). Oil mixture contains different ratio of material 40% oil+60% petrol, 10% oil + 90% diesel, 30% oil + 70% diesel and 50% oil + 50% diesel and various characteristic confirms the presence of pure fuel oil.

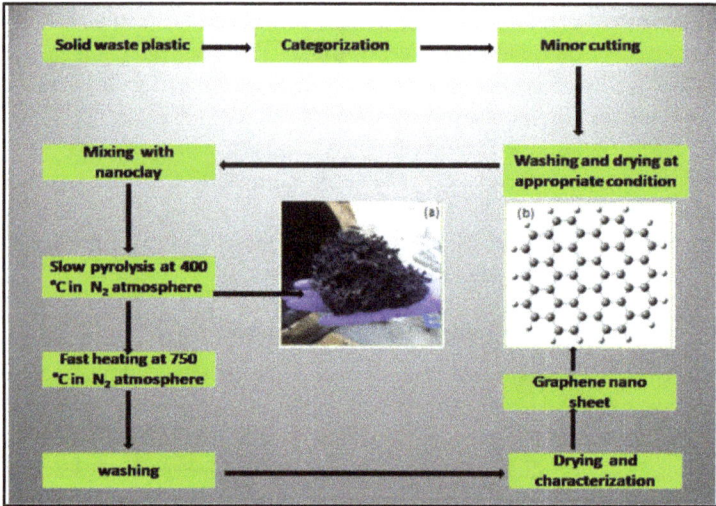

Fig.6 *The methodology flow chart for the production of graphene nano sheets, (a) physical state of carbon residue obtained from slow pyrolysis, (b) Skeleton structure of graphene nano sheets after fast heating. Adapted with permission from Pandey et. al. [127].*

Conclusion and future prospective

The indiscriminate use of plastic and its totally inert behavior to all environmental factors, leads the plastic waste generation all over the world. Waste plastic produces so many harmful effects such as generation of toxic pollutants, spoiling the natural behavior of soil, water and air, creates disturbance to the ecosystem, and causes many diseases in humans and mammals. This inert material is not degrade naturally in the environment or takes nearly 400 years or more to biodegrade. Hence different degradation approaches are currently used to degrade plastic waste. By means of these methods it is possible to manage or overcome this growing problem. However, footprints of plastic in the environment can be reduced by adapting some effective steps such as reduce the use of plastic in our daily life, use biodegradable material for grocery shopping or dispose used plastic in proper manner that it can further be utilized in material processing. The researchers are continuously working to find the alternative for plastic. For the fulfillment of this purpose they are trying to modify synthetic polymers with natural polymers or additives which imitate the biodegradation rate or producing biopolymers with desirable

Degradation of Plastics   Materials Research Forum LLC
Materials Research Foundations **99** (2021) 1-36   https://doi.org/10.21741/9781644901335-1

physical and mechanical properties to deliver real commercial benefits as well as evaluating the resulting polymers for degradation and life span.

Acknowledgements

The work is supported by Department of Science and Technology (Ref No.: DST/TM/WII/WIC/2K17/82(G)), New Delhi, India.

References

[1] H. K. Webb, J. Arnott, R. J. Crawford, E. P. Ivanova, Plastic degradation and its environmental implications with special reference to poly (ethylene terephthalate), Polymers. 5 (2013) 1-18. https://doi.org/10.3390/polym5010001

[2] V. M. Pathak, Navneet, Review on the current status of polymer degradation: a microbial approach, Bioresour. Bioprocess. 4 (2017) 1-31. https://doi.org/10.1186/s40643-017-0145-9

[3] J. R. Fried, Polymer science & technology: Introduction to polymer science, third ed., New Jersey, 1995.

[4] G. Scott, Polymers and the Environment: Polymers in modern life, first ed., Cambridge, UK, 1999.

[5] R. B. Seymour, Polymer science before and after 1899: notable developments during the lifetime of Maurits Dekker, J. Macromol. Sci. Chem. 26 (1989) 1023-1032. https://doi.org/10.1080/00222338908052032

[6] S. Mukherjee, S. Chatterjee, A comparative study of commercially available plastic carry bag biodegradation by microorganisms isolated from hydrocarbon effluent enriched soil. Int. J. Curr. Microbiol. App. Sci. 3(2014) 318-325.

[7] J. G. Speight, Handbook of industrial hydrocarbon processes, first ed., USA,2011

[8] L. Andrady, M. A. Neal, Applications and societal benefits of plastics, Philos. Trans. R. Soc. Lond. B. Biol. Sci. 364(2009) 1977-1984. https://doi.org/10.1098/rstb.2008.0304

[9] M. Tosin, M. Weber, M. Siotto, C.Lott, F. Degli-Innocenti, Laboratory test methods to determine the degradation of plastics in marine environmental conditions, Front. Microbiol. 3 (2012)1-9. https://doi.org/ 10.3389/fmicb.2012.00225

[10] D. W. Laist, Overview of the biological effects of lost and discarded plastic debris in the marine environment, Mar. Pollut. Bull.18 (1987) 319-326. https://doi.org/10.1016/S0025-326X(87)80019-X

[11] A. T. Pruter, Sources, quantities and distribution of persistent plastics in the marine environment. Mar. Pollut. Bill. 18(1987) 305-310. https://doi.org/ 10.1016/S0025-326X(87)80016-4

[12]   C.J. Moore, S.L. Moore, M. K. Leecaster, S.B. Weisberg, A comparison of plastic and plankton in the North Pacific Central Gyre, Mar. Pollut. Bull. 42(2001) 1297–1300. https://doi.org/10.1016/s0025-326x(01)00114-x

[13]   J. G. Derraik, The pollution of the marine environment by plastic debris: a review, Mar. Pollut. Bull. 44(2002) 842-852. https://doi.org/ 10.1016/S0025-326X(02)00220-5

[14]   R. C. Thompson, Y. Olsen, R. P. Mitchell, A. Davis, S. J. Rowland, A. W. G. John, D. McGonigle, A. E. Russell, Lost at sea: where is all the plastic?, Science. 304(2004) 838-838. https://doi.org/10.1126/science.1094559

[15]   C. J. Moore, Synthetic polymers in the marine environment: a rapidly increasing, long-term threat, Environ. Res. 108 (2008)131-139. https://doi.org/10.1016/j.envres.2008.07.025

[16]   D. K. A. Barnes, F. Galgani, R. C. Thompson, M. Barlaz, Accumulation and fragmentation of plastic debris in global environments,Philos. Trans. R. Soc. Lond. B. Biol. Sci. 364 (2009) 1985–1998. https://doi.org/10.1098/rstb.2008.0205

[17]   M. R. Gregory, Environmental implications of plastic debris in marine settings—entanglement, ingestion, smothering, hangers-on, hitch-hiking and alien invasions, Philos. Trans. R. Soc. Lond. B. Biol. Sci. 364 (2009) 2013–2025. https://doi.org/ 10.1098/rstb.2008.0265

[18]   K. L. Law, S. Morét-Ferguson, N.A. Maximenko, G. Proskurowski, E. E. Peacock, J. Hafner , C. M. Reddy, Plastic accumulation in the North Atlantic Subtropical Gyre, Science.329(2010) 1185–1188. https://doi.org/10.1126/science.1192321

[19]   Regional consumption of plastic materials per capita 2015. https://www.statista.com/statistics/270312/consumption-of-plastic-materials-per-capita-since-1980/, 2016 (accessedMarch 2016).

[20]   Plastic pollution- Our world in data. https://ourworldindata.org/plastic-pollution, 2018 (accessed September 2018)

[21]   R. Chandra, Environmental waste management: The role of microbes in plastic degradation. first ed., India, 2015.

[22]   S. Goel, Advances in solid and hazardous waste management: degradation of plastics, first ed.,Capital Publishing Company, New Delhi, India, 2017.

[23]   The plastics industry in India: An overview. https://www.bpf.co.uk/article/the-plastics-industry-in-india-an-overview-446. aspx, 2010 (accessed 2010).

[24]   An Indian consumes 11kg plastic every year and an average American 109 kg. https://www.downtoearth.org.in/news/waste/an-indian-consumes-11-kg-plastic-every-year-and-an-average-american-109-kg-60745, 2018, (accessed June 2018).

[25]   Tackling increasing plastic waste- World Bank Group. http://datatopics.worldbank.org/whatawaste/tackling_increasing_plastic_waste.html, 2018, (accessed 2018).

[26] Beat plastic pollution: This world environment day.
https://www.unenvironment.org/interactive/beat-plastic-pollution/, 2018 (accessed 2018).

[27] M. Eriksen, L. C. Lebreton, H. S. Carson, M. Thiel, C. J. , Moore, J. C. Borerro, F. Galgani, P. G. Ryan, J. Reisser, Plastic pollution in the world's oceans: more than 5 trillion plastic pieces weighing over 250,000 tons afloat at sea, PlOS ONE. 9(2014). https://doi.org/10.1371/journal.pone.0111913

[28] S. Doetsch-Kidder, Social change and intersectional activism: The spirit of social movement, US, 2012.

[29] A.A. Shah, F. Hasan, A. Hameed, S. Ahmed, Biological degradation of plastics: A comprehensive review, Biotechnol. Adv.26 (2008) 246–265. https://doi.org/10.1016/j.biotechadv.2007.12.005

[30] B. Singh, N. Sharma, Mechanistic implications of plastic degradation, Polym. Degrad. Stab., 93(2008) 561-584. https://doi.org/10.1016/j.polymdegradstab.2007.11.008

[31] S. K. Kale, A. G. Deshmukh, M. S. Dudhare, V. B. Patil, Microbial degradation of plastic: a review, J. Biochem. Technol. 6 (2015) 952-961.

[32] Ejim, E. Patrick, J. Eze, Plastic pollution management: a panacea for nigeria's untapped waste to wealth growth; a study of some selected urban cities in south east nigeria-enugu, owerri, awka and umuahia, Adv. J. Man. Acc. Fin. 4 (2019) 21-30.

[33] T. Bond, V. Ferrandiz-Mas, M. Felipe-Sotelo, E. van Sebille, The occurrence and degradation of aquatic plastic litter based on polymer physicochemical properties: a review, Crit. Rev. Env. Sci. Tec. 48 (2018) 685-722. https://doi.org/10.1080/10643389.2018.1483155

[34] R. K. Singh, B. Ruj, Plastic waste management and disposal techniques-Indian scenario, Int. J. Plast. Technol. 19 (2015) 211-226. https://doi.org/10.1007/s12588-015-9120-5

[35] Y. Tokiwa, B. P. Calabia, C. U. Ugwu, S. Aiba, Biodegradability of plastics, Int. J. Mol. Sci. 10 (2009) 3722-3742. https://doi.org/10.3390/ijms10093722

[36] Y. Zheng, E. K. Yanful, A. S. Bassi, A review of plastic waste biodegradation, Crit. Rev. Biotechnol., 25 (2005) 243-250. https://doi.org/10.1080/07388550500346359

[37] A. Kulkarni, H. Dasari, Current status of methods used in degradation of polymers: A review, In. MATEC Web of Conferences. 144 (2018) 02023. https://doi.org/10.1051/matecconf/201814402023

[38] D. Hosler, S. L. Burkett, M. J. Tarkanian, Prehistoric polymers: rubber processing in ancient Mesoamerica, Science. 284 (1999) 1988-1991. https://doi.org/10.1126/science.284.5422.1988

[39] Celluloid. https://www.merriam-webster.com/dictionary/celluloid

[40]   S. Tripathi, A. Yadav, Bioremediation of industrial pollutants: plastic waste: envirnomental pollution, health hazards and biodegradtion strategies, New Delhi, India, 2016.

[41]   M. Raziyafathima, P. K. Praseetha, I. R. S. Rimal, Microbial degradation of plastic waste: a review, J. Pharm. Chem. Biol. Sci. 4 (2016) 231-242. https://doi.org/ 10.17352/ojeb.000010

[42]   R. Premraj, M. Doble, Biodegradation of polymer, Indian J. Biotechnol. 4 (2005) 186-193.

[43]   Plastics-the Facts 2019. An analysis of European plastics production, demand and waste data.

[44]   N. Pandey, C. Tewari, S. Dhali, B. S. Bohra, S. Rana, S. P. S. Mehta, S. Singhal, A. Chaurasia, N. G. Sahoo, Effect of graphene oxide on the mechanical and thermal properties of graphene oxide/hytrel nanocomposites, J. Thermoplast. Compos. Mater. (2019) 1-13. https://doi.org/ 10.1177/0892705719838010.

[45]   I. L. Nerland, C. Halsband, I. Allan, K. V. Thomas, Microplastics in marine environments: Occurrence, distribution and effects, (2014) 1-71.

[46]   D. Danso, J. Chow, W. R. Streita, Plastics: environmental and biotechnological perspectives on microbial degradation, Appl. Environ. Microbiol. 85 (2019) 1-14. https://doi.org/ 10.1128/AEM.01095-19

[47]   R. E. Hester, R. M. Harrison, Marine Pollution and Human Health, first ed., 2011

[48]   V. Koushal, R. Sharma, M. Sharma, R. Sharma, V. Sharma, Plastics: Issues challenges and remediation, Int. J. Waste Resources. 4 (2014) 2-6. https://doi.org/10.4172/2252-5211.1000134

[49]   P. Pavani, T. R. Rajeswari, Impact of plastics on environmental pollution, J. chem. pharm. sci. (2014) 87-93

[50]   R. Proshad, T. Kormoker, M. S. Islam, M. A. Haque, M. M. Rahman, M. M. R. Mithu, Toxic effects of plastic on human health and environment: A consequences of health risk assessment in Bangladesh, Int. J. Health Serv. 6 (2018) 1-5. https://doi.org/10.14419/ijh.v6i1.8655

[51]   R. Mohee, G. D. Unmar, A. Mudhoo, P. Khadoo, Biodegradability of biodegradable/degradable plastic materials under aerobic and an aerobic conditions, Waste Manag. 28 (2008) 1624- 1629. https://doi.org/10.1016/j.wasman.2007.07.003

[52]   J. P. Eubeler, S. Zok, M. Bernhard, T. P. Knepper, Environmental biodegradation of synthetic polymers I. Test methodologies and procedures, Trends Anal. Chem. 28 (2009) 1057-1072. https://doi.org/10.1016/j.trac.2009.06.007

[53]   S. M. Al-Salem, A. Al-Hazza'a, H. J. Karam, M. H. Al-Wadi, A. T. Al-Dhafeeri, A. A. Al-Rowaih, Insights into the evaluation of the abiotic and biotic degradation rate of commercial pro-oxidant filled polyethylene (PE) thin films, J Environ. Manage. 250 (2019). https://doi.org/10.1016/j.jenvman.2019.109475

[54]   J. Pospisil, S. Nespurek, R. Pfaendner, H. Zweifel, Material recycling of plastics waste for demanding applications: upgrading by restabilization and compatibilization. Trends Polym. Sci. 5 (1997) 294-300.

[55]   J. A. Glaser, Biological degradation of polymers in the environment, Intech Open. 2019. https://doi.org/ 10.5772/intechopen.85124

[56]   A. Sivan, New perspectives in plastic biodegradation, Curr. Opin. Biotech. 22(2011) 422-426. https://doi.org/10.1016/j.copbio.2011.01.013

[57]   P. K. Roy, M. Hakkarainen, I. K. Varma, A. C. Albertsson, Degradable Polyethylene: Fantasy or Reality, Environ. Sci. Technol. 45 (2011) 4217–4227. https://doi.org/10.1021/es104042f

[58]   Y. G. Ángeles-López, A. M. Gutiérrez-Mayen, M. Velasco-Pérez, M. Beltrán-Villavicencio, A. Vázquez-Morillas, M. Cano-Blanco, Abiotic degradation of plastic films, J. Phys. Conf. Ser, 792 (2017).

[59]   P. Nayak A. Tiwari, Biodegradability of polythene and plastic by the help of microorganism: A way for brighter future, J. Environ Anal. Toxicol.1 (2011). https://doi.org/10.4172/2161-0525.1000111

[60]   J. D. Gu, Microbiological deterioration and degradation of synthetic polymeric materials: recent research advances, Int. Biodeter. Biodegr. 52 (2003) 69-91. https://doi.org/10.1016/S0964-8305(02)00177-4

[61]   P.S. Chahal, D. S. Chahal, G. André, Cellulase production profile of Trichodermareeseion different cellulosic substrates at various pH levels, JFerment. Bioeng. 74 (1992) 126–128. https://doi.org/10.1016/0922-338X(92)80015-B

[62]   J. D. Gu, R. Mitchell, Biodeterioration, in M. Dworkin, S. Falkow, E. Rosenberg, K. H. Schleifer, E. Stackebrandt (Eds.), The Prokaryotes, Springer, New York, 2006, pp.864-903.

[63]   S. K. Mohan, T. Srivastava ,Microbial deterioration and degradation of polymeric materials, J. Biochem. Technol. 2(2011) 210–215. https://doi.org/10.12691/jaem-5-1-2

[64]   A. Kulkarni, H. Dasari, Current status of methods used in degradation of polymers: A review, MATEC Web of Conf. 144 (2018) 02023. https://doi.org/10.1051/matecconf/201814402023

[65]   I. Kyrikou, B. Briassoulis, Biodegradation of agricultural plastic films: a critical review, J. Polym. Environ.15(2007) 125-150. https://doi.org/10.1007/s10924-007-0053-8

[66]   B. Rånby, Photodegradation and photo-oxidation of synthetic polymers, J. Ana.l Appl. Pyrol. 15 (1989) 237-247. https://doi.org/10.1016/0165-2370(89)85037-5

[67]   J. T. Jensen, J. Kops, Photochemical degradation of blends of polystyrene and poly (2, 6-dimethyl-1, 4-phenylene oxide), J. Polym. Sci. Polym. Chem. 18 (1980) 2737-2746. https://doi.org/10.1002/pol.1980.170180830

[68]   G. E. Sheldrick, O. Vogl, Induced photodegradation of styrene polymers: a survey, J.Polym. Eng. Sci. 16 (2004) 65- 73. https://doi.org/10.1002/pen.760160202

[69]   J. W. Martin, J. W. Chin, T. Nguyen, Reciprocity law experiments in polymeric photodegradation: a critical review, Prog. Org. Coat. 47 (2003) 292-311. https://doi.org/10.1016/j.porgcoat.2003.08.002

[70]   J. Czerný, Thermo-oxidative and photo-oxidative aging of polypropylene under simultaneous tensile stress, J. Appl. Polym. Sci. 16 (1972) 2623-2632. https://doi.org/10.1002/app.1972.070161015

[71]   M. Obadal, R. Čermák, M. Raab, V. Verney, S. Commereuc, F. Fraïsse, Study on photodegradation of injection-moulded β-polypropylenes, Polym. Degrad. Stab. 91 (2006) 459-463. https://doi.org/10.1016/j.polymdegradstab.2005.01.046

[72]   S. H. Hamid, W. H. Prichard, Mathematical modeling of weather-induced degradation of polymer properties, J. Appl. Polym. Sci., 43 (1991) 651-678. https://doi.org/10.1002/app.1991.070430404

[73]   A. Marek, L. Kaprálková, P. Schmidt, J. Pfleger, J. Humlíček, J. Pospíšil, J. Pilař, Spatial resolution of degradation in stabilized polystyrene and polypropylene plaques exposed to accelerated photodegradation or heat aging, Polym. Degrad. Stab. 91(2006) 444-458. https://doi.org/10.1016/j.polymdegradstab. 2005.01.048

[74]   A. L. Andrady, J. E. Pegram, Y. Tropsha, Changes in carbonyl index and average molecular weight on embrittlement of enhanced-photodegradable polyethylenes, J. Environ. Polym. Degrad. 1 (1993) 171-179. https://doi.org/10.1007/BF01458025

[75]   A. Ghaffar, A. Scott, G. Scott, G., The chemical and physical changes occurring during UV degradation of high impact polystyrene, Eur. Polym. J., 11 (1975) 271-275. https://doi.org/10.1016/0014-3057(75)90075-0

[76]   Y. Nagai, T. Ogawa, Y. Nishimoto, F. Ohishi, Analysis of weathering of a thermoplastic polyester elastomer II. Factors affecting weathering of a polyether–polyester elastomer, Polym. Degrad. Stab. 65 (1999) 217-224. https://doi.org/10.1016/S0141-3910(99)00007-5

[77]   A. Torikai, Wavelength sensitivity of photodegradation of polymer, in: S. H. Hamid (Eds.), Handbook of polymer degradation, New York: Markel Dekker; 2000, pp. 573- 604

[78]   K. Yamada-Onodera, H. Mukumoto, Y. Katsuyaya, A. Saiganji, Y. Tani, Degradation of polyethylene by a fungus, Penicilliumsimplicissimum YK, Polym. Degrad. Stab. 72 (2001) 323-327. https://doi.org/10.1016/S0141-3910(01)00027-1

[79]   Y. Zheng, E. K. Yanful, A. S. Bassi, A review of plastic waste biodegradation, Crit. Rev. Biotechnol.,25 (2005) 243-250. https://doi.org/10.1080/07388550500346359

[80]   A. L. Andrady, Microplastics in the marine environment, Mar. Pollut. Bull., 62 (2011) 1596-1605. https://doi.org/10.1016/j.marpolbul.2011.05.030

[81]   J. M. Raquez, A. Bourgeois, H. Jacobs, P. Degée, M. Alexandre, P. Dubois, Oxidative degradations of oxodegradable LDPE enhanced with thermoplastic pea starch: Thermo-mechanical properties, morphology, and UV-ageing studies, J. Appl. Polym. Sci. 122 (2011) 489-496. https://doi.org/10.1002/app.34190

[82]   R. J. Müller, I. Kleeberg, W. D. Deckwer, Biodegradation of polyesters containing aromatic constituents, J. Biotechnol. 86 (2001) 87-95. https://doi.org/10.1016/S0168-1656(00)00407-7

[83]   Y. Nagai, D. Nakamura, T. Miyake, H. Ueno, N. Matsumoto, A. Kaji, Photo degradation mechanisms in poly(2,6-butylenenaphthalate-co tetramethyleneglycol) (PBN–PTMG). II: wavelength sensitivity of the photodegradation, Polym. Degrad. Stab.88 (2005) 251-255

[84]   O. Chiantore, L. Trossarelli, M. Lazzari, Photooxidative degradation of acrylic and methacrylic polymers, Polymer, 41 (2000) 1657-1668. https://doi.org/10.1016/S0032-3861(99)00349-3

[85]   Thermal Degradation of plastic. http://www.appstate.edu/~clementsjs/polymer properties/zeus_thermal_degradation.pdf, 2005 (accessed 2005).

[86]   MSDS- Nylon 6. http://skipper.physics.sunysb.edu/HBD/MSDS/NylonMSDS. pdf, 2001. (accessed 2001).

[87]   D. R. Tayler, Mechanistic aspects of the effect of stress on the rate of photochemical degradation reactions in polymers, J Macromol. Sci. Part C. Polym. Rev. 44 (2004) 351-388. https://doi.org/10.1081/MC-200033682

[88]   F. Khabbaz, A. C. Albertsson, S. Karlsson, Chemical and morphological changes of environmentally degradable polyethylene films exposed to thermo-oxidation. Polym. Degrad. Stab. 63(1999) 127-138. https://doi.org/10.1016/S0141-3910(98)00082-2

[89]   H. V. Boenig, Unsaturated polyesters: structure and properties, London, New York, 1964.

[90]   J. Li, S. Guo, X. Li, Degradation kinetics of polystyrene and EPDM melts under ultrasonic irradiation, Polym. Degrad. Stab. 89 (2006) 6-14. https://doi.org/10.1016/j.polymdegradstab.2004.12.017

[91]   K. Baranwal, Mechanochemical degradation of an EPDM polymer, J. Appl. Polym. Sci. 12 (2003) 1459-1469. https://doi.org/10.1002/app.1968.070120617

[92]   L. C. Bateman, Chemistry and physics of rubber-like substances, first ed., (1963)

[93]   P. Ghosh, Polymer science and technology of plastics, rubbers, blends and composites, third ed.,New Delhi,1990.

[94]   G. Schmidt-Naake, M. Drache, M. Weber, Combination of Mechanochemical Degradation of Polymers with Controlled Free-Radical Polymerization, Macromol. Chem. Phys. 203(2002) 2232-2238. https://doi.org/10.1002/1521-3935(200211)203:15<2232::AID-MACP2232>3.0.CO;2-N

[95]   Y. H. Lin, H. Y. Yen, Fluidised bed pyrolysis of polypropylene over cracking catalysts for producing hydrocarbons, Polym. Degrad. Stab. 89 (2005) 101-108. https://doi.org/10.1016/j.polymdegradstab.2005.01.006

[96]   P. T. Williams, R. Bagri, Hydrocarbon gases and oils from the recycling of polystyrene waste by catalytic pyrolysis, Int. J. Energy Res. 28 (2003) 31-44. https://doi.org/10.1002/er.949

[97]   J. R. Kim, J. H. Van, D. W. Park, M. H. Lee, Catalytic degradation of mixed plastics using natural clinoptilolite catalyst. React. Kinet. Catal. Lett. 81 (2004) 73-81. https://doi.org/10.1023/B:REAC.0000016519.59458.08

[98]   W. Kaminsky, F. Hartmann, New pathways in plastics recycling, Angew. Chem. Int. Ed., 39 (2000) 331-333. https://doi.org/10.1002/(SICI)1521-3773(20000117)39:2%3C331::AID-ANIE331%3E3.0.CO;2-H

[99]   X. Jia, C. Qin, T. Friedberger, Z. Guan, Z. Huang, Efficient and selective degradation of polyethylenes into liquid fuels and waxes under mild conditions, Sci. Adv., 2 (2016) 1501-1591. https://doi.org/ 10.1126/sciadv.1501591

[100]   D. C. Tiwari, E. Ahmad, K. K. Singh, Catalytic degradation of waste plastic into fuel range hydrocarbons, Int. J. Chem. Res. 1 (2009) 31-36. https://doi.org/ 10.9735/0975-3699.1.2.31-36

[101]   Y. Zhang, M. A. Nahil, C. Wu, P. T. Williams, Pyrolysis–catalysis of waste plastic using a nickel–stainless-steel mesh catalyst for high-value carbon products, Environ. Technol. 38 (2017) 2889-2897. https://doi.org/10.1080/09593330.2017.1281351

[102]   V. Sinha, M. R. Patel, J. V. Patel, PET waste management by chemical recycling: a review. J. Polym. Environ.18 (2010) 8-25. https://doi.org/10.1007/s10924-008-0106-7

[103]   D. Simón, A. M. Borreguero, A. De Lucas, and J. F. Rodríguez, Recycling of polyurethanes from laboratory to industry, a journey towards the sustainability, J. Waste Manag.76 (2018) 147-171. https://doi.org/ 10.1016/j.wasman.2018.03.041

[104]   W. Yang, Q. Dong, S. Liu, H. Xie, L. Liu, J. Li, Recycling and disposal methods for polyurethane foam wastes, Procedia Environ. Sci. 16 (2012) 167-175. https://doi.org/10.1016/j.proenv.2012.10.023

[105]   S. Thomas, A. V. Rane, K. Kanny, V. K. Abitha, M. G. Thomas, Recycling of Polyurethane Foams. William Andrew, first ed., 2018

[106]   F. Cataldo, G. Angelini, Some aspects of the ozone degradation of poly (vinyl alcohol), Polym. Degrad. Stab. 91 (2006) 2793-2800. https://doi.org/10.1016/ j.polymdegradstab.2006.02.018.

[107]   S. Chattopadhyay, G. Madras, Kinetics of the enzymatic degradation of poly(vinyl acetate) in solution, J. Appl. Polym. Sci. 89 (2003) 2579-2582. https://doi.org/10.1002/app.12403

[108]   M. Vaverková, D. Adamcová, J. Kotovicová, F. Toman, Evaluation of biodegradability of plastics bags in composting conditions, Ecol. Chem. Eng., S , 21 (2014) 45-57. https://doi.org/10.2478/eces-2014-0004

[109]   Y. Tokiwa, T. Suzuki, Hydrolysis of Polyesters by Rhizopusdelemar Lipase, Agric. Biol. Chem., 42 (1978) 1071- 1072. https://doi.org/10.1271/bbb1961.42.1071

[110]   F. Kawai, M. Watanabe, M. Shibata, S. Yokoyama, Y. Sudate, S. Hayashi, Comparative study on biodegradability of polyethylene wax by bacteria and fungi, Polym. Degrad. Stab. 86 (2004) 105-114. https://doi.org/10.1016/j.polymdegradstab.2004.03.015

[111]   M. Mierzwa-Hersztek, K. Gondek, M. Kopeć, Degradation of polyethylene and biocomponent-derived polymer materials: An Overview, J Polym. Environ. 27 (2012) 600-611. https://doi.org/10.1007/s10924-019-01368-4

[112]   I. Jakubowicz, Evaluation of degradability of biodegradable polyethylene (PE), Polym. Degrad. Stabil. 80 (2003) 39-43. https://doi.org/10.1016/S0141-3910(02)00380-4

[113]   R. K. Singh, B. Ruj, Plasticwaste management and disposal techniques-Indian scenario, Int. J. Plast. Technol. 19 (2015) 211-226. https://doi.org/10.1007/s12588-015-9120-5

[114]   B. Tansel, B.S. Yildiz, Goal-based waste management strategy to reduce persistence of contaminants in leachate at municipal solid waste landfills, Environ. Dev. Sustain. 13(2011). 821-831. https://doi.org/10.1007/s10668-011-9290-z

[115]   E.W. Tollner, P.A. Annis, K.C. Das, Evaluation of strength properties of polypropylene-basedpolymers in simulated landfill and oven conditions,J. Environ. Eng. 137 (2011) 291–296

[116]   N. Miskolczi, L. Bartha, A. Angyal, High energy containing fractions from plastic wastes by their chemical recycling, Macromol. Symp. 245-246 (2006) 599-606. https://doi.org/ 10.1002/masy.200651386

[117]   E. Butler, G. Devlin, K. McDonnell,Waste polyolefins to liquid fuels via pyrolysis: review ofcommercial state-of-the-art and recent laboratory research, Waste Biomass Valor., 2 (2011) 227-255. https://doi.org/10.1007/s12649-011-9067-5

[118]   B. Herman, R. Biczak, P. Rychter, M. Kowalczuk, Degradation of selected synthetic polyesters under industrial composting: impact on compost properties and phytotoxicity, Proceedings of ECOpole, 4 (2010) 133-140

[119]   K. Azim, B. Soudi, S. Boukhari, C. Perissol, S. Roussos, I. T. Alami, Composting parameters and compost quality: a literature review, Org. Arg.8(2018) 141-158. https://doi.org/10.1007/s13165-017-0180-z

[120]   T. Ishigaki, W. Sugano, A. Nakanishi, M. Tateda, M. Ike, M. Fujita, The degradability of biodegradable plastics in aerobic and anaerobic waste landfill model reactors, Chemosphere, 54 (2004) 225-233. https://doi.org/10.1016/S0045-6535(03)00750-1

[121] D. Y. Kim, Y. H. Rhee, Biodegradation of microbial and synthetic polyesters by fungi, Appl. Microbiol. Biotechnol. 61 (2003) 300-308. https://doi.org/10.1007/s00253-002-1205-3

[122] D. Adamcová, M. D.Vaverková, S. Hermanová, and S. Voběrková, Ecotoxicity of composts containing aliphatic-aromatic copolyesters, Pol. J. Environ. Stud. 24 (2015)1497-505. https://doi.org/10.15244/pjoes/31227

[123] V. Koushal, R. Sharma, M. Sharma, R. Sharma, and V. Sharma, Plastics: issues challenges and remediation, International Journal of Waste Resources, 4 (2014) 2-6. https://doi.org/ 10.4172/2252-5211.1000134

[124] M. A. L. Russo, C. O'Sullivan, B. Rounsefell, P. J. Halley, R. Truss, W. P. Clarke, The anaerobic degradability of thermoplastic starch: Polyvinyl alcohol blends: Potential biodegradable food packaging materials, Bioresour. Technol. 100 (2009) 1705–1710. https://doi.org/10.1016/j.biortech.2008.09.026

[125] M. Nitschke, S.G. Costa, and J. Contiero, Rhamnolipids and PHAs: Recent reports on Pseudomonas-derived molecules of increasing industrial interest, Process Biochemistry. 46 (2011) 621-630. https://doi.org/10.1016/j.procbio.2010.12.012

[126] C. Zhuo, Y. A. Levendis, Upcycling waste plastics into carbon nanomaterials: A review, J. Appl. Polym. Sci., 131(2014) 1-14. https://doi.org/10.1002/app.39931

[127] S. Pandey, M. Karakoti, S. Dhali, N. Karki, B. SantiBhushan, C. Tewari, S. Rana, A. Srivastava, A. B. Melkani, N. G. Sahoo, Bulk synthesis of graphene nanosheets from plastic waste: An invinciblemethod of solid waste management for better tomorrow, J. Waste Manag. 88 (2019) 48-55. https://doi.org/ 10.1016/j.wasman.2019.03.023

[128] L. N. Rao, J. L. Jayanthi, D. Kamalakar , Conversion of waste plastics into alternative fuel, International journal of engineering sciences & research technology. (2015) 195-201

Degradation of Plastics   Materials Research Forum LLC
Materials Research Foundations **99** (2021) 37-80   https://doi.org/10.21741/9781644901335-2

Chapter 2

Biodegradable Plastics from Renewable Raw Materials

Pravin D. Patil[1*]; Manishkumar S. Tiwari[2]; Vivek P. Bhange[3]

[1]Department of Basic Science and Humanities, Mukesh Patel School of Technology Management and Engineering, SVKM's NMIMS University, Mumbai, 400056, Maharashtra, India

[2]Department of Chemical Engineering, Mukesh Patel School of Technology Management and Engineering, SVKM's NMIMS University, Mumbai, 400056, Maharashtra, India

[3]Department of Biotechnology, Priyadarshini Institute of Engineering and Technology, Nagpur, 440019, Maharashtra, India

* dr.pravinpatil.ict@gmail.com; pravin.patil@nmims.edu

Abstract

Fossil oil prices are soaring steeply due to the depleting petroleum raw materials. Extensive research has been carried out around the globe to develop efficient processes that can replace oil-derived polymers (conventional plastic) with bio-based polymers that originate from renewable resources. Fossil-oil based plastic products take decades to degrade, leading to the unwanted accumulation of plastic waste that can be seen all around. Further, greenhouse gases emission occurs during the production and destruction of synthetic plastic. Therefore, plastic waste has become a massive threat to the biosphere and needs to be addressed immediately. To overcome this issue, a new type of plastic can be produced from bio-resources that can fulfill even the energy demand in today's world. This new form of plastic must be accommodated fast in daily life, considering the range of applications of plastics. Biodegradable plastics made from renewable raw materials can retain all the benefits of petroleum-based plastic without having any negative impacts on the environment. Bioplastics are not toxic in nature and can easily decay back into carbon dioxide via degradation. The products made from bioplastics may be commercialized, considering their superior properties over conventional plastic. The discovery and implementation of plastic made from renewable raw material resources could be a giant leap into the sustainable future.

Degradation of Plastics
Materials Research Foundations **99** (2021) 37-80

Materials Research Forum LLC
https://doi.org/10.21741/9781644901335-2

Keywords

Bio-Degradable Plastics, Biopolymer, Bio-Based Polymers, Renewable Resources

Contents

Materials Research Forum LLC

https://doi.org/10.21741/9781644901335-2

1.   Introduction

Products made from plastic have been widely used by humankind on a daily basis, serving the non-lasting demands of our fast-paced lifestyle. Society demands a considerable amount of plastic for the manufacturing of millions of products. Though plastic has numerous advantages over conventional materials, it also poses a massive threat to the environment since the degradation period of plastic products can vary from 20 to 500 years. Despite the fact that plastic can be recycled and reused, the high cost of the process and the confined market for recycled products, deposition of solid waste of plastics can be seen all around. Nearly 300 million tons of plastic are produced worldwide every year [1]. The accumulation of synthetic plastic has been continuously rising around the globe and has emerged as a crucial environmental issue that needs to be addressed immediately. Moreover, the production of plastic also consumes a high proportion of oil, a non-renewable resource that is running out of supply. Most of the plastic originates from oil and is composed of synthetic polymers; however, polymer chains can also be obtained from natural resources.

The idea of having material derived from nature with environmental interests of being biobased and/or biodegradable is a lucrative choice to the consumer and industry. From an environmental point of view, natural renewable resources could be employed for the synthesis of plastic while producing plastic material that is biologically degradable at specific conditions of humidity and temperature [1]. On the one hand, it can reduce the consumption of a dwindling supply of fossil resources while reducing the emission of greenhouse gases, providing an eco-friendly alternative to the traditional plastic that can ultimately reduce the negative impact on the environment. Natural renewable resources could be a plant, animal, or microbial biomass that is being used for the production of material or energy apart from food and feed area [2]. Frequent use of natural renewable resources can also boost the field of forest and agriculture while leading to alterations in bio-based chemicals and materials. However, it is imperative that these renewable raw materials must be employed sustainably in a way that they set a minimum impact on the soil, water, air, and climate.

Since the last few decades, bioplastic has been recognized as an essential invention that serves the plastics and chemical industry by presenting numerous opportunities [3]. However, bio-based plastic only endures just over 1% of the total global plastic market, which is still a tiny fraction [4]. Nevertheless, the production of plastic has been rapidly replaced by bio-plastics such as polyethylene, polyhydroxyalkanoate, polylactic acid, etc [5]. Natural renewable resources can be employed in the production of biopolymers that comprise in large part, or even wholly, renewable resources. The biopolymers can be derived from various sources, including vegetable, animal, and microbial biomass/products. Vegetable origin can offer biopolymer synthesis from starch, cellulose, lignin, chitosan, pectin, zein, etc. In animals, whey protein, gelatin, and casein are the most popular sources; whereas, polyhydroxy-butyrate and polyhydroxy-valerate are widely popular in microbial origin [6,7]. Since its discovery, biopolymers have been an integral element of human development and served in the field of food, medical, packaging, food additives, and water treatment, among many others [8,9]. However, the immerging generation of biopolymers is still in its infant stage that needs several developments considering the demand for the environment-friendly alternative of synthetic plastic [10,11].

2. Types of bioplastics

Bioplastics can be a group of different materials; based on biodegradable ability, type of biomass used, or a combination of both (Fig. 1). For instance, there are forms of biobased materials that are not biodegradable and fossil-based materials that are biodegradable. Moreover, bioplastics can also be found in a multi-layered form blended with fossil-based plastics [12]. Bio-based plastics can be classified as novel or drop-in plastic. A few forms of bioplastics are completely novel materials such as polyethylene furanoate (PEF) and polylactic acid (PLA), unlike drop-in plastics that have a similar structure to their fossil-based counterparts [13]. Bioplastics can be further categorized based on their source and degradability. The following are the three categories of bio-plastics: a) bio-based, b) bio-degradable, and c) bio-based and bio-degradable.

2.1 Bio-based plastics

The bio-based plastics originate from natural renewable resources such as agriculture-based raw material, hemp, sugarcane, potatoes, corn starch, etc. These resources are abundant in nature and help to reduce the emission of CO_2 by replacing traditional plastic. Some examples include polyamide (PA), polyethylene terephthalate (PET), and polyethylene (PE) [5].

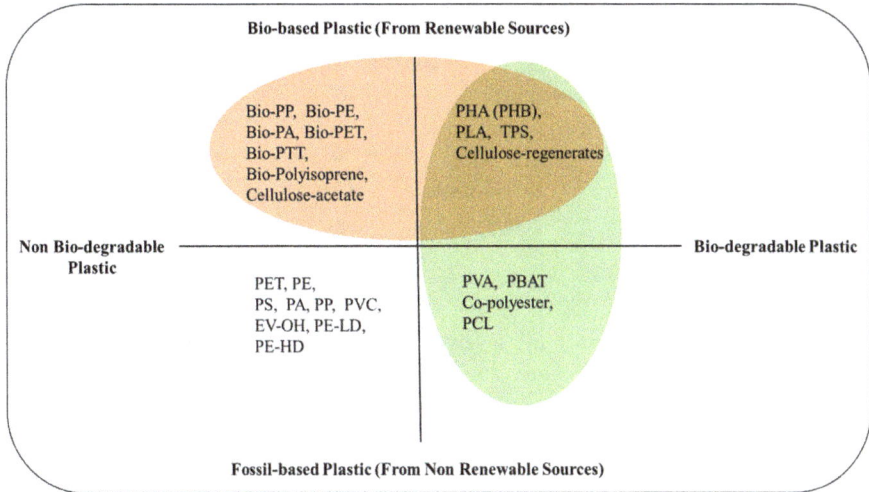

Bio-based Plastic (From Renewable Sources)

Bio-PP, Bio-PE, Bio-PA, Bio-PET, Bio-PTT, Bio-Polyisoprene, Cellulose-acetate

PHA (PHB), PLA, TPS, Cellulose-regenerates

Non Bio-degradable Plastic ——————————————————— **Bio-degradable Plastic**

PET, PE, PS, PA, PP, PVC, EV-OH, PE-LD, PE-HD

PVA, PBAT Co-polyester, PCL

Fossil-based Plastic (From Non Renewable Sources)

Fig. 1 *Material coordinate system of bioplastics*

Polyethylene

Polyethylene (PE) derived from natural renewable resources is flexible and brittle/rigid in nature, showing identical physical and chemical properties when compared to conventional polyethylene. Several agricultural feedstocks can be used in the synthesis of bio-based polyethylene, including corn, sugarcane, etc. It was also observed that the emission of CO_2/kg was far less when compared to the traditional form of polyethylene [14]. Biobased PE can serve a similar purpose by exhibiting similar properties as the conventional form. Additionally, bio-based PE has demonstrated better durability that makes it an obvious choice over conventional PE [5,15].

Polyamide

Vegetable oil is a source to derive another form of bioplastic, polyamide (PA). Castor beans have proved an efficient source for the production of PA; however, it lacks bio-degradability. Biobased PA has shown several better features when compared to conventional PA, such as low cost, superior appearance with better dimensional stability while providing superior water, chemical, and thermal resistance. Moreover, it also records a little environmental impact over conventional form. Bio-based PA is mainly being employed in packaging, automobile, and electrical industries, where durability, safety, and versatility are significant concerns for manufacturers [5,16].

2.2   Bio-degradable plastics

Bio-degradable plastic is a structure that originates from traditional raw materials and can be wholly degraded in the environment without employing any additional additives. This is mostly used as a one-time packaging material that can be disposed of after its use. It may take degradation time ranging from a few days to months, depending on the environmental conditions. Polycaprolactone (PCL) and poly-butyrate adipate-terephthalate (PBAT) are among many of these types of bioplastics available in the market [5,14].

Polycaprolactone (PCL)

Despite originating from crude oil, a few forms of plastic still show biodegradability. PCL is one such type that is not derived from a renewable resource and still demonstrates several attractive properties such as superior oil, water, and solvent resistance. It also has a lower melting point ranging between 58-60°C. PCL is mostly used for shorter life applications, and its lower melting point makes it easily amiable to composting as a means of disposal with a low degradation period [5,17].

2.3   Bio-based and bio-degradable plastics

A bioplastic that possesses properties of both bio-based and biodegradable plastics can also be used as an efficient means of biobased polymers. Polyhydroxylalkanoate (PHA) and poly lactic acid (PLA) are among many types of these plastics.

Poly lactic acid

Poly lactic acid (PLA) is a bio-based polymer mainly derived from agricultural resources (tapioca roots, potato, sugar cane, chips, etc.) and can be biologically degraded in the environment. PLA is the second most consumable plastic in the modern age due to its easy degradability feature that makes it an excellent candidate, considering the unfavorable impact of traditional plastic on the environment. Moreover, it has reduced carbon footprints, where it only produces 1.8 kg CO_2 compared to traditional plastic (6 kg). PLA is mostly used in products meant for longer period usage [5,13].

Polyhydroxylalkanoates

Polyhydroxylalkanoates (PHA) is another type of plastic that exhibits properties of both bio-based and bio-degradable plastic. It is a UV stable linear polyester generated by fermentation of sugar and lipids using bacterial strains. Agricultural raw materials are mainly used during the production of PHA using conventional processing tools. The bio-compostable and mechanical properties of PHA can be altered by blending it with other types of plastics, consequently widening its domain of application [5,16].

3. Advantages of bio-plastics

Bio-plastics have emerged as a promising alternative to petroleum-based plastics due to its remarkable features that favor numerous industrial applications. Global awareness concerning global warming that transpired due to severe pollution has led consumers and industries to reconsider their choices. Bioplastic is an obvious choice since it outperforms traditional plastics in several areas. Several features of bioplastics are as following; (1) Unlimited raw material: agricultural raw materials are abundant in nature and can be obtained on a yearly basis. Unlike the limited supply of fossil oil, natural renewable resources can be used for the production of plastic without worrying about its depletion, (2) Bio-degradable: bio-based polymers require less time to get entirely degraded after being discharged as waste (Fig. 2). Therefore, it does not demand any recycling, leading to less or no requirement of landfilling when compared to conventional plastic, (3) Requires less energy for production: the production process of bioplastic demands less than half of the energy when compared to the conventional form of plastics, (4) Reduces dependency on foreign oil: It also reduces the import of fossil oil that is the primary raw material for the production of conventional plastic, (5) Non-toxic in nature: since conventional plastic is composed of several harmful chemicals [18] and it releases several toxic gases during its breakdown, making it harmful to the environment. On the contrary, bioplastics are entirely safe and do not require or produce hazardous chemicals/gases during their synthesis/break down, and (6) Bio-plastic is eco-friendly in nature: there are very few greenhouse gases, and less carbon emission generated during the production and incineration process of bio-plastics. Table 1 [19-59] depicts the different renewable sources for bioplastic production along with their respective applications.

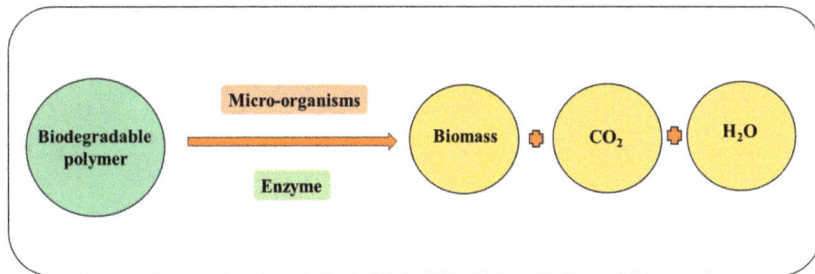

Fig. 2 *Degradation process of bio-based polymers*

Table 1 *Different renewable sources for bioplastic production and their respective applications [19-59].*

Sr. No.	Source material	Material	Techniques	Properties	Applications	Ref.
1.	Potato peels, banana peels, cassava, wheat, Maize	Starch	Polymerizatio, casting,	Short-living, biodegradable and Compostable	Food packaging, medical devices, agriculture foils, textiles, automotive and transport, building and construction	[19–22]
2.	Bunch, citrus waste, corn leaf, oil palm fruit, rice straw	Cellulose	Polymerizatio, casting	Biodegradable, limited compostable,	Car interiors, construction materials, toys, medical applications, sports equipment, decor materials, etc	[23–26]
3.	Plant sources: Rapeseed oil, mesocarp fiber, corn; Animal sources	Protein	Injection moulding, casting	Biodegradable	Toys, sports gear, containers, packaging materials	[27–29]
4.	Wood, straws	Lignin	Injection moulding, plasticizer	Biodegradable		[30,31]
5.	Fungi, yeast	Chitin	Casting, moulding	Biodegradable, bio compostable	Thermoplastic material, Biomedical applications, packaging materials	[32]
6.	Shrimp, crab, mushrooms	Chitosan	Casting, moulding	Biodegradable, bio compostable	Biomedical applications, packaging materials	[33]
7.	Prokaryotic organisms	Polyhydroxy-alkanoates (PHA)	Fermentation, casting, evaporation.	Long-living, depending on composition fast biodegradable to non-biodegradable, limited compostable	Coating, food packaging, medical implant	[34,35]

Materials Research Forum LLC
https://doi.org/10.21741/9781644901335-2

8.	Lactic acid, paper waste, commercial polylactic acid	Polylactic acid (PLA)	Casting	Weatherproof, biodegradable, not compostable	Films, food packaging	[36–38]
9.	Bio-MEG, bagasse, hay, and sugar cane molasses	Polyethylene terephthalate (PET)	Polycondesnation	Long-living, biodegradable	Soda and drinking water bottles, automotive interiors, packaged goods, electronics and construction goods,	[31,39]
10.	Bio-MEG and 2,5-furan dicarboxylic acid (FDCA) from lignocellulose biomass	Polyethyle-ne furanoate (PEF)	Polycondesnation	Non-biodegradable and non-compostable	Packaging industry for milk, soft drink, water, and alcoholic beverages	[31,40]
11.	1,3-Propanediol from glycerol, vegetable oil	Polytrimethyl terephthalate (PTT)	Polycondesnation	Non-biodegradable and Non-compostable	Fiber manufacturing, textiles, and carpet synthesis	[31,41,42]
12	Carboxylic acid from lignocellulose,	Alkyds Resin	Polycondesnation	Biodegradable, biocompatible	Paints, lacquers, production of printing inks, adhesives, insulating	[43,44]
13	Butanediol and succinic acids from starch	Polysuccinate	Polycondesnation	Biodegradable, biocompatible	Fiber spinning, film blowing, and thermoforming	[45,46]
14	Butanediol from starch, vegetable oil	Polyesters	Polycondesnation	Biodegradable, biocompatible	Adhesives, casting material, insulating materials, and printing inks	[47–49]
15	Castor oil	Polyamides	Step-growth polymerization	Depending on composition, biodegradable to non-biodegradable	pneumatic air brake tubes, flexible oil and gas pipes, electrical cable jackets, and automotive fuel lines	[50–52]
16	Plant oils such as soya or castor oil, vegetable oils	Polyurethane	Polycondesnation	Biodegradable	Cushions, coatings, and insulations	[53,54]

17	Glycerol, fatty acids, or sugar	Polyacrylates	Polymerization	Non-biodegradable	Thickener, pigment dispersants	[55,56]
18	Sugarcane, vegetable oil	Polyolefins	Polymerization, cracking	Depending on composition fast biodegradable to non-biodegradable	Bags, pouches, petrol canisters, barrels, packaging, textiles	[57,58]
19	Vinyl chloride, ethylene from fermentation	Polyvinyl Chloride	Casting	Non-biodegradable and non-compostable	Wires, films, cables, pipes, and bottles	[17,58]
20	Vegetable oil, lignin, rosin, furan, sugars	Epoxy Resins	Ring-opening polymerization	Biodegradable	Adhesives, coatings, aerospace, and electronics industry.	[59]

4. Overview of renewable raw materials

Manufacturing of materials, chemicals, and other bio-based products at an industrial scale using renewable raw materials as a feedstock has gained attention since it helps to conserve fossil resources while making the process environmentally friendly. In comparison to the products based on fossil resources, bio-based products have shown excellent greenhouse gas balance over the complete life cycle and have less toxicity while requiring less energy for production as well as for disposal. A limited supply of fossil resources has driven innovation for the use of renewable raw materials as resources. A wide range of renewable raw materials, basically plant-based raw material, can be used for the synthesis of bio-based plastics. These raw materials can be classified as first, second, and third-generation feedstocks. All these resources can be procured, modified, and processed to obtain bio-based plastics. Today, biobased plastics are mainly synthesized from starch, sugar, plant oil belonging to the first-generation feedstock. In recent years, due to the reasonable competition with food and animal feed, lignocellulose feedstock has gathered significant attention towards its utilization to produce fermentable sugar.

First-generation feedstock

The first-generation feedstock includes carbohydrate abundant-plants such as sugarcane or corn that are suitable for human as well as animal consumptions. They are the most efficient feedstock to produce bioplastic since their cultivation needs less land and provides a high amount of yields. The efficiency can be measured based on the annual yield of carbohydrates per hectare and the area (land) used per ton of bioplastics produced. The first-generation feedstocks include crops such as sugarcane, wheat, corn, potato, sugar beet, plant oil, rice, etc.

Second generation feedstock

The plant and crops that are not suitable for animal consumption (feed) and human consumption (food) are considered second-generation feedstocks. They can be non-food crops (cellulose feedstock, chitin, etc.) or waste material from first-generation feedstocks (e.g., waste vegetable oil). Lignocellulose biomass comprising a composition of cellulose hemicellulose and lignin are the most used second-generation feedstock. Lignocellulose biomass includes short -rotation coppice such as willow, miscanthus, poplar, or agricultural waste (by-products). Additionally, by-products from the forest or other biomass waste containing sugar and starch such as corncobs, bagasse, wheat straw, palm fruit bunches, and switchgrass are some of the examples of second-generation feedstock.

Third generation feedstock

Algae-based biomass is considered a third-generation feedstock for bioplastic synthesis. Algae biopolymers were mainly evolved as by-products of algae biofuel production. Algae can be more efficient while obtaining a higher yield of bioplastics when compared to first and second-generation feedstocks and does not require the use of pesticides, herbicides, fertilizer, or land to grow. However, the processing of bioplastic production is comparatively expensive and is still in its infant stage. Algae produce a variety of base materials such as carbohydrates, proteins, and hydrocarbons that can be used in bioplastic synthesis [60]. Different types of bio-plastics prepared by algae feedstock are as following: (1) Hybrid plastics: usually made by the addition of denatured algae biomass to petroleum-based plastics like polyethylene and polyurethane as fillers. Green algae are found to best suit as the hybrid plastic source. (2) Polylactic acid: bacterial fermentation of algae yields lactic acid, a biobased monomer that polymerized to polylactic acid. (3) Polyethylene: fermentation of algae with yeast gives ethanol, which can be further converted to ethylene, a biobased monomer that can be employed in the production of various plastic likes polyethylene. (4) Cellulose-based plastics: after the extraction of algal oil, 30% of algal biomass is left with cellulose, a natural polymer. This type of biomass can be useful for the production of cellulose-based bioplastics.

5.   Production of biobased polymers

Production of bio-based and biodegradable plastics can be grouped according to their source (Fig. 3): (1) Polymers from biomass (agro-resources and agro-polymers) such as starch or lignocellulose, (2) Bioplastic polymers derived employing microbial fermentation such as polyhydroxyalkanoates (PHAs), (3) Bioplastic monomers derived using biotechnology approach comprising conventional synthesis, and (4) Bioplastic from industry-originated wastes.

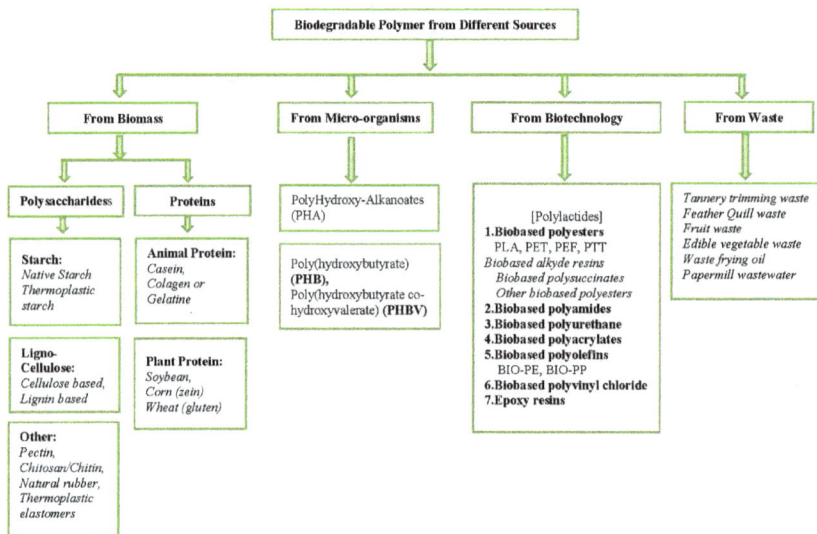

Fig. 3 *Classification of the sources producing biodegradable polymers*

5.1   From biomass (by extraction and separation)

Polysaccharides and proteins are the main agro-polymers that can be used as a raw material for bioplastic synthesis. Several conventional techniques can be employed to process such types of biomass.

Materials Research Forum LLC

https://doi.org/10.21741/9781644901335-2

5.1.1 Polysaccharides

Polysaccharides are macromolecules comprising most of the biosphere. They are mostly found in structures of cellulose, carrageenan, chitin, pectin, etc., forming the main structural elements of animals and plants exoskeleton. They are composed of complex carbohydrates constituting glycosidic bonds. Among many, this chapter focuses on starch, lignocellulosic material, natural rubber, and thermoplastic elastomers.
Starch

Native starch

Starch is the central reserve for polysaccharides, a source of energy in plants. Starch is mainly stocked in seeds or roots of the plants and derived primarily in the form of cereals and tubers (e.g., rice, corn, wheat, manioc, potatoes, etc.) [27,61]. Starch, in its natural form, is composed of crystalline granules and cannot be soluble in water. Starch granules vary in size from 0.5 to 175 μm with various shapes such as polygon, sphere, platelet, etc., depending on their botanical source [62]. These are constituted of two homopolymers of α-D-glucose, amylose, and amylopectin. Amylose is a linear polymer, whereas amylopectin is ramified in nature. Although amylose is a minor component covering only 20-30 % in overall mass [63], its linear nature yields better flowing properties that make amylose-rich starch material, an excellent candidate for bioplastic synthesis [64,65].

Thermoplastic starch

The natural form of starch, chains exhibiting various intermolecular hydrogen bonds, lead to higher melting temperature (T_m) than its degradation temperature [66,67]. Therefore, to obtain plastic-like properties, it is essential to render high water content or/and some non-volatile plasticizers such as sorbitol, glycerol, etc., that help decreases the T_m [68,69]. This plasticized form of starch is known as thermoplastic starch (TPS). To fabricate TPS, the granular structure of starch has to be de-structured. Starch granules can be disrupted by employing the solvent-casting or melting process where starch is blended with plasticizers under the extrusion process (thermo-mechanical treatment) [27,70]. Starch plastification has gained considerable attention from researchers around the globe and is an extensively studied phenomenon [21,22]. The TPS can be obtained by employing a melting process that is often performed using plasticizers to get a homogeneous molten phase. The mechanism of the extrusion (a thermo-mechanical process) is widely studied (Fig. 4) [71], and several related factors can affect the overall water absorption capacity and starch solubility. It can also get highly influenced by several other experimental factors such as pressure, shear, humidity, and temperature [61]. Moreover, it gets affected by the viscosity of the chemicals and amylose/amylopectin ratio [72]. TPS offers desired flowing properties close to the conventional form of thermoplastics originated from petroleum oils such as polyethylene (PE) or polypropylene (PP) [61]. These TPSs are widely being used as a matrix or as co-constituent blended with other forms of thermoplastics, to obtain several commercialized biodegradable plastics such as Biotec®, Novon®, and Mater-Bi® [73]. It can further apply in several biodegradable compositions, including films, injection grades, films, expanded materials, compression grades, etc.

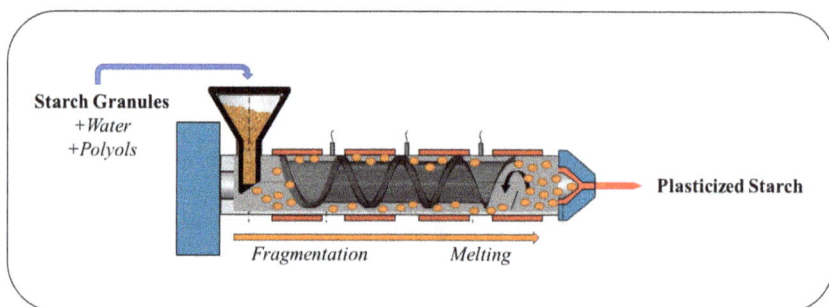

Fig. 4 *Schematic representation of the starch extrusion process*

5.1.2 Ligno-cellulose material

Cellulose-based plastics

Cellulose is the most abundant polysaccharide found in nature that is a primary component of cell walls in higher plants. Around 1.3 billion tons of cellulose is obtained around the globe for numerous technical applications. Cellulose is an unbranched multi-sugar molecule that contains 100-1000 glucose molecules or cellobiose units. The cellulose molecule mainly binds to higher structures that support tear-resistant fibers in plants [31]. The cellulose content varies according to the botanical source of the plant. Hardwood contains 40-75% cellulose, whereas softwood contains 30-50%. In cotton wool, the content of cellulose can reach up to 95%. Several processes have been employed to obtain a robust form of biodegradable materials. A group of researchers has produced bioplastic from the cellulose-rich waste of rice, rice straw. A particular type of industrial extractor called Naviglio was used to pre-treat rice straw that was further treated with trifluoroacetic acid (TFA). TFA is capable of solubilizing cellulose with other organic matter in rice straws and forms a bioplastic material [75]. In another study, corn leaf biomass was heated at high temperatures (150°C) and further treated with acetic acid and sodium chlorite (NaClO₂) solution to obtain nano-coating biomaterial that can be used in biopolymers [76]. A group reported biofilm production from milled orange waste by employing a casting method where the gelling ability of pectin associated with cellulosic fibers was found to be significant on the biofilm production [25]. Similarly, cellulose-based bioplastics from oil palm empty fruit bunch have been reported where obtained cellulose with cassava starch matrix was blended with glycerol as plasticizer [23].

Cellulose-based materials are mainly classified in two distinct forms: cellulose regenerates and derivatives. Cellulose in dissolved form using chemicals to restructure into film or fibers are known as cellulose regenerates. This cellulose regenerates are used to make several groups of textile materials such as artificial silk, viscose silk, and many others. However, only cellulose derivatives are being used for the production of bioplastic. Cellulose derivatives are mainly classified into cellulose ethers and esters [31]. When compared, cellulose esters are significantly more important for plastic manufactures. Cellulose derivatives can be obtained via the esterification of cellulose using organic acids [77]. Researchers made celluloid, the first thermoplastic material, by blending camphor (25 %) and cellulose nitrate 75 % (synthesized using cellulose and nitric acid) [78]. From a technical point of view, the most significant forms of cellulose esters are cellulose propionate (CP combined with propionic acid), cellulose acetate (CA combined with acetic acid), and cellulose butyrate (CB combined with botanic acid). Cellulose-based plastics are widely used in car interiors, construction materials, toys, medical applications, sports equipment, decor materials, etc.

Lignin based plastics

Changing the chemical and physical properties of lignin polymer is conceivable via several modifications. Advanced research in lignin has led to a new generation of lignin-rich plastics featuring superior properties to conventional plastic polystyrene and PMMA (polymethyl methacrylate). The innovative structure offering promising tensile strength can accommodate more than 80% lignin comprehending lignin sulfonate in the methylated or non-methylated form [79]. Moreover, the blending of these materials with plasticizers or other polymers can further improve their mechanical properties. Also, lignin can be used as a blending companion for several conventional plastics and natural fiber-reinforced plastics. Liquid wood is the most popular lignin-based bioplastic in the market that can be easily processed in injection moulding machines [31]. The strength of the liquid wood can be enhanced by blending with flax and hemp (natural fibers). Further, advanced thermosets are being formulated, employing lignin for epoxy, polyurethane, and phenolic resins.

Wood-plastic composites

Wood-plastic composites (WPC) are a special type of natural fiber-reinforced polymer (NFRP) that is manufactured by blending plastic and additives with wooden raw material. WPCs are thermoplastic processible compounds and can be processed by employing extrusion and press or injection moulding [2]. WPC's 3-D plasticity and greater stiffness, unlike its counterparts from conventional plastics, making it a favorable candidate. Typically, WPCs are used in handrails, fences, noise protection walls, harbor docks, etc. Newly developed WPC material comprising polyvinyl chloride and wood fibers in the ratio of 1:1 is used for the manufacturing of door/window frames, indoor furniture, and construction material [2].

5.1.3 Other material

Chitosan/Chitin

After cellulose, chitin is the second most abundant agro-originated polymer and can be found in the exoskeleton of fungi or arthropods, conferring crystalline microfibrils to the cell wall [32]. Similar to cellulose, whisker-shaped nanofillers can be obtained by acidic treatment of the semi-crystalline structure of chitin microfibrils and can be further combined into polymer form to develop nano-hybrid materials. Chitosan can be derived from chitin employing the deacetylation process. Amino groups of chitosan offer some specific properties when compared to chitin. When chitosan is exposed to the acidic treatment, amino groups get protonated, which makes it a water-soluble polycation. Carrageenan is one such polysaccharide that shows polyelectrolyte function; however, polyanions are widely produced from these agro-polymers [80]. Glycerol is among many plasticizers that can be blended with chitosan to obtain thermoplastic material similar to plasticized starch [81].

Natural rubber and thermoplastic elastomers

Rubber can be found in the form of caoutchouc, natural latex, and elastomers; that possess a common relation to plastics. Though synthetic rubber has been demanded by most of the world, 40% of the market is accommodated by caoutchouc [31]. Caoutchouc is a plant-based polymer that is a type of latex, which has a primary role in protecting the wound from bacterial contamination. However, by making deliberate slits in cultivated plantations, crude rubber can be obtained that can be furthered hardened by employing vulcanization. Similar to natural rubber, thermoplastic elastomers (TPE) are biological materials that can be used in plastic manufacturing. TPE has better elasticity and is not cross-linked in natural form, making it favorable for re-melting and can be used as thermoplastics. They can be found in both complete and partial biobased forms.

Thermoplastic polyurethanes (TPE-U) is one such candidate that is being widely used in several industries, including the shoe and tooth-brush industry, where hard-soft bonded properties are mainly required to produce desirable products. Similarly, thermoplastic ether-ester elastomer (TPC- ET) is produced in a hard-soft form, where petrochemical polybutylene terephthalate (PBT) extends hardening whereas polyether from 1,3 propanediol extends softening. These blends are mainly used for the synthesis of airbags in the automobile industry. Moreover, a complete bio-based TPE derived from bio-based polyamide 11 and polyether was reported in the form of polyether block amide that was used in the manufacturing of ski boots [31].

6.   Proteins

Proteins are agro-polymers and crucial renewable resources produced by plants, animals, and micro-organisms. Though several proteins have been reported for the production of biodegradable polymers, only a few made it to actual industrial-scale production due to the high production and downstream processing cost.   Plant proteins derived from soybean, corn (zein), wheat (gluten) are extensively studied. Proteins, including casein, collagen, or gelatin, and keratin, are among many well-established animal proteins. Similarly, fumarase, chymotrypsin, and lactate dehydrogenase are commonly used as bacterial proteins in bioplastic synthesis [27].

6.1   Plant protein

Soybean

Soybean proteins were the first agro-biopolymers that were extensively researched and used in the manufacturing of moulded materials. During 1930-40, it was being used in the manufacturing of parts of automobile industries where a substance composed of phenol-formaldehyde/soybean flour mixture was used [29]. However, it was discontinued due to the comparatively lower costs of synthetic plastics. Soya proteins are still being used in the manufacturing of bioplastic due to the demand for biodegradable products. Typically, a dried soya bean contains crude protein (38-42 %), carbohydrates (33 %), and triglycerides (16-20 %) [27]. The soya proteins can be processed via physical or chemical treatments. For polymer processing, several processes can be employed, such as casting, designing, extraction, injection moulding, etc. Reports suggest that the soya proteins blended with starch can favor the superior synthesis of bioplastic products such as toys, sports gear, containers, packaging materials, etc. Plastic obtained through injection moulding has shown superior water-resistant properties while making these products easy to recycle and reuse. Similarly, films can be obtained using soy proteins that can be used as UV protective material making them applicable in packaging material. They can also be used for land fillings since they quickly degrade in the environment. If processed properly, soy proteins can also be used to obtain foam products as well as insulating materials having superior thermal properties. Moreover, non-electrostatics, non-flammability, and efficient biodegradability make it an excellent candidate for the bioplastic industry.

Corn (Zein)

Corn accommodates about 9% protein and is mainly composed of zein, albumins, glutelin, and globulins. Zein is a highly hydrophobic protein that is soluble in alcohol [82]. It is typically extracted from corn employing the wet-milling process [83]. It is a by-product of ethanol production. Zein's hydrophobic properties make it water-resistant, a desirable characteristic in bioplastic production. Moreover, it is entirely biodegradable and also contains amino acids that are rich in nitrogen; therefore, it can be used as a fertilizer in agriculture after being used since it supports and promotes plant growth [84].

Wheat (Gluten)

Gluten, a wheat protein, has successfully attracted researchers from the bioplastic field during the last few decades, and a few materials produced from wheat proteins have been reported as well [85]. Typically, the processing of protein-based material requires the following steps: 1. breaking of intermolecular bonds using chemical or physical rupturing agents that stabilize polymers. 2. Rearrangement or re-orientation of polymer chains in the desired size and shape. 3. Lastly, allowing the formation of intermolecular bonds that maintains the 3-D network [86]. This can be attained using the casting method or physio-chemical method via employing chemical reactants in the process. A group of researchers has successfully prepared bioplastic from wheat gluten by using a casting method while employing plasticizers in the process [28]. Some specific applications of wheat protein-based products are developed by manufacturers of toys, office products, leather imitation, design, and furniture, etc.

6.2   Animal protein

Casein

Casein is an animal protein that can be used for the production of bioplastics. It is usually found in mammalian milk, such as cow milk (making 80 % of total protein content) and human milk (making 60-65 % of total protein content) [61]. Casein is an open and random coil structure of predominant phosphoprotein. Acid-precipitation can be used to obtain caseinates casein with an alkali solution. It can be found in various forms such as insoluble casein, particle calcium salts, or entirely soluble sodium or potassium caseinates [87]. Both caseins and caseinates can be used to obtain aqueous solutions that can further be applied to produce transparent films without treating hydrogen bonds [88]. Similarly, to build a casein-based plastic, plasticized casein from skimmed milk can be treated with formaldehyde to obtain a cross-linked plastic followed by the removal of water. Casein-based plastic can be used in several areas to produce flower pots [61], where lignin can be blended with the products to increase their strength and stiffness. Furthermore, caseinates can be polymerized by gamma radiations by the creation of bityrosine covalent bonds [89] and employed in the production of biobased materials. However, due to its slightly substandard properties compared to the competitors, casein plastics are only being used in small niche markets [31].

Gelatine or Collagen

Gelatin is an animal protein derived from collagen. Collagen is elongated fibrils often located in fibrous tissues such as the cornea, blood vessels, bones, cartilage, skin, tendon and ligaments, and intervertebral disc. It is primarily used as a building material for capsules/tablets in the pharmaceutical industry. Moreover, it is also used as a nutritional supplement. Gelatin-based biofilms can be fabricated by solution casting method, and several products have been developed in the last few years. The production of bioplastics from a rapeseed meal by using injection moulding was investigated at varying mould temperatures [90].

7. From micro-organisms (fermentation)

7.1 Polyhydroxyalkanoates

Polyhydroxyalkanoates (PHAs) are aliphatic polyesters obtained via bacterial (sometimes yeast or plants) fermentation processes of lipids and sugars. PHAs can potentially replace traditional polymers (hydrocarbon-based) and are considered to be 100% bio-based and biodegradable in a varied range of environments, whether it is fresh/seawater or home/industrial composters [13]. PHAs can combine >150 monomers while producing a material with specific desirable properties of bioplastic. Moreover, they can alter the biological/mechanical compatibility employing enzymes while blending PHA with other polymers or inorganic materials, leading to a wide range of applications. Several bacterial strains can utilize sucrose, vegetable oils, glucose, glycerin, etc., as a source of carbon and energy where cells are exposed to the stressed conditions to make them produce PHA by expressing secondary metabolites. PHAs are a family of assorted plastics such as polyhydroxy-valerate (PHV), polyhydroxybutyrate (PHB), polyhydroxybutyrate-polyhydroxyvalerate (PHBV), and polyhydroxyhexanoate (PHH). Among several types of PHAs, PHB is a potential candidate to replace conventional polypropylene (PP) and polyethylene terephthalate (PET) due to its similar barrier, thermal, and mechanical properties. PHB from the component of the bacterium *Bacillus megaterium* was reported in 1923. During the 1990s PHA has emerged as a potential candidate and received extensive attention around the globe; however, due to high processing cost when compared to oil-based plastic, it could not cover the anticipated portion of the overall plastic market.

A range of bacterial strains has been reported for the accumulation of PHA that comprises 30-80% of total dry cell biomass [91,92]. Bacteria are allowed to grow in a suitable medium provided with an adequate amount of nutrients so that they divide and grow at a fast pace. Further, they are supplied with excess carbon along with limited essential growth nutrients (deficiency of certain micro-elements: phosphorus, nitrogen, elements in traces, or lack of oxygen), encouraging them to synthesize PHA [93]. To isolate and purify PHA produced during the fermentation process, cells are concentrated, followed by drying. Further, the extraction of PHA can be carried out using solvents (chloroform/acetone). The solid-liquid separation process can be exercised to remove residual cell debris from the solvent containing PHA extract. Finally, the PHA extract can be precipitated using methanol and recovered by the precipitation process [94].

PHA synthesis in the soil environment can be beneficial due to the lack of phosphorus or nitrogen. The types of monomers and co-polymers produced are directly influenced by the type of carbon substrate supplied and the metabolism of microorganisms [35]. The most common substrates for PHA are fatty acids and glucose [95] that can be obtained from a wide range of agro-based biomass. Sugarcane and corn (glucose source) along with soybean and rapeseed oil (fatty acid source) are among some common raw materials that can be fermented by bacterial strains for PHA synthesis. PHAs are biocompatible and biodegradable, making them the right candidate for several industries. The non-toxic nature of PHAs favors medicinal applications where the implants need not be removed later. Similarly, PHAs can be used in the packaging industry for coating purposes as well as in consumer/household products. PHA growth has been increasing in the past few years; however, it is expected to grow steadily in the coming years. Researchers are also trying to produce the production of PHA from methane in low gravity environments. If it works, it could substantially push the growth for the PHA industry due to the abundant availability of methane[13].

From biotechnology (by conventional synthesis from bio-derived monomers)

The molecules that can be polymerized to obtain polymers and are derived from several renewable resources are called bio-monomers. These bio-monomers are polymerized to synthesize various types of bioplastics (e.g., polyesters, polyamide, polyurethane, polyacrylates, etc.).

7.2   Biobased polyesters (PLA, PET, PEF, PTT)

Polylactic acid

Polylactic acid or polylactide (PLA) is one of the most critical bioplastics in the market due to its abundant availability and relatively low price in comparison to other bioplastics [96]. PLA consists of natural acid, i.e., lactic acid obtained from various renewable resources such as corn starch, tapioca roots, and sugarcane, etc. [97]. Lactic acid is a chiral molecule with two different stereoisomers, i.e., L and D-lactic acid, and prepared by either chemical or biological methods [98]. The fermentation of carbohydrates produces lactic acid with the help of microorganisms such as lactic bacteria belonging to the genus lactobacillus, or a few strains of fungi [97,99]. The process requires carbon sources such as carbohydrates, nitrogen sources such as yeast extracts or peptides and minerals element for the growth of bacteria while producing lactic acid. The polycondensation of lactic acid generally leads to the formation of low molecular weight polylactic acid. However, Moon et al. [100,101] have proposed an enhanced melt-polycondensation method using Sn (II) catalyst for the synthesis of high molecular weight. At present, ~30% lactic acid is only used for PLA synthesis, while the consumption of PLA is about 200,000 tons/year, thus presents a high potential for development [98]. The properties of PLA depend on their molecular characteristics and ordered structures such as crystallinity, morphology, degree of chain orientation, and spherulite size. PLA plastics have the advantage of film transparency, high rigidity, good thermoplasticity, and excellent processing performance on existing equipment [31]. However, PLA has a low softening temperature around 60°C, which limits its applications at higher temperatures [102]. The processed PLA is not a final plastic product; however, it can be used as a raw polymer for the specific application by copolymerization or compounding with suitable additives and can be further blended with other bioplastics. PLA can be plasticized by using citrate esters [103], low molecular weight polyethylene glycol [104], or lactic acid [105]. Using proper combinations of L- and D- lactides and additives, an enhanced version of modified PLA can be produced. This modified PLA can withstand high-temperature [37]. In comparison to polystyrene, PLA has a medium level of water and oxygen permeability and makes it suitable for different packaging applications [106,107]. PLA and PLA-blends are available in various grades and granulate forms that can be used for the manufacturing of drink containers, cups, film, moulded parts, bottles, and other everyday items [108,109]. Owing to their short life, the PLA has great potential for the manufacturing of items such as packaging films [38]. PLA can also be used in the manufacturing of desktop accessories, a casing of mobile phones, lipstick tubes, dashboards, door tread plates, and in textile applications depending on the fibers spun from PLA [37,39,110].

Polyethylene terephthalate

Polyethylene terephthalate (PET) is a thermoplastic polymer and one of the mass-produced plastic since the second half of the 20th century due to its extensive use in beverage bottle manufacturing. PET is generally produced by polycondensation of terephthalic acid/dimethyl terephthalate, and monoethylene glycol (or ethylene glycol, a diol or bivalent alcohol) having 70:30 composition ratio by weight [36,111]. Both the raw materials can be produced from bioresources, and hence the petroleum-based PET is now being replaced by biobased PET called Bio-PET. The petroleum route usually prepares the first component of PET, i.e., terephthalic acid, and its bio routs are still under development phase [112,113]. Bio-PET is a partially biobased product produced by using bio-derived mono ethylene glycol (Bio-MEG) [31]. Bio-MEG is produced from agro-based resources and includes bagasse, hay, and sugar cane molasses [114]. Bio-PET-based products have similar quality as that of regular PET in terms of weight, appearance, etc., and they can be recycled and reused. Today, Bio-PET is mainly used in the manufacturing of soda and drinking water bottles, automotive interiors, packaged goods, electronics, and construction goods, which makes them an environmentally friendly alternative to traditional plastic.

Polyethylene furanoate

As discussed in the above section, Bio-PET is not 100 % biobased as the significant component of the synthesis, terephthalic acid, is obtained from petroleum resources. As an alternate, polyethylene furanoate (PEF) is a type of thermoplastic, that is 100% biobased and has the potential to replace the PET completely. PEF is generally synthesized by polymerization of ethylene glycol, i.e., Bio-MEG and a 2,5-furan dicarboxylic acid (FDCA); both are biobased chemicals. FDCA is produced from the oxidation of 5-hydroxymethyl furfural, which is a platform molecule derived from the lignocellulosic biomass. PEF has a high glass transition temperature, lower melting point, and excellent barrier properties to water and oxygen in comparison to PET, which makes the PEF not only 100% biobased but also a superior competitor to PET [40]. PEF can be employed in the packaging industry for milk, soft drink, water, and alcoholic beverages. PEF based products are not biodegradable but can be easily recycled or incinerated to convert back into CO_2.

Polytrimethylterephthalate

Polytrimethylterephthalate (PTT) is a partially biobased plastic made by the condensation polymerization or esterification of terephthalic acid and 1, 3-propanediol [115]. 1,3-propanediol is a biobased molecule (referred to as Bio-PDO) derived from hydrogenation of glycerol or by fermentation reaction. PTT has similar physical properties, i.e., heat resistance, strength, toughness, and stiffness as that of PET. PTT is used in fiber manufacturing, textiles, and carpet synthesis for the domestic market and automobile industry since they are soft in texture but can bear heavy wear [41,42,116]. Due to similar processing properties to polybutylene terephthalate (PBT), PTT can also be used in injection moulding. PTT-based plastics are resistant to shrinking and deformation and can be used in electric and electronic components, air breath outlets, and car instrument panels due to their high-quality surface texture [31,117].

8.   Biobased alkyds resin

Alkyd resins are produced by polycondensation of polyvalent alcohol such as sorbitol, glycerin or glycol with a carboxylic acid such as phthalic acid, succinic acid, maleic acid, adipic acid, or their anhydride. Some of these polyvalent alcohols and acids can be produced from renewable resources; hence the alkyd resins can be biobased or partly biobased [43]. Succinic acid, adipic acid, or maleic acid can be produced from the platform molecules derived from biomass resources such as lignocellulose biomass. Similarly, acids, such as fatty acids derived from vegetable oil, are biobased and can be used for the synthesis of various alkyd resins [44]. Combining both biobased derived carboxylic acid and polyvalent alcohol, 100% biobased alkyd resins can be synthesized (e.g., Voxtar M100 synthesized by PersorpAB, Sweden) [74]. Alkyd resins are used as a raw material in paints, lacquers, and also as filler compounds. Additionally, they find applications in the production of printing inks, adhesives, insulating material, floor covering material, and textile enhancers [118].

Biobased polysuccinate

Polysuccinate can be synthesized either by polycondensation of succinic acid with the diols or by ring-opening polymerization of succinic anhydride with ethylene oxide [46]. Depending on the sources of diols and acid, the polysuccinates can be partly or 100% biobased. Polybutylene succinate (PBS) is a biodegradable plastic that is synthesized by using succinic acid and butanediol (BDO) [119]. Bio-based butanediol, i.e., Bio-BDO and succinic acid, are produced from the sugar derived from starch hydrolysis in a single step fermentation by *E. coli* [120,121]. PBS can also have more than one acid to get the diacid polymers suitable for different applications. Polybutylene succinate adipate (PBSA) is a derived product of PBS wherein addition to succinic acid, adipic acid is polymerized within the compound [122]. PBS has a high melting temperature and a wide processing method that is suitable in extrusion, injection moulding, fiber spinning, film blowing, and thermoforming [31,45].

9. Other biobased polyesters

Other partly or wholly biobased polyesters are polybutylene terephthalate (PBT), polybutylene adipate terephthalate (PBAT), and vegetable oil-based polyesters. PBTs are produced from the polycondensation of terephthalic acid and butanediol (BDO) [123]. Similar to the above plastic material, PBT can be fully biobased, but as discussed earlier, the cost of acid is very high; hence at present, it is partly based on bioresources. Similar to PBT, PBAT has adipic acid as additional acid, which gets polymerized along with other acids to obtain the final product. In 2009, BASF, a German chemical company, synthesized fully biodegradable PBAT from renewable resources [124]. These polyesters have similar properties to that of PET and can be employed in injection moulding to produce automotive exterior/interior parts and compostable plastic bags for gardening and agricultural uses [47]. Vegetable oil-based polyesters are resin-type polyester made by the polycondensation of acid or ester with alcohol in which one of the monomers should be derived from the vegetable oil. The monoglyceride method is the most commonly used for polyesters preparation. In this process, oil is processed to get the mono or diglycerides, which are further converted into polyesters by employing polycondensation with anhydrides such as succinic, phthalic, or maleic anhydride [125]. Depending on the anhydride source, it can be partly or 100% biobased plastic. These alkyds can be used for the manufacturing of adhesives, casting material, insulating materials, and printing inks [48,49]. They can also be used as textile finishing agents. Linoleum is one such product produced from linseed oil and is used as a floor covering material [126].

9.1 Biobased polyamides

Polyamides are an essential class of polymer having a wide range of applications. The production and development of polyamides are continuously growing since the establishment of nylon and perlon in 1930 [50]. Polyamides were particularly suitable for the synthesis of fibers and few technical applications. It is still being used in the production of extruded products, hollowware, textiles for the manufacture of decorative material, clothing, and technical fabrics [50]. Polyamides are produced via polycondensation of amino acids or ring-opening polymerization of lactams (cyclic amides), resulting in AABB-types and AB-type polyamides, respectively. In these polymers, the monomers are connected by amide bonds. Bio-polyamides can be wholly or partly biobased depending on the source of dicarboxylic acid (C10) and diamine. The majority of these bio-amides are based on sebacic acid, synthesized from ricinoleic acid derived from castor oil [127]. Apart from this, undecenoic acid can also be prepared from the same ricinoleic acid via pyrolysis and used for the synthesis of bio polyamides named

PA11 that is available in the market since1940 [128]. It is produced via polycondensation of amino acid, i.e., 11-amino undecanoic acid. Using sebacic acid as a monomer, various partially biobased polyamides such as PA4.10, PA6.10 are produced with "10" components as their biobased part. PA6.10 is prepared by the step-growth polymerization of diacid, which is biobased with hexamethylenediamine as diamine source, making it partly biobased polyamides. Similarly, PA10.10 is produced from the polymerization of sebacic acid and decamethylenediamine (C10) [51]. In this, both components are biobased derived from castor oil. These polyamides have excellent mechanical properties such as high flexibility, tensile strength, toughness, resilience, and abrasion resistance while having lower moisture absorption ability with improved chemical resistance when compared to traditional PA6.6 polyamides [52,129,130]. Due to their superior properties, bio polyamides are used in various applications such as pneumatic air brake tubes, flexible oil and gas pipes, electrical cable jackets, and automotive fuel lines [52].

9.2   Biobased polyurethane

Polyurethanes (PUs) are a versatile class of plastic having a wide range of applications in the coating, foam, adhesive, and fiber industry [131]. PUs are prepared by carrying out a reaction between polyols and diisocynates, forming urethane linkage, which is a carbamate ester linkage generated during the reaction between the isocyanate and alcohol group [54]. They may be used as a thermosetting or thermoplastic form. Bio-polyurethane (Bio-PUs) can be partly or wholly biobased while utilizing biobased polyols and petroleum-based or biobased isocyanates. Most of the Bio-PUs are partly biobased and derived using biobased polyols obtained from plant oils such as soyor castor oil [53]. Castor oil already contains OH groups, while polyols derived from vegetable oils (sunflower, rapeseed, or soya) are produced by epoxidation of unsaturated fatty acid followed by the addition of multiple alcohols in the epoxide opening reaction. Another route of synthesis of bio polyols is based on starch liquefaction, liquified lignin, and nutshell liquid [132]. Recently, biobased diisocyanates such as pentamethylenediisocyanates have been commercialized that can produce a pure form of Bio-PU. Bio-PUs are biodegradable foams having high thermal and dimensional stability than petroleum-based PUs. Bio-PUs have been used to replace conventional PUs in various applications, particularly for the production of cushions, coatings, and insulations.

9.3   Biobased polyacrylates

Polymerization of various acrylic esters results in the formation of a resin-type structure know as polyacrylates. The commonly used polyacrylates are polyethyl acrylates and polymethyl acrylates. These acrylic esters are made by esterification of acrylic acid, which is generally produced from petroleum resources [55]. The biobased polyacrylates can be synthesized by using bio-based materials, e.g., glycerol, fatty acids, or sugar. The dehydration of glycerol gives acrolein, which is further oxidized to biobased acrylic acid, which then gets transformed into ester by the reaction with alcohol, ultimately polymerizing to form the polyacrylate. Depending on the source of alcohol and acrylic acid, the polyacrylates can be complete or partly biobased plastic. Currently, several efforts are being made to use a biobased version of platform molecule, e.g., 3-hydroxy propionic acid, lactic acid, etc. [133], for the synthesis of raw materials of plastic. The bio polyacrylates have similar properties to that of the conventional ones. Polyacrylates find application as superabsorbent in consumer products as a thickener and are mainly used in the salt form (sodium polyacrylate) in pigment dispersants [56].

9.4 Biobased polyolefins

Polyolefins (polyethylene and polypropylene) are the most common and essential plastics used in various applications. They have very low density ($>1g/cm^3$) and hence can easily float in water. Both types of this category can be produced from renewable resources [57].

Biobased polyethylene (Bio-PE)

Polyethylene (PE) is the most used global plastic today, produced by the polymerization of ethylene gas. Ethylene can be obtained from petroleum resources or ethanol dehydration [134]. The latter method was used in the early 20th century before the availability of direct ethylene gas for industrial production from petroleum resources. Biobased polyethylene (Bio-PE) can be obtained from renewable resources such as sugarcane and has a similar characteristic to that of PE obtained from fossil oil. The production of Bio-PE is based on the ethylene produced from ethanol (bioethanol obtained from sugarcane) [31,57]. The process includes the fermentation of sugarcane juice to ethanol, which is dehydrated to get the ethylene, and after polymerization, the formed product is Bio-PE (Fig. 5). Bio-PE is mostly used in the packaging industry [135]. It is an important intermediate for the synthesis of other polymers like PET and polyols for polyurethanes [136]. It can also be used in the production of bags, pouches, petrol canisters, barrels, automobile fuel tanks through blow moulding and injection moulding [58].

Sugar cane $\xrightarrow[\text{Distillation}]{\text{Fermentation}}$ Ethanol $\xrightarrow{\text{Dehydration}}$ Ethylene $\xrightarrow{\text{Polymerization}}$ Bio-Polyethylene

Fig. 5 *Bio-polyethylene production from renewable sugar cane biomass*

Bio-polypropylene

Like Bio-PE, Bio-polypropylene (Bio-PP) can also be synthesized from bioethanol using more sophisticated methods [137]. The part of ethylene obtained, as mentioned above, will dimerize to butenes, which will react with remaining ethylene to produce Bio-PP. One more way to it is via vegetable oil cracking in the fluid catalytic cracker (FCC) [138,139]. Bio-PP is generally used in injections, packaging, textiles, and due to its light-weight feature, it can also be used in automobile industries for the production of seat-backs, shelving, interior storage bins, and load floors [17,140].

9.5 Biobased polyvinyl chloride

Polyvinyl chloride (PVC) is the polymerized product of the vinyl chloride (VC) monomer. VC is synthesized from ethylene gas via oxychlorination methods. As discussed in the section of bio propylene synthesis, the fermentation of sugarcane yields ethanol, which can be converted to bio ethylene gas. The bio ethylene gas can be used as a starting material for the synthesis of biobased VC (Bio-VC) that can be further polymerized to obtain Bio-PVC [31,141]. They have similar properties to that of general PVC obtained from petroleum resources. Bio-PVC is used in the manufacturing of wires, films, cables, pipes, and bottles with end-use in various industries, including transportation, packaging, electrical and electronic instrument, and construction [17,58].

9.6 Epoxy resins

Epoxy resins are thermosetting polymers having several desirable properties such as low shrinking with strong adhesion and hardness property while offering excellent stability for many solvents and alkali solutions [59,142]. Most of the epoxy pre-polymers are synthesized by condensation of bisphenol A and epichlorohydrin that produces diglyceride ether of bisphenol A, a most commercially available form of epoxy resin. Bio-based epoxy resins can be prepared using renewable resources such as vegetable oil, lignin, rosin, furan, sugars, and itaconic acid. Biobased molecules (e.g., FDCA from furan, eugenol from lignin, etc.) react with the epichlorohydrin (another biobased compound) to obtain the 100 % biobased epoxy resins [59]. These biobased epoxy resins have similar properties to petroleum-based epoxy resins and find applications in the synthesis of structural polymers, epoxy foams, adhesives, coatings. Moreover, it has also been explored in several applications in the aerospace and electronics industry.

9.7   From waste (by processing industrial bio-waste)

Bio-based industrial waste is another huge opportunity to develop bioplastic materials. Researchers are trying to find means to use waste materials for the production of bioplastics. A potato industry (for chips) in the Netherlands, trying to use wastewater produced during the process of peeling and slicing the potatoes. This large amount of processed water contains a high percentage of starch that can be further processed and used. These industries have successfully developed methods to obtain bioplastic from the waste produced. Similar approaches to use processed wastewater containing starch and associated waste are also being applied in several other parts of the world [31].

Tannery trimming waste

Trimming hydrolysate comprising bio-crosslinkers (non-toxic) can be obtained from tannery solid waste for the production of bioplastic material. Trimming hydrolysate powder can be blended with polyvinyl alcohol to obtain transparent and highly flexible films where a solution casting process can be employed. Using citric acid as a plasticizer, the resultant bioplastic films can lead to a smooth, uniform, and defect-free form of plastic with a better tensile strength (>20 Mpa). Moreover, it also showed a notably high elongation capacity (break value >343 %). Following the soil burial test, the bioplastic degraded up to 62 % within 70 days. Also, transparency and the anti-microbial properties of the films were found to be superior due to the presence of citric acid interactions. Therefore, bioplastics derived from trimming hydrolysate could potentially replace fossil-based plastics while having several applications in the field of medicine (wound healing), packaging, among many others [143].

Feather quill waste

The poultry industry generated about 3-4 billion pounds of feathers each year as a by-product in the United States only [144] and more than 157 million pounds in Canada. Feathers comprise around 90 % of protein called keratin, are sent to landfills, which increases environmental impacts and health hazards [145,146]. Several attempts were carried out to use this waste to produce bioplastic. A twin-screw extruder was used for suspending poultry feather quills while using sodium sulfite as a reducing agent. Ethylene glycol as plasticizers was able to mix more effectively with quill keratin at the molecular level, manifesting only one sharp glass transition with excellent mechanical and transparent properties related to other plasticized resins. The melting temperature (T_m)

and the glass transition temperature of quill material were reduced by the addition of plasticizers. The mechanical properties of the materials were also seen to be reliant on the kind of plasticizer used. It was also noted that propylene glycol and diethyl tartrate were able to transform quill material into comparatively hard and brittle polymer related to ethylene glycol and glycerol. However, the material with superior mechanical properties along with better clearness, flowability, and processability was observed by the ethylene glycol as a plasticized resin compared to others [147].

Fruit waste

Industrial fruit juice production generates a large amount of fruit waste, which holds about 60% of fibers in dried biomass. Bioplastics prepared using the fruit waste can also serve as a possible option for conventional plastic materials. Bioplastic can be produced by using fruit waste such as potato and banana peel, among many others. Banana peel blended with glycerol could help generate a polymer that can be used in bioplastic production. The plastic produced could have characteristics of better flexibility and strength. However, potato peels can outperform banana peels since they contain more amounts of starch and polymer chains that are required for the synthesis of good quality plastic [148].

Edible vegetable waste

The agro-food industry produces huge amounts of inedible debris, originated from processed edible vegetables and cereals. More than 24 million tons of processed vegetable waste is produced per annum in Europe alone [149]. Edible cereals and vegetable waste from industries that are rich in cellulose can be converted into bioplastics by just aging them in trifluoroacetic acid (TFA) solutions regardless of their bio-origin. Biopolymers produced from these wastes were found to exhibit distinct mechanical properties varying from brittle and hard to soft and stretchable, depending on the bio-source used in the process. All-natural plasticization of amorphous cellulose is obtained by combining these vegetable waste solutions with TFA solutions of pure cellulose. Bioplastics produced from edible vegetable waste can replace many non-degrading plastics, conserving the environment while being employed in the packaging and biomedicine industry [150].

Waste frying oil (WFO)

Rapeseed oils for the purpose of frying are commonly used in Europe. In other countries of the world, the type of frying oil can differ. The Polyhydroxyalkonates (PHAs) with valerate monomer have been produced successfully from rapeseed oil waste [151]. In PHA production, it can be used without filtration and could produce more biopolymer when compared to pure vegetable oil as a source of production. Waste frying oils from the food industry can also be employed for the production of PHB. While waste frying oil is plentiful, the main obstacle is to put the collection arrangements in place to collect this waste resource. However, if processed properly, waste frying oil could establish as a cost-effective and environment-friendly alternative to traditional plastic [152].

Papermill wastewater

Production of Polyhydroxyalkonates (PHA) was investigated by treating wastewater of paper mills with a microbial community (*Plasticicumulansacidivorans*). The overall process can be carried out as follows: (1) acidogenic fermentation of paper mill wastewater was carried out in a simple batch process, (2) a feast-famine regime was employed in a sequencing batch mode to provide enrichment of PHA-producing microbial culture, and (3) finally, cellular PHA content was enriched and accumulated in a fed-batch system. This study unveils the potential of wastewater produced in the paper industry to render PHA-based materials. Similarly, these PHA-producing microbial communities can be further explored for the treatment of other types of wastewater while obtaining bioplastic in the process [153].

Conclusion and future prospects

Owing to the obstacles induced by petroleum-based plastic, along with their high cost due to the fast-paced depletion of fossil fuel, they have attracted the use of renewable biomass for the synthesis of bio-based and bio-degradable plastic. Several renewable resources have been identified and employed in the last few decades, and a few are still under development to make them efficiently utilized while attaining a better quality of the bioplastic material. The bioplastic can be biodegradable or non-biodegradable, depending on the monomer type present in the structure. The renewable feedstocks are further classified based on their use in the food chain. So far, the most utilized feedstocks are the carbohydrates and starch originating from plants, i.e., first-generation feedstocks; however, the food crisis in 2008 has stirred the research towards the synthesis of bioplastic from the biomass (derived from the non-food chain), mainly lignocellulose and industrial waste. Bioplastics can be derived from biomass by employing different methods, depending on the type of biomass used in the process. The bioplastics have almost similar properties to that of petroleum-based plastic and can be applied to a related application similar to traditional plastics. Lignocellulose has gained lots of attention in recent years as it contains lignin and cellulose as the main constituent, which is a natural polymer and can also be utilized to generate bio monomer (raw material for polymers) via different treatment methods. Since the large-scale utilization of lignocellulose and waste from plant or biobased industrial waste are still not economically established, more research is required before moving entirely away from a first-generation feedstock for bioplastic synthesis. Some of the major concerns about existing renewable resources are to have low efficiency as their production drops due to several factors such as weather conditions, land quality, and available area. Algae could be a potential alternative that has the advantage of no land requirement and can be produced in a large amount with the use of a simple photosynthesis process. They can be produced using wastewater from different sources where CO_2 gets utilized for their growth, which makes algae biomass a very attractive option. Moreover, it serves the dual purpose where it can be used as a feedstock while controlling the CO_2 in the environment. However, the high cost of production of bioplastic from algae and several other biomasses has been a concern among the research community, and the broader adoption of bioplastic at the industrial-scale is yet to be seen. Nevertheless, bioplastic could be a savior of the planet earth since the biosphere is on the verge of a global environmental crisis.

References

[1]   J. Farrin, Biodegradable plastics from natural resources, Rochester Institute of Technology report for the Institute of Packaging Professionals, (2005).

[2]   A. Jering, J. Günther, A. Raschka, M. Carus, Use of renewable raw materials with special emphasis on chemical industry, ETC/SCP Rep. (2010) 1–58.

[3]   N. Cioica, C.Ń. Co, M. Nagy, G. Fodorean, Plastics made from renewable sources – potential and perspectives for the environment and agriculture of the third millenium, Bull. Univ. Agric. Sci. Vet. Med. Cluj-Napoca - Agric. 65 (2008) 23–28. https://doi.org/10.15835/buasvmcn-agr:1083

[4]   Global Bioplastics Market Research & Industry Trends Report, Available at:https://www.bccresearch.com/market-research/plastics/global-markets-and-technologies-for-bioplastics.html (accessed January 19, 2020).

[5]   P. Kumar, Sonia, Green Plastic: A new plastic for packaging, Int. J. Eng. Sci. Res. Technol. 5 (2016) 778–781. https://doi.org/10.5281/zenodo.61482

[6]   M.G.A. Vieira, M.A. Da Silva, L.O. Dos Santos, M.M. Beppu, Natural-based plasticizers and biopolymer films: A review, Eur. Polym. J. 47 (2011) 254–263. https://doi.org/10.1016/j.eurpolymj.2010.12.011

[7]   R.N. Tharanathan, Biodegradable films and composite coatings: past, present and future, Trends Food Sci. Technol. 14 (2003) 71–78. https://doi.org/10.1016/S0924-2244(02)00280-7

[8]   J.F. Martucci, R.A. Ruseckaite, Biodegradable bovine gelatin/Na$^+$-montmorillonite nanocomposite films, structure, barrier and dynamic mechanical properties, Polym. Plast. Technol. Eng. 49 (2010) 581–588. https://doi.org/10.1080/03602551003652730

[9]   M. Flieger, M. Kantorová, A. Prell, T. Řezanka, J. Votruba, Biodegradable plastics from renewable sources, Folia Microbiol. (Praha). 48 (2003) 27–44. https://doi.org/10.1007/BF02931273

[10]   J.W.Hill, T.W. McCreary, Chemistry for changing times, Forteenth ed., Prentice Hall, New Jersey, 2015

[11]   K. Marsh, B. Bugusu, Food packaging-roles, materials, and environmental issues, J. Food Sci. 72 (2007) 39–55. https://doi.org/10.1111/j.1750-3841.2007.00301.x

[12]   I. Odegard, S. Nusselder, E. Lindgreen, G. Bergsma, L. Graaff, Biobased plastics in a circular economy, (2017) 1–136. https://www.cedelft.eu/publicatie/biobased_plastics_in_a_circular_economy/2022

[13]   D. Verma, E. Fortunati, Biobased and biodegradable plastics, Handb. Ecomater. (2018) 1–23. https://doi.org/10.1007/978-3-319-48281-1_103-1

[14]   B. Choi, S. Yoo, S. Il Park, Carbon footprint of packaging films made from LDPE, PLA, and PLA/PBAT blends in South Korea, Sustainability 10 (2018), 2369. https://doi.org/10.3390/su10072369

[15]   F. Gu, J. Guo, W. Zhang, P.A. Summers, P. Hall, From waste plastics to industrial raw materials: a life cycle assessment of mechanical plastic recycling practice based on a real-world case study, Sci. Total Environ. 601–602 (2017) 1192–1207. https://doi.org/10.1016/j.scitotenv.2017.05.278

[16]   D. Jyoti Sen, P.N. Patel, K.G. Parmar, A.N. Nakum, M.N. Patel, P.R. Patel, V.R. Patel, Biodegradable polymers: an ecofriendly approach in newer millenium, Asian J. Biomed. Pharm. Sci. 1 (2011) 23–39.

[17]   A. Rudin, P. Choi, Chapter 13 - Biopolymers, in: A. Rudin, P.B.T.-T.E. of P.S.& E. (Third E. Choi (Eds.), Academic Press, Boston, 2013: pp. 521–535. https://doi.org/https://doi.org/10.1016/B978-0-12-382178-2.00013-4

[18]   R.C. Thompson, C.J. Moore, F.S.V. Saal, S.H. Swan, Plastics, the environment and human health: current consensus and future trends, Philos. Trans. R. Soc. B Biol. Sci. 364 (2009) 2153–2166. https://doi.org/10.1098/rstb.2009.0053

[19]   S. Kumar, K. Thakur, Bioplastics - classification, production and their potential food applications, J. Hill Agric. 8 (2017) 118. https://doi.org/10.5958/2230-7338.2017.00024.6

[20]   R. Porta, Plastic Pollution and the Challenge of Bioplastics, J. Appl. Biotechnol. Bioeng. 2 (2017). https://doi.org/10.15406/jabb.2017.02.00033

[21]   R.F.T. Stepto, Thermoplastic starch, in: Macromol. Symp., John Wiley and Sons Ltd, 2000: pp. 73–82. https://doi.org/10.1002/1521-3900(200003)152:1<73::AID-MASY73>3.0.CO;2-1

[22]   G. Della Valle, A. Buleon, P.J. Carreau, P.A. Lavoie, B. Vergnes, Relationship between structure and viscoelastic behavior of plasticized starch, J. Rheol. 42 (1998) 507–525. https://doi.org/10.1122/1.550900

[23]   Isroi, A. Rahman, K. Syamsu, Biodegradability of oil palm cellulose-based bioplastics, IOP Conf. Ser. Earth Environ. Sci. 183 (2018). https://doi.org/10.1088/1755-1315/183/1/012012

[24]   W.J. Orts, J. Shey, S.H. Imam, G.M. Glenn, M.E. Guttman, J.F. Revol, Application of cellulose microfibrils in polymer nanocomposites, J. Polym. Environ. 13 (2005) 301–306. https://doi.org/10.1007/s10924-005-5514-3

[25]   V. Bátori, M. Jabbari, D. Åkesson, P.R. Lennartsson, M.J. Taherzadeh, A. Zamani, Production of pectin-cellulose biofilms: a new approach for citrus waste recycling, Int. J. Polym. Sci. 2017 (2017). https://doi.org/10.1155/2017/9732329

[26]   S. Kumar, K. Thakur, Bioplastics - classification, production and their potential food applications, J. Hill Agric. 8 (2017) 118. https://doi.org/10.5958/2230-7338.2017.00024.6

[27]   L. Avérous, E. Pollet, Environmental silicate nano-biocomposites, Green Energy Technol. 50 (2012). https://doi.org/10.1007/978-1-4471-4108-2

[28]   A. Jerez, P. Partal, I. Martínez, C. Gallegos, A. Guerrero, Protein-based bioplastics: effect of thermo-mechanical processing, Rheol. Acta. 46 (2007) 711–720. https://doi.org/10.1007/s00397-007-0165-z

[29]   B. Vergnes, G. Della Valle, J. Tayeb, A specific slit die rheometer for extruded starchy products. Design, validation and application to maize starch, Rheol. Acta. 32 (1993) 465–476. https://doi.org/10.1007/BF00396177

[30]   R. Gu, M. Sain, Green polyurethanes and bio-fiber-based products and processes, in : Z. Liu, G. Kraus (Eds.), Green materials from plant oils, RSC publishing, Cambridge, 2014. 77–87.

[31]   M.Thielen, Bioplastics MAGAZINE In: Bioplastics – Plants and crops, raw materials. Published in Berlin: Fachagentur Nachwachsende Rohstoffe e.V. (FNR). Order No. 237, 2014.

[32]   M. Rinaudo, Chitin and chitosan: properties and applications, Prog. Polym. Sci. 31 (2006) 603–632. https://doi.org/10.1016/j.progpolymsci.2006.06.001

[33]   Y.J. Chen, Bioplastics and their role in achieving global sustainability, J. Chem. Pharm. Res. 6 (2014) 226–231.

[34]   M. Brodin, M. Vallejos, M.T. Opedal, M.C. Area, G. Chinga-Carrasco, Lignocellulosics as sustainable resources for production of bioplastics – A review, J. Clean. Prod. 162 (2017) 646–664. https://doi.org/10.1016/j.jclepro.2017.05.209

[35]   M. Zinn, B. Witholt, T. Egli, Occurrence, synthesis and medical application of bacterial polyhydroxyalkanoate, Adv. Drug Deliv. Rev. 53 (2001) 5–21. https://doi.org/10.1016/S0169-409X(01)00218-6

[36]   L. Chen, R.E.O. Pelton, T.M. Smith, Comparative life cycle assessment of fossil and bio-based polyethylene terephthalate (PET) bottles, J. Clean. Prod. 137 (2016) 667–676. https://doi.org/10.1016/j.jclepro.2016.07.094

[37]   S.deVos, Improving heat-resistance of PLA using poly (D-lactide), Bioplastics Mag. 3 (2008) 21–25.

[38]   E. Castro-Aguirre, F. Iñiguez-Franco, H. Samsudin, X. Fang, R. Auras, Poly(lactic acid)—Mass production, processing, industrial applications, and end of life, Adv. Drug Deliv. Rev. 107 (2016) 333–366. https://doi.org/10.1016/j.addr.2016.03.010

[39]   Y. Chen, L.M. Geever, J.A. Killion, J.G. Lyons, C.L. Higginbotham, D.M. Devine, Review of multifarious applications of poly (lactic acid), Polym. Plast. Technol. Eng. 55 (2016) 1057–1075.

[40]   J. Gotro, Polyethylene Furanoate (PEF): 100% Biobased Polymer to Compete with PET, (2013). Available at: https://polymerinnovationblog.com/polyethylene-furanoate-pef-100-biobased-polymer-to-compete-with-pet/.(accessed January 15, 2020)

[41]   C. Sarathchandran, C. Chan, S.R. Karim, Poly(Trimethylene Terephthalate)—The new generation of engineering thermoplastic polyester, in: Phys. Chem. Macromol., Apple Academic Press, 2014: pp. 573–617. https://doi.org/10.1201/b16706-22

[42]   Li Zhao, Hong Hu, S. Wang, Fuzzy-integrative judgment on the end-use performance of knitted fabrics made with polytrimethylene terephthalate blended yarns, Text. Res. J. 81 (2011) 1739–1747. https://doi.org/10.1177/0040517511410103

[43]   J. van Haveren, E.A. Oostveen, F. Miccichè, B.A.J. Noordover, C.E. Koning, R.A.T.M. van Benthem, A.E. Frissen, J.G.J. Weijnen, Resins and additives for powder coatings and alkyd paints, based on renewable resources, J. Coatings Technol. Res. 4 (2007) 177–186. https://doi.org/10.1007/s11998-007-9020-5

Materials Research Forum LLC
https://doi.org/10.21741/9781644901335-2

[44]   J.S. Ling, I. Ahmed Mohammed, A. Ghazali, M. Khairuddean, Novel poly(alkyd-urethane)from vegetable oils: Synthesis and properties, Ind. Crops Prod. 52 (2014) 74–84. https://doi.org/10.1016/j.indcrop.2013.10.002

[45]   J. Xu, B.H. Guo, Poly(butylene succinate) and its copolymers: research, development and industrialization, Biotechnol. J. 5 (2010) 1149–1163. https://doi.org/10.1002/biot.201000136

[46]   A. Oishi, M. Zhang, K. Nakayama, T. Masuda, Y. Taguchi, Synthesis of poly(butylene succinate) and poly(ethylene succinate) including diglycollate moiety, Polym. J. 38 (2006) 710–715. https://doi.org/10.1295/polymj.PJ2005206

[47]   H. Nakajima, P. Dijkstra, K. Loos, The recent developments in biobased polymers toward general and engineering applications: polymers that are upgraded from biodegradable polymers, analogous to petroleum-derived polymers, and newly developed, Polymers (Basel). 9 (2017) 523. https://doi.org/10.3390/polym9100523

[48]   Y. Xia, R.C. Larock, Vegetable oil-based polymeric materials: synthesis, properties, and applications, Green Chem. 12 (2010) 1893. https://doi.org/10.1039/c0gc00264j

[49]   N. Karak, Vegetable oil-based polymers: Properties, Processing and Applications, Woodhead Publishing Limited, Sawston, 2012. https://doi.org/10.1533/9780857097149

[50]   M. Winnacker, B. Rieger, Biobased polyamides: recent advances in basic and applied research, macromol. Rapid Commun. 37 (2016) 1391–1413. https://doi.org/10.1002/marc.201600181

[51]   M. Kyulavska, N. Toncheva-Moncheva, J. Rydz, Biobased Polyamide Ecomaterials and Their Susceptibility to Biodegradation BT - Handbook of Ecomaterials, in: L.M.T. Martínez, O.V. Kharissova, B.I. Kharisov (Eds.), Springer International Publishing, Cham, 2019: pp. 2901–2934. https://doi.org/10.1007/978-3-319-68255-6_126

[52]   I.B. Page, Polyamides as engineering thermoplastic materials, Smithers Rapra Publishing, Shropshire, 2000.

[53]   A. Noreen, K.M. Zia, M. Zuber, S. Tabasum, A.F. Zahoor, Bio-based polyurethane: an efficient and environment friendly coating systems: A review, Prog. Org. Coatings. 91 (2016) 25–32. https://doi.org/10.1016/j.porgcoat.2015.11.018

[54]   M. Alinejad, C. Henry, S. Nikafshar, A. Gondaliya, S. Bagheri, N. Chen, S.K. Singh, D.B. Hodge, M. Nejad, Lignin-based polyurethanes: opportunities for bio-

based foams, elastomers, coatings and adhesives, Polymers (Basel). 11 (2019), 1202. https://doi.org/10.3390/polym11071202

[55]   V. Grimm, M. Braun, O. Teichert, A. Zweck, Biomasse–Rohstoff der Zukunft für die chemische Industrie, Zukünftige Technol. (2011).

[56]   P. Gontia, M. Janssen, Life cycle assessment of bio-based sodium polyacrylate production from pulp mill side streams: case study of thermo-mechanical and sulfite pulp mills, J. Clean. Prod. 131 (2016) 475–484. https://doi.org/10.1016/j.jclepro.2016.04.155

[57]   A. Morschbacker, Basics of bio-polyolefins, Bioplastics. 5 (2010) 52–55.

[58]   R.P. Babu, K. O'Connor, R. Seeram, Current progress on bio-based polymers and their future trends, Prog. Biomater. 2 (2013) 8. https://doi.org/10.1186/2194-0517-2-8

[59]   S. Kumar, S.K. Samal, S. Mohanty, S.K. Nayak, Recent development of biobased epoxy resins: a review, Polym. - Plast. Technol. Eng. 57 (2018) 133–155. https://doi.org/10.1080/03602559.2016.1253742

[60]   P. Geada, V. Vasconcelos, A. Vicente, B. Fernandes, Chapter 13 - Microalgal Biomass Cultivation A2 - Rastogi, Rajesh Prasad, in: D. Madamwar, A.B.T.-A.G.C. Pandey (Eds.), Elsevier, Amsterdam, 2017, 257–284. https://doi.org/https://doi.org/10.1016/B978-0-444-63784-0.00013-8

[61]   A. Rouilly, L. Rigal, Agro-materials: a bibliographic review, J. Macromol. Sci. - Polym. Rev. 42 (2002) 441–479. https://doi.org/10.1081/MC-120015987

[62]   G.O. Aspinall, eds, The Polysaccharides, Elsevier, Amsterdam, 1983. https://doi.org/10.1016/C2013-0-10317-0

[63]   A. Dufresne, S. Thomas, L.A. Pothen, eds., Biopolymer Nanocomposites, John Wiley & Sons, Inc., Hoboken, New Jersey, USA, 2013. https://doi.org/10.1002/9781118609958

[64]   N.L. Lacourse, P.A. Altieri, Biodegradable shaped products and the method of preparation thereof, US Patent No. 5043196, 1991.

[65]   C. Bastioli, Properties and applications of mater-Bi starch-based materials, Polym. Degrad. Stab. 59 (1998) 263–272. https://doi.org/10.1016/S0141-3910(97)00156-0

[66]   J.K. Jang, Y.R. Pyun, Effect of moisture content on the melting of wheat starch, Starch - Starke. 48 (1996) 48–51. https://doi.org/10.1002/star.19960480204

[67]   R.L. Shogren, Effect of moisture content on the melting and subsequent physical aging of cornstarch, Carbohydr. Polym. 19 (1992) 83–90. https://doi.org/10.1016/0144-8617(92)90117-9

[68]   C.L. Swanson, R.L. Shogren, G.F. Fanta, S.H. Imam, Starch-plastic materials-preparation, physical properties, and biodegradability (a review of recent USDA research), J. Environ. Polym. Degrad. 1 (1993) 155–166. https://doi.org/10.1007/BF01418208

[69]   I. Tomka, Thermoplastic starch, in: Adv. Exp. Med. Biol., 1991: pp. 627–637. https://doi.org/10.1007/978-1-4899-0664-9_34

[70]   Bio-based products - overview of standards, (2011). https://doi.org/10.31030/1775170

[71]   L. Avérous, Biodegradable multiphase systems based on plasticized starch: a review, J. Macromol. Sci. - Polym. Rev. 44 (2004) 231–274. https://doi.org/10.1081/MC-200029326

[72]   N.L. Lacourse, P.A. Altieri, Biodegradable packaging material and the method of preparation thereof, US Patent No. 4863655, 1989

[73]   A.K. Mohanty, M. Misra, G. Hinrichsen, Biofibres, biodegradable polymers and biocomposites: an overview, Macromol. Mater. Eng. 276–277 (2000) 1–24. https://doi.org/10.1002/(SICI)1439-2054(20000301)276:1<1::AID-MAME1>3.0.CO;2-W

[74]   A. Chaudhari, R. Kulkarni, P. Mahulikar, D. Sohn, V. Gite, Development of PU coatings from neem oil based alkyds prepared by the monoglyceride route, JAOCS, J. Am. Oil Chem. Soc. 92 (2015) 733–741. https://doi.org/10.1007/s11746-015-2642-3

[75]   F. Bilo, S. Pandini, L. Sartore, L.E. Depero, G. Gargiulo, A. Bonassi, S. Federici, E. Bontempi, A sustainable bioplastic obtained from rice straw, J. Clean. Prod. 200 (2018) 357–368. https://doi.org/10.1016/j.jclepro.2018.07.252

[76]   A.B.M. Sharif Hossain, M.M. Uddin, V.N. Veettil, M. Fawzi, Nano-cellulose based nano-coating biomaterial dataset using corn leaf biomass: an innovative biodegradable plant biomaterial, Data Br. 17 (2018) 162–168. https://doi.org/10.1016/j.dib.2017.12.046

[77]   C.R. Fordyce, Cellulose Esters of Organic Acids, Adv. Carbohydr. Chem. 1 (1945) 309–327. https://doi.org/10.1016/S0096-5332(08)60413-0

[78]   J.A. Reilly, Celluloid objects: their chemistry and preservation, J. Am. Inst. Conserv. 30 (1991) 145. https://doi.org/10.2307/3179527

[79]   S. Sarkanen, Y. Chen, Y.Y. Wang, Journey to polymeric materials composed exclusively of simple lignin derivatives, ACS Sustain. Chem. Eng. 4 (2016) 5223–5229. https://doi.org/10.1021/acssuschemeng.6b01700

[80]   H. Chen , J. Shu, P. Li, B. Chen , N. Li, L. Li, Application of coating chitosan film-forming solution combined β-cd-citral inclusion complex on beef fillet, J. Food Nutr. Res. 2 (2014) 692–697. https://doi.org/10.12691/jfnr-2-10-7

[81]   V. Epure, M. Griffon, E. Pollet, L. Avérous, Structure and properties of glycerol-plasticized chitosan obtained by mechanical kneading, Carbohydr. Polym. 83 (2011) 947–952. https://doi.org/10.1016/j.carbpol.2010.09.003

[82]   A. Prinz, K. Koch, A. Górak, T. Zeiner, Multi-stage laccase extraction and separation using aqueous two-phase systems: experiment and model, Process Biochem. 49 (2014) 1020–1031. https://doi.org/10.1016/j.procbio.2014.03.011

[83]   R. Shukla, M. Cheryan, Zein: The industrial protein from corn, Ind. Crops Prod. 13 (2001) 171–192. https://doi.org/10.1016/S0926-6690(00)00064-9

[84]   M.S. Helgeson, "Horticultural evaluation of zein-based bioplastic containers" (2009). Graduate Theses and Dissertations.10554. https://lib.dr.iastate.edu/etd/10554

[85]   S. Guilbert, C. Guillaume, N. Gontard, New Packaging Materials Based on Renewable Resources: Properties, Applications, and Prospects, in: J.M. Aguilera, R. Simpson, J. Welti-Chanes, D. Bermudez-Aguirre, G. Barbosa-Canovas, eds., Food Engineering Interfaces, Springer, New York, 2011 pp. 619–630. https://doi.org/10.1007/978-1-4419-7475-4_26

[86]   A. Gennadios,eds, Protein-based films and coatings, CRC Press, Boca Raton, 2002.

[87]   S. Nakai, Structure-function relationships of food proteins: with an emphasis on the importance of protein hydrophobicity, J. Agric. Food Chem. 31 (1983) 676–683. https://doi.org/10.1021/jf00118a001

[88]   I. Arvanitoyannis, E. Psomiadou, A. Nakayama, Edible films made from sodium caseinate, starches, sugars or glycerol. Part 1, Carbohydr. Polym. 31 (1996) 179–192. https://doi.org/10.1016/S0144-8617(96)00123-3

[89]   D. Brault, M. LaCroix, M. Ressouany, Biodegradable films containing caseinate and their method of manufacture by irradiation,US Patent No. 6120592, 2000 .

[90]   M. Delgado, M. Felix, C. Bengoechea, Development of bioplastic materials: from rapeseed oil industry by products to added-value biodegradable biocomposite

materials, Ind. Crops Prod. 125 (2018) 401–407.
https://doi.org/10.1016/j.indcrop.2018.09.013

[91] L.L. Madison, G.W. Huisman, Metabolic engineering of poly(3-hydroxyalkanoates): from DNA to plastic., Microbiol. Mol. Biol. Rev. 63 (1999) 21–53.

[92] G.J.. De Koning, Prospects of bacterial poly[(R)-3-(hydroxyalkanoates)], Eindhoven: Technische Universiteit Eindhoven. 1993.
https://doi.org/10.6100/IR403691

[93] L.G. Donaruma, Microbial polyesters, by Yoshiharu Doi, VCH, New York, 1990, 156 pp., J. Polym. Sci. Part A Polym. Chem. 29 (1991) 1365–1365.
https://doi.org/10.1002/pola.1991.080290916

[94] Y. Kathiraser, M.K. Aroua, K.B. Ramachandran, I.K.P. Tan, Chemical characterization of medium-chain-length polyhydroxyalkanoates (PHAs) recovered by enzymatic treatment and ultrafiltration, J. Chem. Technol. Biotechnol. 82 (2007) 847–855. https://doi.org/10.1002/jctb.1751

[95] G.Q. Chen, M.K. Patel, Plastics derived from biological sources: present and future: a technical and environmental review, Chem. Rev. 112 (2012) 2082–2099.
https://doi.org/10.1021/cr200162d

[96] H.J. Endres, A. Siebert-Raths, Engineering Biopolymers, Carl Hanser Verlag GmbH & Co. KG, München, 2011. https://doi.org/10.3139/9783446430020

[97] D. Garlotta, A Literature Review of Poly(Lactic Acid), J. Polym. Environ. 9 (2001) 63–84. https://doi.org/10.1023/A:1020200822435

[98] L. Avérous, E. Pollet, Biodegradable Polymers. In: L. Avérous, E. Pollet, (eds) Environmental Silicate Nano-Biocomposites. Green Energy and Technology. Springer, London, 2012, 13-39

[99] Y.J. Wee, J.N. Kim, H.W. Ryu, Biotechnological production of lactic acid and its recent applications, Food Technol. Biotechnol. 44 (2006) 163–172.

[100] S. Il Moon, C.W. Lee, M. Miyamoto, Y. Kimura, Melt polycondensation of L-lactic acid with Sn(II) catalysts activated by various proton acids: A direct manufacturing route to high molecular weight Poly(L-lactic acid), J. Polym. Sci. Part A Polym. Chem. 38 (2000) 1673–1679. https://doi.org/10.1002/(SICI)1099-0518(20000501)38:9<1673::AID-POLA33>3.0.CO;2-T

[101] S.Il. Moon, C.W. Lee, I. Taniguchi, M. Miyamoto, Y. Kimura, Melt/solid polycondensation of l -lactic acid: an alternative route to poly(l -lactic acid) with high

molecular weight, Polymer (Guildf). 42 (2001) 5059–5062.
https://doi.org/10.1016/S0032-3861(00)00889-2

[102] M. Jamshidian, E.A. Tehrany, M. Imran, M. Jacquot, S. Desobry, Poly-Lactic acid: production, applications, nanocomposites, and release studies, Compr. Rev. Food Sci. Food Saf. 9 (2010) 552–571. https://doi.org/10.1111/j.1541-4337.2010.00126.x

[103] L. V. Labrecque, R.A. Kumar, V. Dav, R.A. Gross, S.P. McCarthy, Citrate esters as plasticizers for poly(lactic acid), J. Appl. Polym. Sci. 66 (1997) 1507–1513. https://doi.org/10.1002/(SICI)1097-4628(19971121)66:8<1507::AID-APP11>3.0.CO;2-0

[104] S. Jacobsen, H.G. Fritz, Plasticizing polylactide-the effect of different plasticizers on the mechanical properties, Polym. Eng. Sci. 39 (1999) 1303–1310. https://doi.org/10.1002/pen.11517

[105] O. Martin, L. Avérous, Poly(lactic acid): plasticization and properties of biodegradable multiphase systems, Polymer (Guildf). 42 (2001) 6209–6219. https://doi.org/10.1016/S0032-3861(01)00086-6

[106] A. Steinbüchel, eds, Biopolymers, General Aspects and Special Applications v. 10,Wiley-Vch, Weinheim, 2003.

[107] A. Albertsson, U. Edlund, I.K. Varma, Synthesis, chemistry and properties of hemicelluloses, Biopolym. Mater. Sustain. Film. Coatings. (2011) 133–150.

[108] R.G. Sinclair, The Case for Polylactic Acid as a Commodity Packaging Plastic, J. Macromol. Sci. Part A. 33 (1996) 585–597. https://doi.org/10.1080/10601329608010880

[109] G. Kale, R. Auras, S.P. Singh, Degradation of commercial biodegradable packages under real composting and ambient exposure conditions, J. Polym. Environ. 14 (2006) 317–334. https://doi.org/10.1007/s10924-006-0015-6

[110] E.S. Stevens, Green plastics: an introduction to the new science of biodegradable plastics, Princeton University Press, New Jersey, 2002.

[111] T. Rieckmann, S. Völker, Micro-kinetics and mass transfer in poly(ethylene terephthalate) synthesis, Chem. Eng. Sci. 56 (2001) 945–953. https://doi.org/10.1016/S0009-2509(00)00309-2

[112] D. Komula, Completing the puzzle: 100% plant-derived PET, Bioplastics Mag. 6 (2011) 14–17.

[113] D.I. Collias, A.M. Harris, V. Nagpal, I.W. Cottrell, M.W. Schultheis, Biobased terephthalic acid technologies: a literature review, Ind. Biotechnol. 10 (2014) 91–105.

[114] A. Bušić, N. Marđetko, S. Kundas, G. Morzak, H. Belskaya, M. Ivančić Šantek, D. Komes, S. Novak, B. Šantek, Bioethanol production from renewable raw materials and its separation and purification: a review, Food Technol. Biotechnol. 56 (2018). https://doi.org/10.17113/ftb.56.03.18.5546

[115] Q. Xie, X. Hu, T. Hu, P. Xiao, Y. Xu, K.W. Leffew, Polytrimethylene terephthalate: an example of an industrial polymer platform development in China, Macromol. React. Eng. 9 (2015) 401–408. https://doi.org/10.1002/mren.201400070

[116] C. Saricam, N. Okur, Polyester Usage for Automotive Applications, in: Polyest. - Prod. Charact. Innov. Appl., InTech, London, 2018. https://doi.org/10.5772/intechopen.74206

[117] C. Sarathchandran, C. Chan, S.R. Karim, Poly(Trimethylene Terephthalate)-The New Generation of Engineering Thermoplastic Polyester, in: Phys. Chem. Macromol., Apple Academic Press, New Jersey, 2014: pp. 573–617. https://doi.org/10.1201/b16706-22

[118] J. van Haveren, E.A. Oostveen, F. Miccichè, B.A.J. Noordover, C.E. Koning, R.A.T.M. van Benthem, A.E. Frissen, J.G.J. Weijnen, Resins and additives for powder coatings and alkyd paints, based on renewable resources, J. Coatings Technol. Res. 4 (2007) 177–186. https://doi.org/10.1007/s11998-007-9020-5

[119] M. Bautista, A. de Ilarduya, A. Alla, S. Muñoz-Guerra, Poly(butylene succinate) Ionomers with enhanced hydrodegradability, Polymers (Basel). 7 (2015) 1232–1247. https://doi.org/10.3390/polym7071232

[120] J. Xu, B.H. Guo, Microbial Succinic Acid, Its Polymer Poly(butylene succinate), and Applications, in: Chen GQ. (eds) Plastics from Bacteria. Microbiology Monographs, vol 14. Springer, Berlin, 2010: pp. 347–388. https://doi.org/10.1007/978-3-642-03287-5_14

[121] B. Bai, J. Zhou, M. Yang, Y. Liu, X. Xu, J. Xing, Efficient production of succinic acid from macroalgae hydrolysate by metabolically engineered *Escherichia coli*, Bioresour. Technol. 185 (2015) 56–61. https://doi.org/10.1016/j.biortech.2015.02.081

[122] C.T. Brunner, E.T. Baran, E.D. Pinho, R.L. Reis, N.M. Neves, Performance of biodegradable microcapsules of poly(butylene succinate), poly(butylene succinate-co-adipate) and poly(butylene terephthalate-co-adipate) as drug encapsulation systems,

Colloids Surfaces B Biointerfaces. 84 (2011) 498–507.
https://doi.org/10.1016/j.colsurfb.2011.02.005

[123] C. Lavilla, A. Alla, A. Martínez de Ilarduya, E. Benito, M.G. García-Martín, J.A.
Galbis, S. Muñoz-Guerra, Bio-based poly(butylene terephthalate) copolyesters
containing bicyclic diacetalized galactitol and galactaric acid: Influence of
composition on properties, Polymer (Guildf). 53 (2012) 3432–3445.
https://doi.org/10.1016/j.polymer.2012.05.048

[124] A. Künkel, J. Becker, L. Börger, J. Hamprecht, S. Koltzenburg, R. Loos, M.B.
Schick, K. Schlegel, C. Sinkel, G. Skupin, M. Yamamoto, Polymers, Biodegradable,
in: Ullmann's Encycl. Ind. Chem., Wiley-VCH Verlag GmbH & Co. KGaA,
Weinheim, Germany, 2016: pp. 1–29. https://doi.org/10.1002/14356007.n21_n01.pub2

[125] M. Gobin, P. Loulergue, J.-L. Audic, L. Lemiègre, Synthesis and characterisation
of bio-based polyester materials from vegetable oil and short to long chain
dicarboxylic acids, Ind. Crops Prod. 70 (2015) 213–220.
https://doi.org/10.1016/j.indcrop.2015.03.041

[126] S. Miao, P. Wang, Z. Su, S. Zhang, Vegetable-oil-based polymers as future
polymeric biomaterials, Acta Biomater. 10 (2014) 1692–1704.
https://doi.org/10.1016/j.actbio.2013.08.040

[127] D. Ogunniyi, Castor oil: a vital industrial raw material, Bioresour. Technol. 97
(2006) 1086–1091. https://doi.org/10.1016/j.biortech.2005.03.028

[128] M. Genas, Rilsan (Polyamid 11), Synthese und Eigenschaften, Angew. Chemie. 74
(1962) 535–540. https://doi.org/10.1002/ange.19620741504

[129] M. Kyulavska, N. Toncheva-Moncheva, J. Rydz, Biobased Polyamide
Ecomaterials and Their Susceptibility to Biodegradation, in: Handb. Ecomater.,
Springer International Publishing, Cham, 2017: pp. 1–34. https://doi.org/10.1007/978-
3-319-48281-1_126-1

[130] R. Holsti-Miettinen, J. Seppälä, O.T. Ikkala, Effects of compatibilizers on the
properties of polyamide/polypropylene blends, Polym. Eng. Sci. 32 (1992) 868–877.
https://doi.org/10.1002/pen.760321306

[131] F.E. Golling, R. Pires, A. Hecking, J. Weikard, F. Richter, K. Danielmeier, D.
Dijkstra, Polyurethanes for coatings and adhesives – chemistry and applications,
Polym. Int. 68 (2019) 848–855. https://doi.org/10.1002/pi.5665

[132] J. D'Souza, N. Yan, Producing bark-based polyols through liquefaction: effect of liquefaction temperature, ACS Sustain. Chem. Eng. 1 (2013) 534–540. https://doi.org/10.1021/sc400013e

[133] G.M. Yee, M.A. Hillmyer, I.A. Tonks, Bioderived acrylates from alkyl lactates via pd-catalyzed hydroesterification, ACS Sustain. Chem. Eng. 6 (2018) 9579–9584. https://doi.org/10.1021/acssuschemeng.8b02359

[134] A. Morschbacker, Bio-ethanol based ethylene, J. Macromol. Sci. Part C Polym. Rev. 49 (2009) 79–84.

[135] J. Jane, Starch properties, modifications, and applications, J. Macromol. Sci. Part A Pure Appl. Chem. 32 (1995) 751–757.

[136] H. Eslami, M.R. Kamal, Elongational rheology of biodegradable poly (lactic acid)/poly [(butylene succinate)-co-adipate] binary blends and poly (lactic acid)/poly [(butylene succinate)-co-adipate]/clay ternary nanocomposites, J. Appl. Polym. Sci. 127 (2013) 2290–2306.

[137] K. Bula, Ł. Klapiszewski, T. Jesionowski, A novel functional silica/lignin hybrid material as a potential bio-based polypropylene filler, Polym. Compos. 36 (2015) 913–922. https://doi.org/10.1002/pc.23011

[138] V. Koncar, Composites and hybrid structures, in: Smart Text. Situ Monit. Compos., Elsevier, Amsterdam, 2019: pp. 153–215. https://doi.org/10.1016/B978-0-08-102308-2.00002-4

[139] A. Rudin, P. Choi, Chapter 13 - Biopolymers, in: A. Rudin, P.B.T.-T.E. of P.S.& E. (Third E. Choi (Eds.), Academic Press, Boston, 2013: pp. 521–535. https://doi.org/https://doi.org/10.1016/B978-0-12-382178-2.00013-4

[140] C. Smith, Braskem commits to producing bio-based polypropylene, Plast. News. 28 (2010).

[141] S.A. Solvay, Solvay Indupa will produce bioethanol-based vinyl in Brasil & considers state-of-the-art power generation in Argentina, Brussels, Belgium, December. (2007).Avilable at : https://www.chemeurope.com/en/news/75840/solvay-indupa-will-produce-bioethanol-based-vinyl-in-brasil-considers-state-of-the-art-power-generation-in-argentina.html (accessed January 15, 2020)

[142] F.L. Jin, X. Li, S.J. Park, Synthesis and application of epoxy resins: A review, J. Ind. Eng. Chem. 29 (2015) 1–11. https://doi.org/10.1016/j.jiec.2015.03.026

[143] V. Muralidharan, M.S. Arokianathan, M. Balaraman, S. Palanivel, Tannery trimming waste based biodegradable bioplastic: facile synthesis and characterization

of properties, Polym. Test. 81 (2020).
https://doi.org/10.1016/j.polymertesting.2019.106250

[144] S. Huda, Y. Yang, Feather fiber reinforced light-weight composites with good acoustic properties, J. Polym. Environ. 17 (2009) 131–142.
https://doi.org/10.1007/s10924-009-0130-2

[145] N. Reddy, Y. Yang, Structure and properties of chicken feather barbs as natural structure and properties of chicken feather barbs as natural protein fibers protein fibers, J. Polym. Environ. 15 (2007) 81–87. https://doi.org/10.1007/s10924-007-0054-7

[146] W.I.A. Saber, M.M. El-Metwally, M.S. El-Hersh, Keratinase production and biodegradation of some keratinous wastes by alternaria tenuissima and *Aspergillus nidulans*, Res. J. Microbiol. 5 (2010) 21–35. https://doi.org/10.3923/jm.2010.21.35

[147] A. Ullah, T. Vasanthan, D. Bressler, A.L. Elias, J. Wu, Bioplastics from feather quill, Biomacromolecules. 12 (2011) 3826–3832. https://doi.org/10.1021/bm201112n

[148] J. Yaradoddi, V. Patil, S. Ganachari, N. Banapurmath, A. Hunashyal, A. Shettar, J.S. Yaradoddi, Biodegradable plastic production from fruit waste material and its sustainable use for green application ,Int. J. Pharm. Res. Allied Sci. 5 (2016) 56–66.

[149] L De las Fuentes "AWARENET: Agro-food wastes minimization and reduction network." *WIT Transactions on Ecology and the Environment* 56 (2002).

[150] I.S. Bayer, S. Guzman-Puyol, J.A. Heredia-Guerrero, L. Ceseracciu, F. Pignatelli, R. Ruffilli, R. Cingolani, A. Athanassiou, Direct transformation of edible vegetable waste into bioplastics, Macromolecules. 47 (2014) 5135–5143.
https://doi.org/10.1021/ma5008557

[151] S. Obruca, I. Marova, O. Snajdar, L. Mravcova, Z. Svoboda, Production of poly(3-hydroxybutyrate-co-3-hydroxyvalerate) by Cupriavidus necator from waste rapeseed oil using propanol as a precursor of 3-hydroxyvalerate, Biotechnol. Lett. 32 (2010) 1925–1932. https://doi.org/10.1007/s10529-010-0376-8

[152] R.A.J. Verlinden, D.J. Hill, M.A. Kenward, C.D. Williams, Z. Piotrowska-Seget, I.K. Radecka, Production of polyhydroxyalkanoates from waste frying oil by cupriavidus necator, AMB Express. 1 (2011) 1–8. https://doi.org/10.1186/2191-0855-1-11

[153] Y. Jiang, L. Marang, J. Tamis, M.C.M. van Loosdrecht, H. Dijkman, R. Kleerebezem, Waste to resource: converting paper mill wastewater to bioplastic, Water Res. 46 (2012) 5517–5530. https://doi.org/10.1016/j.watres.2012.07.028

Chapter 3

Degradable Plastic Recycling

Nadia Akram[1*], Khalid Mahmood Zia[1], Asim Mansha[1]

[1] Department of Chemistry, Government College University Faisalabad, Faisalabad-38000.Pakistan

* nadiaakram@gcuf.edu.pk

Abstract

The public demand of plastics for food, drinks, consumable and packaging is increasing enormously all over the world. Due to limited available plastic resources, it is challenging to meet the stipulation of the massive population. The contribution of the synthetic plastic industry is encouraging to cope with these challenges. However, it is not only restricted towards production, but the degradation of its waste is also equally arduous and even more complicated to a large extent. A useful solution to this problem is recycling instead of degradation. In order to optimize the utility of recycling, various techniques are in progress. Plastic recycling is an acceptable technique to keep the economy in circulation. Moreover, it is an effective way to reduce the environmental pollution and to promote green environment.

Keywords

Plastic Industry, Global Warming, Economy Circulation, Environment, Reusability

Contents

1.   Introduction to degradable & biodegradable plastics

Twentieth century is marked with one of the greatest invention of plastics, consisting of large number of synthetic or semi synthetic organic materials. Plastics are high molecular weight polymers conventionally derived from petrochemical resources but not limited to such materials only. A large number of products are derived from renewable resources preferably termed as "bioplastics" extracted from biobased, refined materials which are biodegradable in nature. Undoubtedly, plastics are very convenient to use in food, packaging and all sort of consumables. The clean and lightweight material can not only be used conveniently, but it can also be thrown away easily along with other debris [1]. As the concerns towards a clean and safe planet grew, society became more vigilant towards disposal modes of plastics. Soon it was realized, that life without plastic is unimaginable but what sort of safety measures can be adopted? This was a very important question which lead scientists to make a clear distinction among the sources and products of the plastic materials. With the enormous efforts of chemists, biochemists, environmentalists, scientists were able to figure out that all the plastics can ultimately be disposed of but at the cost of our health and environment. Plastics and their toxicity have endangered the planet making its proper disposal indispensable. In order to develop a safe method of disposal, the plastics have been categorized into degradable and biodegradable plastics, it is mere effort to evade the severe misperception amongst the community surrounding the difference between degradable and biodegradable plastics (Fig. 1). The word degrades refers to "break down" and essentially, all plastics are degradable. The degradable plastics are also known as "oxo degradable plastics". Polyethene bags or commonly known as traditional shopping bags are the product of a non-renewable resource; these bags are exceptionally injurious to the environment as by no means they are biodegraded. Instead, they can only disintegrate into tiny pieces over hundreds of years, inflicting indescribable damage to the natural ecosystem. Conspicuously, degradable plastics do not possess living organisms as a vital part of the degradation process. Instead, chemical constituents used in the plastic allow it to disintegrate faster

Materials Research Forum LLC
https://doi.org/10.21741/9781644901335-3

than a standard plastic usually would. Principally, the plastics advertised as "degradable"' are undeniably not advantageous, they can even be *"worst"* for the environment! The degradable plastics can adopt various modes of degradation including, aerobic, aquatic, thermal and chemical (Fig. 2a). Hence, degradable plastic will find multiple environments to degrade, however it is not as simple. As a matter of fact it takes time if just dumped on land. The process is usually speedup when burned. However, it hazardously affects the environment in all its disposal modes. In every mode of its disposal, degradation of plastics converts them into smaller fragments and then poses serious threats to ecosystem. Alarmingly, even the microplastics enter the food chain, get eaten by smaller species and then continue to make their way through the food chain when these smaller species are consumed in the ecosystem [2]

On the other hand, biodegradables are differently identified as compared to degradable plastics. These materials can be "biobased plastics" or "petroleum based plastics". Biodegradable biobased plastics are the product of plant-based materials like corn, soy, sugarcane, potato, or other renewable material source. However, the biodegradable petroleum plastics can be the product of synthetic polymers as well. It is convenient to acclimate the composition of biodegradable plastic to degrade it at a faster rate. The resultant fragments are generally less toxic to the environment based on its composition. The life cycle of biodegradable plastic is quite simple as shown in (Fig. 2b). Biodegradable plastics are considered superior over petrochemical-based plastic due to their higher proportion of natural sources which greatly diminishes their probability of eventually converting into a horrid plastic pollutant. Specific environmental factors including temperature, pressure, humidity, natural gases are required for the process of biodegradation. Biodegradable plastics are transformed to biomass, CH_4, CO_2, and H_2O and other organic residues by a thermochemical protocol in a certain time and environment. Numerous biodegradable plastics provides the solution to many environmental problems and it is the need of the hour to develop products and techniques to enhance the functionality of biodegradable plastics so that they can replace the traditional plastics [3].

Materials Research Forum LLC

https://doi.org/10.21741/9781644901335-3

Fig. 1 Degradable and non degradable plastics.

Fig. 2 (a): Degradation modes of plastics: (A) Aerobic degradation ;(B);Aquatic degradation(C)Thermal degradation(D);Chemical degradation.

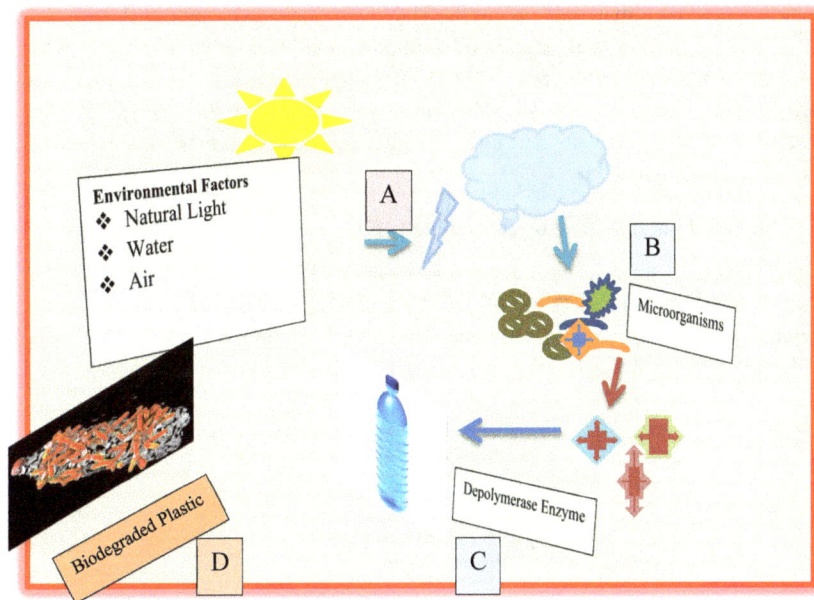

Fig. 2(b): Life cycle of Biodegradable Plastics. (A); Environmental factors of natural heat, wind and water under suitable conditions nourish micro-organisms(B); Micro-organism boost the enzyme activity (C); Depolymerase enzyme attacks the plastic substrate (D); Enzymatic reaction engulf biodegradable plastic.

Currently, there are two eminent forms of plastics which are of great importance. One is the petrochemicals polymers like polyethylene (PE) and polypropylene (PP). On the other side is the starch based biobased plastics such as Polylactic acid (PLA). [4]

2. Petroleum based plastics

This category utilizes simple and conventional methods of drying and processing and is basically known to develop plastics from petrochemicals sources. However, biobased petrochemical plastics are also available.

2.1 Biobased Polypropylene

Typically, the polypropylene (PP) (Fig. 3) is petroleum based synthesized from natural gas and petroleum products however, it can also be produced from sugarcane and other

Degradation of Plastics   Materials Research Forum LLC
Materials Research Foundations **99** (2021) 81-93   https://doi.org/10.21741/9781644901335-3

organic plant-based materials as well. The biobased polypropylene has the same composition as conventional petroleum based polypropylene. Biobased polypropylene is produced from ethanol by the process of fermentation which converts the sugar of sugar cane, corn or potato to ethanol. The ethanol is converted into polypropylene via formation of butylene and ethylene as an intermediate step, known as metathesis reaction of butylene and ethylene [5].

2.2 Biobased polyethylene terephthalate (PET)

Another important form of bio based polyethylene is PET (Fig. 3),which is made from plant sources of monoethylene glycol (MEG) and terephthalic acid. There is usually no difference in the properties of petroleum based PET or biobased. Only 30% biobased PET is currently synthesized which can be extended to 100% in near future [6].

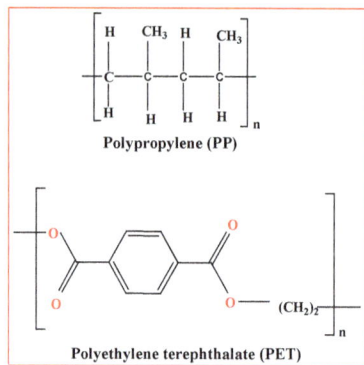

Fig. 3 Chemical structures of polyprolene (PP) and polyethelene terephthalate (PET).

3. Recycling

Recycling is the process of conversion of a waste material into a new product. It is usually performed with the intention to attain the original properties of the material. The concept gained popularity as an alternative to the conventional attempt to reduce the greenhouse gas emission from waste materials. However, currently recycling is more useful than the previous practices. Now a days it is the prevention of disposal of useful materials and also an attempt to diminish the consumption of new raw materials, thus reducing not only the energy usage but also the pollution. The two important categories of recycling are the closed-loop recycling and cradle to cradle recycling; the former describe a process where the recycled material is converted to original material without giving it a new form. Hence polypropylene will essentially remain polypropylene even after recycling. Whereas, the later term promotes the concept in which the industrial practices imitate nature in a closed loop by recycling feedstock materials in an endless loop, the waste produced during this process is converted into feedstock for a succeeding process. The recycling is based on a very systematic approach of "Life cycle assessment" popularly known as (LCA), it is based on the principle of product's life from "cradle to grave," It describes how a material is produced, how its raw materials are extracted, which kind of process are involved in its manufacturing, how the product is manufactured, where it was distributed, how it was used, when it was expired, disposed of or recycle. So LCA is a complete guide of any product. Internationally the recycling is represented with a "chasing arrow triangle" of arrows as shown in (Fig. 4).

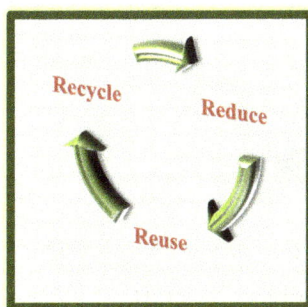

Fig. 4 Conventional symbol of Recycling.

The process of recycling is not complicated as shown in Fig. 5. The recycling of a plastic bottle can be carried out in either chemical or biochemical way ending up in the useful intermediates and final products. However, the cross-contamination during the recycling can make it a real burden on the recycling industry. Prolific recycling typically requires separation and cleaning of materials before they are transported to manufacturers to launch it as a new brand, nevertheless inducing discordant materials makes it challenging to process occasionally unsuccessful recycling trials, resulting in yield loss and end up disposing of the material. Recycling process varies for each materials and different protocols are developed for better recycling yield. Recycling is also dependent on the

Materials Research Forum LLC

https://doi.org/10.21741/9781644901335-3

design of new product and the techniques of assembling various parts. The recycling automobile industry is considered difficult and for the reproduction of various vehicles mostly biodegrable materials is preferred in various parts [7-9].

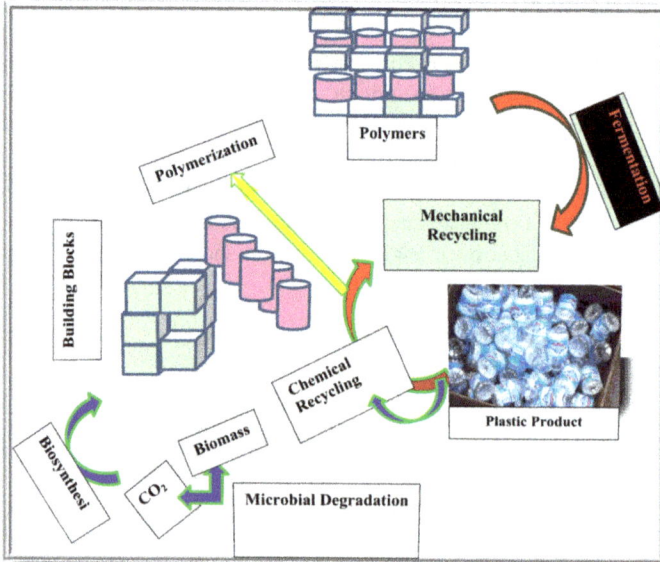

Fig. 5 Biobased plastic recycling Process.

Recycling is not devoid of issues. There are many concerns related to recycling such as which items can be recycled? How many times an item can be recycled? In order to answer the first question for example we assume that plastic and paper can be recycled. Paper can be recycled easily whereas the situation for plastic is very complexed. The one thing which is most important in this regard is the lack of public awareness. Many of us do not know that either a useable item can be recycled or not. Primarily, it is the duty of production agencies and manufacturing companies to write the information guidelines regarding the recycling on packaging. One another way to promote recycling is the promotion of reusable material. It will keep the product in circulation for a longer time for example; packaging bags, boxes, envelops can be reused several times by the proper awareness campaign [10-12]. Another important issue related with recycling is to discard the materials right in the litter, but lots of our recyclable materials actually end up in landfills or oceans, it makes a real mess to the ecosystems. As a matter of fact recycling forces customers to take the responsibility to manage the waste, as a result the customer sometimes look for the alternate available resources instead of using such materials. Consequently the, recycling industry is working effectively but the piles of waste are increasing as well. It has resulted in the proper legislation of the recycling material and to dispose of the solid waste material.

Plastic waste management prohibits the dumping of plastic into the sea which was used as a major dumping area due to lack of awareness regarding marine life and without considering a major part of the ecosystem. The plastic waste is one of the major causes of marine pollution. A movement aroused throughout the USA to question the ability of plastic to degrade which finally resulted into the spread of more public awareness regarding the use of plastic bag. On the other continents, European union made it requirements that all automobiles companies must use 85 % of material which can be reused [9].

3.1 Recyclable bioplastics

Bioplastics made from renewable resources possess the natural tendency to be recycled by biological methods using limited resources and reduced greenhouse gas emission. However, the third generation bioplastics are bio-based which has dominated the plastic market of – polyethylene (PE), polypropylene (PP) and polyethylene terephthalate (PET). It is expected that in near future all this market will be replaced by bioplastics.

3.2 Reusable plastic bag

Plastic bags are best example of reusable materials; these are available in both biodegradable and non-biodegradable materials. The biodegradable plastic bags are major contributors to reduce the environmental pollution. The reuse of plastic bags overcome number of problems such as; cost, greenhouse gas emission, multiple type of pollution including land and water pollution. Plastic bags can usually be washed which reduces the chances of infection of bacteria [13].

3.3 Recycling issues related to biodegradable plastic

The recycling of biodegradable plastics is very challenging due to removal of macro and micro contaminations. The biodegradable plastic displays a very unique affinity with the microbes and becomes polluted instantly. Hence the convenience in the biodegradation is actually a great hurdle in recycling. However, all the impurities cannot be identified during this process. For example polylactic acid (PLA) and polyethylene terephthalate

(PET) has the same density, but after the recycling process PLA exhibits lower quantity of impurities as compared to PET which not only seriously impact the quality of recycling process but also the quality of the product as well. The situation is the same for some other materials where contamination cannot be avoided despite taking all the necessary measures. The problem becomes sever in case of multilayers due to greater chances of contaminants to be entrapped in between layers. Research show that 10 % contamination of biodegradable plastics does not show any negative impact and the quality of recycled material can be ensured below this range. For optimal recycling ability the mechanical system of recycling should be adopted which is useful for both biodegradable and non-biodegradable plastics [14,15].

3.4   Marine plastic debris recycling

Plastic pollution is a major threat to the oceans. The enormous usage of plastics poses a serious threat to the aquatic environments, the marine debris includes both degradable and nondegradable plastics however, the nondegradable severely affects the marine life. Plastics disintegrate into smaller and smaller pieces, unfortunately these smaller parts never settle down, nor are they absorbed in the aquatic system, resultantly they are always disguised until engulfed by some aquatic life. Marine debris acts as sponges absorbing toxins, hence any impurities entrapped in these materials becomes the part of the ocean, when any marine animal try to digest it as a food it becomes a part of food chain. Ocean gyre is very important part of marine system. It is a large system of ocean currents, these currents are usually generated by wind blowing. The mass of ocean gyre varies from 1000 miles to 5000 miles in the south pacific. These gyres are known for collecting marine waste, while in the north pacific region it is titled as "Garbage patch" because it is one of the major area of marine waste collection. This gyre absorbs excessive amount of plastic which is broken because of water current or due to wave movement, the floating plastic debris produce rings of gyre. A similar example is observed in south pacific region of Japan where the quantity of plastic debris has increased to a large extent during the last 20 years. According to Marine research foundation the plastic creates naval litter and contaminate the water of beach, surface and even in the depth of ocean. The large amount of suspended plastic ultimately settles down in the bottom. The seashores and beaches are the major sites of sea litter, it is harmful to human beings and marine creatures. Sea birds that swallow the plastic floating on sea surface often feed themselves and their children, the sea litter, plastic bottles, caps buttons party balloons are swallowed by the birds resultantly they often die of these remains in stomach [15, 16].

Degradation of Plastics

Materials Research Foundations **99** (2021) 81-93

Materials Research Forum LLC

https://doi.org/10.21741/9781644901335-3

The most feasible solution to the marine debris is to treat it with plastic recycling technology which is employed for industrial and domestic plastic recycling. Plastic recycling technologies are classified into mechanical recycling, chemeical recycling and thermal recycling. However, there are some factors which make the recycling difficult due to degradation of materials such as:

1. Decomposing via fungi, this leads to of fiber swelling and deteriorates the plastic swiftly. 2. Availability of polar and non-polar groups, making the recycling difficult once the plastic is in water, 3. The inconsistency in the temperature in different zones of the ocean, especially the upper limit of temperature. In order to reduce the natural moisture ability of fibers hydrothermal method is applied. Similarly, chemical treatment leads to the disintegration of structural bonds in non cellulosic material such as; acrylates [17].

3.5   Recycling of packaging plastics

As a matter of fact the 7.8 billion population of the world is using the plastic material in any form, even the demand of developing plastic goods is increasing continuously. It only gives us a clue about how gigantic plastic is! Day by day, the concern over the problem of environmental waste is increasing and undoubtedly plastic industry is the major contributor. It produces 20 to 30% of whole waste. There is no solution of waste problem except recycling. Indeed, recycling rates remain small (approx. 14%) in the plastic packaging field on a global scale. Even in Europe, the recycling rate is approx. 32.5 wt%. However, these values only deal with the collected waste instead of total circulated plastic. Europe is dealing with the waste problem by introducing the circular economy policy. Indeed circular economy is badly affected by the principle of "take, make and dispose:, it is now facing the consequences of decrease in natural resources [18-20].

Conclusions

Plastic is a giant industry producing a vast range of products for over 7 billion people of the world, majority of these plastic products are of daily use. The huge pile of waste given to the earth is indescribable. This waste is dumped on land, in the air and in water. The disposal of this plastic waste is expensive, tedious and an endless task, even the resources of advance countries never allow them to invest only to dispose of the material. In this situation we are left with no option but to recycle the degradable plastics. After a long time, the world has recognized the dire need and the potential challenges of recycling. People are more interested to use recyclable products to promote ecofriendly environment. However, prioritized reduction of plastic consumption and turning towards

Materials Research Forum LLC
https://doi.org/10.21741/9781644901335-3

recycling is the only safe route to meet the demands of the growing population worldwide.

Challenges and future perspective

The future of degradable plastic recycling industry is very challenging. Great efforts will be required to make people realize that use of plastic is associated with hidden hazards. Humans have already started to face the some consequences of the use of plastics but the future is more polluted by the heaps of unrecycled plastics. They are not going to dispose of easily, and now we cannot eliminate it from our lives as well. Hence, it will need a common legislation to be adopted by the world to conserve the natural resources and to control the pollution caused by plastic litter. It will not be easy to bring 195 countries of the world on a single page. However, by adopting the right strategies this industry has the potential to be emerged equally important as plastics are today. It will need multisector collaboration to be emerged and accepted.

References

[1] P. Anastas, J. Zimmerman, Design through the twelve principles of green engineering, Environ. Sci. Technol. 37 (2003) 94A–101A. https://doi.org/10.1021/es032373g

[2] G. Atkinson, K. Hamilton. Savings, growth, and the resource cure hypothesis. World. Dev. 31 (2003) 1893–1807. https://doi.org/10.1016/j.worlddev.2003.05.001

[3] F. Esteves, J. Santos, P. Anunciacao, Sustainability in the information society a proposal of information systems requirements in view of the DPOBE model for organizational sustainability. *Procedia. Technol.* 5 (2012) 599–606. https://doi.org/10.1016/j.protcy.2012.09.066

[4] F. Granek, M. Hassanali. The toronto region sustainability program insights on the adoption of pollution prevention practices by small to medium-sized manufacturers in the Greater Toronto Area (GTA). J. Clean. Prod. 14 (2006) 572–579. https://doi.org/10.1016/j.jclepro.2005.07.008

[5] A Kulig, H. Kolfoort, R. Hoekstra, The case for the hybrid capital approach for the measurement of the welfare and sustainability. Ecol. Indic. 10 (2010) 118–128. https://doi.org/10.1016/j.ecolind.2009.07.014

[6] W.McDonough, M.Braungart, P.T. Anastas, J.B. Zimmerman. Applying the principles of green engineering to cradle-to-cradle design. Environ. Sci. Technol. 37 (2003) 434A–441A. https://doi.org/10.1021/es0326322

[7] L. White, B. Noble. Strategic environmental assessment for sustainability, A review of a decade of academic research. Environ. Impact Assess. Rev. (2013)4260–66. https://doi.org/10.1016/j.eiar.2012.10.003

[8] H. Beltrami, E. Bourlon. Ground waring patterns in the Northern Hemisphere during the last five centuries. Earth Planet. Sci. Lett. 227 (2004) 169–177. https://doi.org/10.1016/j.epsl.2004.09.014

[9] E. Berthier, Y. Arnaud, R. Kumar, S. Ahmad, P. Wagnon, P. Chevallier. Remote sensing estimates of glacier mass balances in the Himachal Pradesh (Western Himalaya, India). Remote. Sens. Environ. 108 (2007) 27–338. https://doi.org/10.1016/j.rse.2006.11.017

[10] E. J.Carpenter, K. L. Smith Jr. Plastics on the Sargasso Sea surface. J. Sci 175 (1972) 1240–1241. https://doi.org/10.1126/science.175.4027.1240.

[11] E. Carpenter, S. Anderson, G. Harvey, H. Miklas, B. Peck. Polystyrene spherules in coastal waters. J. Sci. 178 (1972) 749–750. https://doi.org/10.1126/science.178.4062.749

[12] A. Cazenave, A. Lombard, W. Llovel. Present-day sea level rise: A synthesisHausse actuelle du niveau de la mer: synthèse .340 (2008) 761–770. https://doi.org/10.1016/j.crte.2008.07.008

[13] M. Akiyama, T. Tsuge, Y. Doi. Environmental life cycle comparison of polyhydroxyalkanoates produced from renewable carbon resources by bacterial fermentation. Polym. Degrad. Stabil. 80 (2003) 183–194. https://doi.org/10.1016/S0141-3910(02)00400-7

[14] F. Aouada, L. Mattoso, E. Longo. New strategies in the preparation of exfoliated thermoplastic starch–montmorillonite nanocomposites. Ind. Crops.Prod. 34 (2011). 1502–1508. https://doi.org/10.1016/j.indcrop.2011.05.003

[15] P. Bordes, E. Pollet, L. Avernous. Nano-biocomposites biodegradable polyester/nanoclay systems. Prog. Polym.Sci. 34 (2009) 125–134. https://doi.org/10.1016/j.progpolymsci.2008.10.002

[16] J. Greene. PHA biodegradable blow-molded bottles compounding and performance. Plast. Eng. 69 (2013) 16–21. https://doi.org/10.1002/j.1941-9635.2013.tb00940.x

[17] K. Hofvendahl, B. Hahn-Hagerdal. Factor affecting the formative lactic acid production from renewable resources. Enzyme. Microb. Technol. 26 (2000) 87–107. https://doi.org/10.1016/s0141-0229(99)00155-6

Materials Research Forum LLC

https://doi.org/10.21741/9781644901335-3

[18] K. Kim, S. Woo. Synthesis of high molecular weight poly (L-lactic acid) by direct polycondensation. Macromol. Chem. Phys. 203 (2002) 2245–2250. https://doi.org/10.1002/1521-3935(200211)203:15<2245::AID-MACP2245>3.0.CO;2-3

[19] D. Kint, S. Munoz-Guerra. A review on the potential biodegradability of poly(ethylene terephthalate. Polym. Int. 44 3 (1999) 46–352. https://doi.org/10.1002/(SICI)1097-0126(199905)48:5<346::AID-PI156>3.0.CO;2-N

[20] P. Lescher, K. Jayaraman, D, Bhattacharyya. Characterization of water-free thermoplastic starch blends for manufacturing processes, Mat. Sci. Eng. 532 (2012).178–189. https://doi.org/10.1016/j.msea.2011.10.079

Degradation of Plastics
Materials Research Foundations **99** (2021) 95-110

Materials Research Forum LLC
https://doi.org/10.21741/9781644901335-4

Chapter 4

Enzymes Involved in Plastic Degradation

S.Z.Z. Cobongela[1]*

[1]Nanotechnology Innovation Centre, Advanced Materials Division, Mintek, Randburg, South Africa

snazozezethu@gmail.com

Abstract

The global increase in production of plastic and accumulation in the environment is becoming a major concern especially to the aquatic life. This is due to the natural resistance of plastic to both physical and chemical degradation. Lack of biodegradability of plastic polymers is linked to, amongst other factors, the mobility of the polymers in the crystalline part of the polyesters as they are responsible for enzyme interaction. There are significantly few catabolic enzymes that are active in breaking down polyesters which are the constituents of plastic. The synthetic polymers widely used in petroleum-based plastics include polyethylene (PE), polypropylene (PP), polyvinylchloride (PVC), polyurethane (PUR), polystyrene (PS), polyamide (PA) and polyethylene terephthalate (PET) being the ones used mostly. Polymers with heteroatomic backbone such as PET and PUR are easier to degrade than the straight carbon-carbon backbone polymers such as PE, PP, PS and PVC.

Keywords

Polymers, Lipase, Cutinase, Esterase, Polyethylene Terephthalate, Polyurethane, Active Site, Protease, Enzymatic Degradation

Contents

1.   Introduction

Polymeric material can be degraded by enzymes secreted by microorganisms. They are degraded either by hydrolases or oxidation enzymes. The struggle in degrading these polymers is due to their low hydrophilicity. Their hydrophobic nature makes it challenging to wet them as they hinder water from penetrating the pores of the material. Alkali treatment is a conventional treatment commonly used by industries to hydrolyse the polyester bonds which requires high amounts of sodium hydroxide and high temperatures which in turn are a disadvantage in the environment. Therefore, there is a need for substances that improve surface hydrophilicity and increase hydrolysis of these polymers. The enzymes enhance water absorption ability of the synthetic fibres through increase in surface hydrophilicity as seen on treatments with alkaline treatment [1].

Enzymes have shown a potential in combating this problem. Examples of these surface modifying enzymes are several hydrolases, such as lipases, carboxylesterases, cutinases, and proteases. Some of these enzymes are selected based on their known primary activities e.g. cutinase from *Aspergillusoryzae* (*A. oryzae*) which hydrolyses a protective cutin lipid polyester matric in plants [2]. The enzyme reaction with the polymers occur under mild conditions which require no additional chemicals or complex machinery. Temperature plays a huge role in the performance of the hydrolases. Most of the enzymes seem to have optimum activity between 50-80°C [3]. An ideal hydrolase should possess high hydrolytic activity to plastic polymers, show specificity and thermostability. One way that has been proven to increase thermostability is the presence of disulphide bond between cysteine residues [2,4]. Some of these enzymes are genetically modified to enhance their enzymatic activity while achieving industrial requirements [5]. Cutinase belongs to a group of enzymes that catalyse the hydrolysis of esters and triglycerides and has been ascertained as a biocatalyst for biotechnological and industrial bioprocesses. Cutinases generally display their active site forming a pocket-like structure rather than having the gate structure as observed in most lipases [6]. Polymerases prefer to degrade the amorphous region [3].

Most of these enzymes have a catalytic triad made up of serine, glutamic acid and histidine with an exception of carboxyesterases that replace glutamic acid with aspartic acid [6,7]. These amino acid residues have shown to play a major role in hydrolysis of the ester bond. Billing et al. [6] (2010) confirmed this by using some of the inhibitors that bind irreversible to these amino acids. Phenylmethylsulfonyl fluoride and tosyl-L-phenylalanine-chloromethyl ketone permanently bind and inactivate serine and histidine, respectively, on the active site. The enzymes lose catalytic activity in the presence of either of these inhibitors. In addition, substitution mutagenesis of glutamic acid leads to extensive decrease of the enzyme activity [6]. These observations suggest a major role played by the amino acid triad in catalytic mechanism of enzymes. Usually these enzymes contain a specific sequence around the serine characterised by G-X-S-X-G (G- glycine, X- any polar amino acid, S- serine) motif [7].

2. Enzymatic degradation of polyethylene terephthalate

Polyethylene terephthalate (PET) is used in the manufacturing of bottles, foil and fibers used in the textile industry. It is characterized by polar and linear polymer with aromatic terepthalic acid and ethylene glycol. PET is a thermoplastic polymer and is partly crystalline which is made up of bis(2-hydroxyethyl) terephthalate (BHET) monomer [8,9]. There are relatively few known enzymes that degrade PET to oligomers or monomersand have all shown low turnover rates [10]. PET degrading enzymes are classified as serine hydrolases, e.g., cutinases, lipases and esterases [11]. Most of these enzymes are from a Gram-positive phylum *Actinobacteria* [12], possessing α/β hydrolase folds housing a catalytic triad that is composed of serine, histidine and aspartate residues [13,14]. Some of the enzymes contain cysteine residues which encourages disulfide bonds with the sulphur groups and that in turn promotes thermal stability and specific binding to PET [15]. The cutinase partially hydrolyzes the ester bond of PET on the surface to much more sympler components that are water soluble. The preference for the *A. oryzae* enzyme to hydrolyze longer chain substrates can be explained by the deep continuous groove extending across the active site, while that of *Fusariumsolani* (*F. solani*) pisi favors short chain substrates due to the shallow and interrupted groove. This is due to the arrangement of the key hydrolysing residues on the active site. The distance between the "gatekeeper" residues determines the groove of the active site being narrow or wise. Wider opening of the active site is assumed to be related to rapid hydrolysis of longer chains such as PCL while the narrow active site result in slow hydrolysis and prefers shorter polymer chains [1]. The by products (oligomers and monomers) include terephthalic acid (TA), mono(2-hydroxyethyl) terephthalate (MHET), bis(2-hydroxyethyl) terephthalate (BHET), benzoic acid (BA) and 2-hydroxyethylbenzoate (HEB) [16,17].

Degradation of Plastics   Materials Research Forum LLC
Materials Research Foundations **99** (2021) 95-110   https://doi.org/10.21741/9781644901335-4

An example of such enzymes is a PET-digesting enzyme (PETase) from *Ideonellasakaiensis* (*I. sakaiensis*) which also uses PET as the major source of carbon and energy source. The PETase from *I. sakaiensis* is assumed to be a lipase that shares about 51% amino acid residues with a hydrolase isolated from the actinomycete *Thermobifidafusca* (*T. fusca*) that also exhibit activity in PET hydrolysis [12]. In comparison with other hydrolases such as lipases, hydrolase from *T. fusca*known to be a cutinase depolymerizes the aromatic polyesters of PET at a higher rate [18]. Other cutinases are produced by *Humicola insolens*, *F.solani* pisi, *Fusarium oxysporum*, *Penicillium citrinum*, *Pseudomonas mendocina* and different species of *Thermobifida*. The genus *Thermobifida* has few other known and characterized cutinases that hydrolyzes PET.

Herrero-Acero [12] working with Ribitsch and others in 2011 compared two closely related cutinases from *Thermobifida Cellulosilytica*(*T. Cellulosilytica*) and *T. fusca* about the structure: function relationship. Their distinctive catalytic properties for both cutinases were distinctively different due to their structural difference. The study shows that all the cutinases cloned and tested hydrolyzed substrate bis(benzoyloxyethyl) terephthalate (3PET) more soluble products such as MHET, TA, benzoic acid (BA) and 2-hydroxyethyl benzoate (HEB) *Thermobifida Cellulosilytica* (*T. Cellulosilytica*) cutinase (Thc_Cut1) released more of the soluble products compared to the *T. fusca* (Tf42_Cut1) and *T. Cellulosilytica* (Thc_Cut2) [12]. The release of the PET oligomers and monomers is confirmation that all three enzymes were able to cleave the ester bonds. A linear increase of the by products from 3PET substrate hydrolyzed by cutinases from *T. Fusca* and *F. solani* pisi was observed on the study done by Eberl et al. [19] (2009). Both cutinases, like others, also proved to catalyze the conversion of HEB to BA as the increase in concentration of BA was observed while there was a decline in HEB. The same phenomenon was observeved with the increase of TA against MHET. Other PET degrading enzymes such as lipases and polyesterases are not able to degrade the secondary components of PET [19,20].

In 2012, Ribitsch et al. [21] cloned a cutinase (Tha_Cut1) from *Thermobifida alba* (*T. alba*) for hydrolysis of PET. Data obtained from modeling reveals that Tha_Cut1 is almost identical to a cutinase produced by *T. cellulosilytica* with only four different amino acids found outside the active site [21]. Despite the identical active site, Tha_Cut1 had higher activity on PET surface hydrolisis compared to *T. Cellulosilytica* cutinase due to the ability of the cutinase to increase the hydrophilicity of the polymer. Hydrophilicity is measured by the variation in water contact angle such that decrease in water contact angle equates increase hydrophilicity and that in turn increases polymer hydrolysis. Treatment with cutinase results in a large number of hydroxyl end groups at the surface of the polymer [22]. The process of hydrolysing the ester bond of PET to polyethylene glycol and

Materials Research Forum LLC
https://doi.org/10.21741/9781644901335-4

terephthalic generates hydroxyl and carbonyl groups [23,24]. Interestingly, an esterase isolated from *Thermobifida halotolerans* (*T. halotolerans*) is a 262 amino acid with a mass of 28.7 kDa sharing about 85–87% homology to an esterase from *T. alba*. It hydrolyses PET and polylactic Acid (PLA) by increasing their hydrophilicity measured by water contact angle from 90.8°and 75.5° to 50.4°, respectively [21].

Another Actinomycete*Thermomonospora curvata* (*T.* curvata) is phylogenetically related and sharing about 61% sequence identity with *T. Fusca*enzymes, produces extracellular hydrolases which can degrade synthetic polymers [13]. Tcur1278 and Tcur0390 are two polyester serine hydrolases coded by *T. curvata*. The study compared the two hydrolases catalytic activity on PET nanoparticles with varying temperatures. Tcur0390 showed higher activity at temperatures lower than 50°C while Tcur1278 showed to strive in higher temperatures between 55° C 60° C. It is suggested that the diffence in thermal stability and hydrolytic activity is due to the difference in their molecular dynamics simulation. Tcur1278 would be ideal in codegradation with the thermal degradation of PET process. Agitation and pH control between pH6 to 8 during reaction is necessary to increase the enzyme catalyses hydrolysis [23,25]. Calcium ions and has been shown to activate cutinases [26]. Table 1 [1–3, 6, 12, 17,19–31] contains a list of enzymes involved in PET degradation.

Table 1   Polyethylene terephthalate degrading enzymes [1–3,6,12,17,19–31]

Enzyme	Species	Substrate	Optimum tempeature	References
Lipases	*Candida antarctica*	PET		[27]
	Thermomyces lanuginosus	PET		[19]
	Burkholderia spp.	PET		[28]
	Triticum aestivum	PET		[29]
Cutinases from fungi	*Aspergillus oryzae*			[3]
	Humicola insolens	PET to terephthalic acid and ethylene glycol	70-89 °C	[3]
	Aspergillus oryzae	PET and poly(ε-caprolactone) (PCL)		[2]

	Penicillium citrinum			[20]
	Fusarium solani pisi	PET	50°C	[1,23]
Cutinases from actinomycetes	*Thermomonospora fusca*	PET		[17,22]
	Thermobifida cellulosilytica	PET		[12]
	Thermobifida alba	PET		[21]
	Saccharomonospora viridis	PET	50°C	[24]
Cutinase	Unidentified	Cyclic PET	60°C	[25]
Esteraseserine hydrolases	*Thermobifida halotolerans*	PET and PLA to lactic acid		[30]
Carboxylesterases		PET	60°C	[6]
Protease		Improves hydrophilicity of the polyester fabrics		[31]
Polyesterase	*Saccharomonospora viridis*	PET		[26]

3. Enzymatic degradation of polyurethane

Polyurethane(PUR)is a polymer made up of polyether/ polyester (polyisocyanate) polyols connected by intra-molecular carbamate ester bond. Correspondingly, both fungi and bacteria have been shown to secrete enzymes capable of degrading PUR. The enzymatic degradation of polyether PUR and polyester PUR is quite distinct and specific to different enzymes [32]. PUR is degraded by oxidative enzymes (oxidases) such as horseradish peroxidase, catalase, and xanthine oxidase [33].

One of polyurethase (PURase) is a 48 kDa esterase, extracted from *Pseudomonas fluorescens* (*P. Fluorescens*)in 1998, cloned and expressed in *Escherichia coli*. The *P. fluorescens*utilises PUR as sole carbon source. This enzymes resembles a typical serine protease catalytic triad with G-X-S-X-G motif [34]. Stern and Howard cloned a gene (pueA) that codes for a polyester PURase (lipase) from *Pseudomonas chlororaphis* (*P. chlororaphis*) in the year 2000. The active site for polyurethases is conserved like the other hydrolases mentioned in this text [35]. PUR esterase is highly hydrophobic which is thought to play a huge role on the catalytic activity. The hydrophobicity increases the chance of the enzyme binding to PUR surface domain while helping with PUR binding on the catalytic site. There are two types of PURase, membrane bound and extracellular, degrading PUR in two distinct ways. Memrane bound PURase encourages cell mediated contact with the subtrate [36] while the extracellular PURase binds to the surface of the substrate [34]. PUR esterase degrades PUR by hydrolysing the ester bonds releasing diethylene glycol and adipic acidat neutral pH [36]. In 2017, Schmidt and colleagues [32] tested a few bacterial polyester hydrolase on degradation of polyester PUR [32]. Some of these actinomycete enzymes were previously reported to be hydrolysing PET. These enzymes include hydrolases from *T. Alba* [21], *T. Fusca* [19,24], *Saccharomonospora viridis* (*S. viridis*) [24] and others. These enzymes proved to be hydrolysing the PUR ester bonds while the urethane bond resisted hydrolysis [32]. This is in agreement with what was reported by Akutsu et al. [36], 1998 and Nakajima-Kambe et al. [37] 1995 that degradation of PUR is due to cleaveage of ester bonds. *Pestalotiopsis microspora*, a fungus species has also shown to secrete a metallo-hydrolase that robustly degrade PUR under anaerobic conditions using PUR as the main carbon source [38].

Table 2   Polyurethane degrading enzymes [34–36,38,40–43]

Enzyme	Species	Optimum Tempeature	References
Esterase	*Curvularia Senegalensis*	100°C	[42]
	Pseudomonas Fluorescens		[34,43]
	Comamonas acidovorans TB -35	45°C	[36]
Esterase/ Protease	*Pseudomonas chlororaphis*	100°C	[40]
Lipase	*Pseudomonas chlororaphis* (*pueA*)		[35]
	Pseudomonas chlororaphis (pueB)		[41]
Metallo-hydrolase	Pestalotiopsis microspora	60-80°C	[38]

Cellulase in enzyme that has the hydrophobic PUR surface binding domain containing a hydrophobic domain, a flexible hinge region and most importantly the C-terminus involved in substrate binding [39]. Cellulases are widely known to hydrolyze the internal bond (β-1,4glycosidic) and terminal ends of cellulose. PURases show to be hydrolysing the ester bond; however, there is no clear view on how these enzymes breakdown the carbamate bond. On the other hand, it is unclear whether the urethane bond is hydrolyzed by the PUR esterase, because there is no evidence of the degradation products derived from polyisocyanate segments of the PUR were not detected [36]. The list of enzymes responsible for the degradation of PUR are listed in table 2 [34–36,38,40–43].

4.   Enzymatic degradation of polyethylene

Polyethylene (PE) is considered the most durable polymer consisting of carbon-carbon backbone long chain of ethylene consisting of low density or high density. There is very limited research done on the enzymes capable of degrading PE. Some of the PE degrading enzymes are screened by degrading a protective polymer layers that are also made up of carbon-carbon bond linkages found in plant cell walls [44–46]. One of these enzymes is laccase produced by an Actinomycete *Rhodococcus ruber* (*R. ruber*) [47,48]. It is the first enzyme known to degrade PE. The activity of laccase on utilizing and degrading the PE is greatly induced by copper. These was a 13-fold increase when comparind the copper treated laccase degradation and the untreated reaction [47]. Laccase is an oxidizing enzyme with four binding sites breaking down non-phenolic aromatic rings, carbon-carbon and other bonds [49]. This enzyme plays a huge role in oxidation of thr carbon-carbon backbone on the PE proven by the decrease in molecular weight of the PE and increase in carbonyl index [47].

Manganese peroxidase enzyme has also been confirmed to degrade PE in the presence of malonate buffer and Tween 80 [50]. It was previously assumed by Moen et al. [51] that the degradation is due to manganese peroxidase-lipid peroxidation system generating free radicals that in turn reacts with the carbon-carbon backbone of the PE [51]. Ehara and colleagues [50] concluded that the manganese (III) produced by manganese peroxidase is highly responsible for the PE gradation [50]. On the other hand, an enzyme alkane hydroxylase responsible for PE degradation acts through mineralization of low molecular weight PE into CO_2 [52]. These enzymes modify the PE surface by hydrolysis of the carbon-carbon backbone and most importantly introducing hydrophilic groups thus making the PE surface more hydrophilic and high energy improving adhesion to other material [53]. Some of the enzymes ivoled in PE degradation are listed in Table 3 [47,48,50,52,54,55].

Table 3   *Polyethylene degrading enzymes [47,48,50,52,54,55]*

Enzyme	Species	Temperature	References
Laccase	*Rhodococcus Ruber*	30-70°C	[47]
	Trametes versicolor		[48]
	Aspergillus fumigatus		[54]
Manganese peroxidase	*Phanerochaete chrysosporium*		[50]
Alkane hydroxylase	*Pseudomonas aeruginosa*	37°C	[52,58]

5. Enzymatic degradation of polystyrene

An alternative name for polystyrene (PS) is poly-1-phenylethene, a synthetic aromatic hydrocarbon polymer consisting of monomer styrene. It is a high molecular weight hydrocarbon compared to PE which in turn makes it durable and resistant to degradation. Promising enzymatic degradation has been with the help of chemical additives breaking down the higher molecular weight to low molecular weight PS. This phenomenon was confirmed by Nakamiya et al. [56], in 1997 where they used dichloromethane to break PS to small water soluble molecules that were further broken down by hydroquinone peroxidase enzyme isolated from *Azotobacter beijerinckii* (*A. beijerinckii*) [56]. Besides the pre-chemical treatment, a physicochemical process has been observed using mealworms (*Tenebriomolitor* Linnaeus). The mealworms physically chew and ingest PS and with the help ofunidentified enzymes produced by the mealworm helps in depolymerization the polymer into CO_2 [57,58]. These enzymes might include the previously mentioned.

6. Enzymatic degradation of polyvinylchloride (PVC)

Like other plastic polymers, polyvinylchloride (PVC) is known for phenominal resistance to degradation. It is made up of a chain of vinyl chloride. PVC is highly hydrophobic mostly resistan to abraision by chemical [59] which makes it a good candidate to be used in construction industries. Low-molecular weight PVC has beenshown to be viable degraded by some fungi species such as *As. fumigatus, Phanerochaete chrysosporium, Lentinus tigrinus, Aspergillus niger, and Aspergillus sydowii* although the mchanism of action is not fully understood [60,61]. In 2016, Sumathi and colleagues [62] finally isolated *Cochliobolus sp.* a fungi species from plastic dump soil [62]. This fungi species uses PVC as the sole carbon source and is capable of producing laccase enzyme which is assumed to be responsible for the degradation of low-molecular weight PVC. There is no reported ezymatic degradation on high-molecular PVC.

7. Enzymatic degradation of polyvinylchloride polyamide (PA)

The polymer chain of polyvinylchloride polyamide (PA) is joined together by amide bonds (CO-NH). This bond is generally resistant to degradation. Most common synthetic polyamaides are nylon 6 (PA 6) and nylon 6,6 (PA 6,6). Proteolytic enzymes such as proteases showed hydrolysis on amide oligomers [63,64]. Recently,a protease from *Bacillus sp.* has been shown to hydolyze both PA 6 and PA 6,6 [65]. This particular protease functions optimally between 25-30°C at pH 8 under agitatiom.However, protease and lipase enzymes are normally used to modify and improve PA physical and chemical properties such as hydrophilicity while smoothing the surface [65,66].

Conclusions

Although plastics play a key role in society, their fate has become a major management problem on our environment. The major degradation techniques currently used include thermal and chemical treatments. Enzymatic plastic degradation remains a topic to be explored further with major improvements. Literature shows there has been a lot of research done in PET and PUR compared to other plastic polymers on enzymatic degradation. Occasionally, enzymes are used mainly for modification of plastic such as improving elasticity, hydrophilicity and other important functional physiognomies. Some of the enzymes require a pretreatment either by heat or chemical treatment for maximum degradation. There is very minute research done on the enzyme-substrate interaction, hydrolysis of bonds and enzyme kinetics.

References

[1]   M. Alisch-Mark, A. Herrmann, W. Zimmermann, Increase of the hydrophilicity of polyethylene terephthalate fibres by hydrolases from Thermomonospora fusca and Fusarium solani f. sp. pisi, Biotechnol Lett. 28 (2006) 681–685. https://doi.org/10.1007/s10529-006-9041-7

[2]   Z. Liu, Y. Gosser, P.J. Baker, Y. Ravee, Z. Lu, G. Alemu, H. Li, G.L. Butterfoss, X.-P. Kong, R. Gross, J.K. Montclare, Structural and functional studies of A. oryzae Cutinase: enhanced thermostability and hydrolytic activity of synthetic ester and polyester degradation, J. Am. Chem. Soc. 131 (2009) 15711–15716. https://doi.org/10.1021/ja9046697

[3]   A.M. Ronkvist, W. Xie, W. Lu, R.A. Gross, Cutinase-catalyzed hydrolysis of poly(ethylene terephthalate), Macromolecules. 42 (2009) 5128–5138. https://doi.org/10.1021/ma9005318

[4]   A.J. Doig, D.H. Williams, Is the hydrophobic effect stabilizing or destabilizing in proteins? The contribution of disulphide bonds to protein stability, J. Mol. Biol. 217 (1991) 389–398. https://doi.org/10.1016/0022-2836(91)90551-g

[5]   I.S. Chin, A.M.A. Murad, N.M. Mahadi, S. Nathan, F.D.A. Bakar, Thermal stability engineering of Glomerella cingulata cutinase, Protein. Eng. Des. Sel. 26 (2013) 369–375. https://doi.org/10.1093/protein/gzt007

[6]   S. Billig, T. Oeser, C. Birkemeyer, W. Zimmermann, Hydrolysis of cyclic poly(ethylene terephthalate) trimers by a carboxylesterase from Thermobifida fusca KW3, Applied Microbiology and Biotechnology. 87 (2010) 1753–1764. https://doi.org/10.1007/s00253-010-2635-y

[7]   U.T. Bornscheuer, Microbial carboxyl esterases: classification, properties and application in biocatalysis, FEMS Microbiology Reviews. 26 (2002) 73–81. https://doi.org/10.1016/S0168-6445(01)00075-4

[8]   S. Amin, M. Amin, Thermoplastic elastomeric (TPE) materials and their use in outdoor electrical insulation, Rev. Adv. Mater. Sci.29 (2011) 15-30.

[9]   B. DemiRel, A. Yara, H. Elç, Crystallization behavior of PET Materials, BAÜ Fen Bil. Enst. Dergisi Cilt 13(1) (2011)26-35.

[10]   R. Wei, W. Zimmermann, Microbial enzymes for the recycling of recalcitrant petroleum-based plastics: how far are we?, Microb Biotechnol. 10 (2017) 1308–1322. https://doi.org/10.1111/1751-7915.12710

[11]   S. Heumann, A. Eberl, H. Pobeheim, S. Liebminger, G. Fischer-Colbrie, E. Almansa, A. Cavaco-Paulo, G. Guebitz, New model substrates for enzymes hydrolysing polyethyleneterephthalate and polyamide fibres, Journal of Biochemical and Biophysical Methods. 69 (2006) 89–99. https://doi.org/10.1016/j.jbbm.2006.02.005

[12]   E. Herrero Acero, D. Ribitsch, G. Steinkellner, K. Gruber, K. Greimel, I. Eiteljoerg, E. Trotscha, R. Wei, W. Zimmermann, M. Zinn, A. Cavaco-Paulo, G. Freddi, H. Schwab, G. Guebitz, Enzymatic surface hydrolysis of PET: Effect of structural diversity on kinetic properties of cutinases from Thermobifida, Macromolecules. 44 (2011) 4632–4640. https://doi.org/10.1021/ma200949p

[13]   R. Wei, T. Oeser, J. Then, N. Kühn, M. Barth, J. Schmidt, W. Zimmermann, Functional characterization and structural modeling of synthetic polyester-degrading hydrolases from Thermomonospora curvata, AMB Express. 4 (2014) 44. https://doi.org/10.1186/s13568-014-0044-9

[14]   D.L. Ollis, E.Cheah, M.Cygler, B. Dijkstra, F. Frolow, S.M. Franken, M. Harel, S. Jamse-Remington, I. Silman, J. Schrag, J.L. Sussman, K.H.G. Verschueren, A. Goldman,The alpha/beta hydrolase fold. Protein Eng. 5 (1992) 197-211. https://doi.org10.1093/protein/5.3.197

Materials Research Forum LLC
https://doi.org/10.21741/9781644901335-4

[15]   S. Yoshida, K. Hiraga, T. Takehana, I. Taniguchi, H. Yamaji, Y. Maeda, K. Toyohara, K. Miyamoto, Y. Kimura, K. Oda, A bacterium that degrades and assimilates poly(ethylene terephthalate), Science. 351 (2016) 1196–1199. https://doi.org/10.1126/science.aad6359

[16]   J.T. Ranz-Korpecka, S. Heumann, G. Gübitz, S. Billig, W. Zimmermann, M. Zinn, J. Ihssen, A. Cavaco-Paulo, Cutinase activity of PET-hydrolases., Macromolecular Symposia. 296 (2010) 342–346.

[17]   T. Brueckner, A. Eberl, S. Heumann, M. Rabe, G.M. Guebitz, Enzymatic and chemical hydrolysis of poly(ethylene terephthalate) fabrics, Journal of Polymer Science Part A: Polymer Chemistry. 46 (2008) 6435–6443. https://doi.org/10.1002/pola.22952

[18]   I. Kleeberg, C. Hetz, R.M. Kroppenstedt, R.-J. Müller, W.D. Deckwer, Biodegradation of aliphatic-aromatic copolyesters by Thermomonospora fusca and other thermophilic compost isolates, Appl Environ Microbiol. 64 (1998) 1731–1735.

[19]   A. Eberl, S. Heumann, T. Brückner, R. Araujo, A. Cavaco-Paulo, F. Kaufmann, W. Kroutil, G.M. Guebitz, Enzymatic surface hydrolysis of poly(ethylene terephthalate) and bis(benzoyloxyethyl) terephthalate by lipase and cutinase in the presence of surface active molecules, J. Biotechnol. 143 (2009) 207–212. https://doi.org/10.1016/j.jbiotec.2009.07.008

[20]   S. Liebminger, A. Eberl, F. Sousa, S. Heumann, G. Fischer-Colbrie, A. Cavaco-Paulo, G.M. Guebitz, Hydrolysis of PET and bis-(benzoyloxyethyl) terephthalate with a new polyesterase from Penicillium citrinum, Biocatalysis and Biotransformation. 25 (2007) 171–177. https://doi.org/10.1080/10242420701379734

[21]   D. Ribitsch, E.H. Acero, K. Greimel, I. Eiteljoerg, E. Trotscha, G. Freddi, H. Schwab, G.M. Guebitz, Characterization of a new cutinase from Thermobifida alba for PET-surface hydrolysis, Biocatalysis and Biotransformation. 30 (2012) 2–9. https://doi.org/10.3109/10242422.2012.644435

[22]   M. Alisch, A. Feuerhack, H. Müller, B. Mensak, J. Andreaus, W. Zimmermann, Biocatalytic modification of polyethylene terephthalate fibres by esterases from actinomycete isolates, Biocatalysis and Biotransformation. 22 (2004) 347–351. https://doi.org/10.1080/10242420400025877

[23]   A. O'Neill, R. Araújo, M. Casal, G. Guebitz, A. Cavaco-Paulo, Effect of the agitation on the adsorption and hydrolytic efficiency of cutinases on polyethylene terephthalate fibres, Enzyme and Microbial Technology. 40 (2007) 1801–1805. https://doi.org/10.1016/j.enzmictec.2007.02.012

[24]   F. Kawai, T. Kawase, T. Shiono, H. Urakawa, S. Sukigara, C. Tu, M. Yamamoto, Enzymatic hydrophilization of polyester fabrics using a recombinant cutinase Cut 190 and their surface characterization, Journal of Fiber Science and Technology. 73 (2017) 8–18. https://doi.org/10.2115/fiberst.fiberst.2017-0002

Materials Research Forum LLC
https://doi.org/10.21741/9781644901335-4

[25] J. Hooker, D. Hinks, G. Montero, M. Icherenska, Enzyme-catalyzed hydrolysis of poly(ethylene terephthalate) cyclic trimer, J. App. Polymer Sci. 89 (2003) 2545–2552. https://doi.org/10.1002/app.11963

[26] F. Kawai, M. Oda, T. Tamashiro, T. Waku, N. Tanaka, M. Yamamoto, H. Mizushima, T. Miyakawa, M. Tanokura, A novel Ca^{2+}-activated, thermostabilized polyesterase capable of hydrolyzing polyethylene terephthalate from Saccharomonospora viridis AHK190, Applied Microbiology and Biotechnology. 98 (2014) 10053–10064. https://doi.org/10.1007/s00253-014-5860-y

[27] M.A.M.E. Vertommen, V.A. Nierstrasz, M. van der Veer, M.M.C.G. Warmoeskerken, Enzymatic surface modification of poly(ethylene terephthalate), Journal of Biotechnology. 120 (2005) 376–386. https://doi.org/10.1016/j.jbiotec.2005.06.015

[28] C.W. Lee, J.D. Chung, Synthesis and Biodegradation behavior of poly(ethylene terephthalate) oligomers, Polymer Korea. 33 (2009)198-202.

[29] A. Nechwatal, A. Blokesch, M. Nicolai, M. Krieg, A. Kolbe, M. Wolf, M. Gerhardt, A contribution to the investigation of enzyme-catalysed hydrolysis of poly(ethylene terephthalate) oligomers, Macromol. Mater. Eng. 291 (2006) 1486–1494. https://doi.org/10.1002/mame.200600204

[30] D. Ribitsch, E. Herrero Acero, K. Greimel, A. Dellacher, S. Zitzenbacher, A. Marold, R.D. Rodriguez, G. Steinkellner, K. Gruber, H. Schwab, G.M. Guebitz, A new esterase from thermobifida halotolerans hydrolyses polyethylene terephthalate (PET) and polylactic acid (PLA), Polymers. 4 (2012) 617–629. https://doi.org/10.3390/polym4010617

[31] H.R. Kim, W.S. Song, Optimization of papain treatment for improving the hydrophilicity of polyester fabrics, Fibers and Polymers. 11 (2010) 67–71. https://doi.org/10.1007/s12221-010-0067-z

[32] J. Schmidt, R. Wei, T. Oeser, L.A. Dedavid e Silva, D. Breite, A. Schulze, W. Zimmermann, degradation of polyester polyurethane by bacterial polyester hydrolases, Polymers. 9 (2) (2017) 65. https://doi.org/10.3390/polym9020065

[33] J.P. Santerre, R.S. Labow, D.G. Duguay, D. Erfle, G.A. Adams, Biodegradation evaluation of polyether and polyester-urethanes with oxidative and hydrolytic enzymes, J. Biomed. Mater. Res. 28 (1994) 1187–1199. https://doi.org/10.1002/jbm.820281009

[34] RE Vega, T. Main, G.T. Howard, Cloning and expression in Escherichia coli of a polyurethane!degrading enzyme from Pseudomonas ~uorescens, International Biodeterioration & Biodegradation. 43 (1999) 49–55.

[35] R.V. Stern, G.T. Howard, The polyester polyurethanase gene (pueA) from Pseudomonas chlororaphis encodes a lipase, FEMS Microbiol. Lett. 185 (2000) 163–168. https://doi.org/10.1111/j.1574-6968.2000.tb09056.x

[36] Y. Akutsu, T. Nakajima-Kambe, N. Nomura, T. Nakahara, Purification and properties of a polyester polyurethane-degrading enzyme from Comamonas acidovorans TB-35, Appl Environ Microbiol. 64 (1998) 62–67.

[37] T. Nakajima-Kambe, F. Onuma, N. Kimpara, T. Nakahara, Isolation and characterization of a bacterium which utilizes polyester polyurethane as a sole carbon and nitrogen source, FEMS Microbiol. Lett. 129 (1995) 39–42. https://doi.org/10.1016/0378-1097(95)00131-N

[38] J.R. Russell, J. Huang, P. Anand, K. Kucera, A.G. Sandoval, K.W. Dantzler, D. Hickman, J. Jee, F.M. Kimovec, D. Koppstein, D.H. Marks, P.A. Mittermiller, S.J. Núñez, M. Santiago, M.A. Townes, M. Vishnevetsky, N.E. Williams, M.P.N. Vargas, L.-A. Boulanger, C. Bascom-Slack, S.A. Strobel, Biodegradation of polyester polyurethane by endophytic fungi ᵛ, Appl. Environ. Microbiol. 77 (2011) 6076–6084. https://doi.org/10.1128/AEM.00521-11

[39] J. Knowles, P. Lehtovaara, T. Teeri, Cellulase families and their genes, Trends Biotechnol. 5 (1987) 255–261. https://doi.org/10.1016/0167-7799(87)90102-8

[40] C. Ruiz, T. Main, N.P. Hilliard, G.T. Howard, Purification and characterization of twopolyurethanase enzymes from Pseudomonas chlororaphis, International Biodeterioration & Biodegradation. 43 (1999) 43–47. https://doi.org/10.1016/S0964-8305(98)00067-5

[41] G.T. Howard, B. Crother, J. Vicknair, Cloning, nucleotide sequencing and characterization of a polyurethanase gene (pueB) from Pseudomonas chlororaphis, International Biodeterioration & Biodegradation. 47 (2001) 141–149. https://doi.org/10.1016/S0964-8305(01)00042-7

[42] J.R. Crabbe, J.R. Campbell, L. Thompson, S.L. Walz, W.W. Schultz, Biodegradation of a colloidal ester-based polyurethane by soil fungi, International Biodeterioration & Biodegradation. 33 (1994) 103–113. https://doi.org/10.1016/0964-8305(94)90030-2

[43] G.T. Howard, R.C. Blake, Growth of Pseudomonas fluorescens on a polyester–polyurethane and the purification and characterization of a polyurethanase–protease enzyme, International Biodeterioration & Biodegradation. 42 (1998) 213–220. https://doi.org/10.1016/S0964-8305(98)00051-1

[44] Suhas, P.J.M. Carrott, M.M.L. Ribeiro Carrott, Lignin – from natural adsorbent to activated carbon: A review, Bioresource Technology. 98 (2007) 2301–2312. https://doi.org/10.1016/j.biortech.2006.08.008

[45] J.-M. Restrepo-Flórez, A. Bassi, M.R. Thompson, Microbial degradation and deterioration of polyethylene – A review, International Biodeterioration & Biodegradation. 88 (2014) 83–90. https://doi.org/10.1016/j.ibiod.2013.12.014

Materials Research Forum LLC

https://doi.org/10.21741/9781644901335-4

[46] M.C. Krueger, H. Harms, D. Schlosser, Prospects for microbiological solutions to environmental pollution with plastics, Appl. Microbiol. Biotechnol. 99 (2015) 8857–8874. https://doi.org/10.1007/s00253-015-6879-4

[47] M. Santo, R. Weitsman, A. Sivan, The role of the copper-binding enzyme – laccase – in the biodegradation of polyethylene by the actinomycete Rhodococcus ruber, International Biodeterioration & Biodegradation. 84 (2013) 204–210. https://doi.org/10.1016/j.ibiod.2012.03.001

[48] M. Fujisawa, H. Hirai, T. Nishida, Degradation of polyethylene and Nylon-66 by the Laccase-Mediator System, 9 (2001) 103-108.

[49] S. Kim, S.C. Chmely, M.R. Nimlos, Y.J. Bomble, T.D. Foust, R.S. Paton, G.T. Beckham, Computational Study of Bond Dissociation Enthalpies for a Large Range of Native and Modified Lignins, J. Phys. Chem. Lett. 2 (2011) 2846–2852. https://doi.org/10.1021/jz201182w

[50] K. Ehara, Y. Iiyoshi, Y. Tsutsumi, T. Nishida, Polyethylene degradation by manganese peroxidase in the absence of hydrogen peroxide, J. Wood Sci. 46 (2000) 180–183. https://doi.org/10.1007/BF00777369

[51] M.A. Moen, K.E. Hammel, Lipid peroxidation by the manganese peroxidase of Phanerochaete chrysosporium is the basis for phenanthrene oxidation by the Intact Fungus, Appl. Environ. Microbiol. 60 (1994) 1956–1961. https://doi.org/10.1128/AEM.60.6.1956-1961.1994

[52] M.G. Yoon, H.J. Jeon, M.N. Kim, Biodegradation of Polyethylene by a soil bacterium and AlkB Cloned Recombinant Cell, Journal of Bioremediation & Biodegradation. 03 (2012) undefined-undefined.

[53] J. Zhao, Z. Guo, X. Ma, G. Liang, J. Wang, Novel surface modification of high-density polyethylene films by using enzymatic catalysis, Journal of Applied Polymer Science. 91 (2004) 3673–3678. https://doi.org/10.1002/app.13619

[54] C. Ndahebwa Muhonja, G. Magoma, M. Imbuga, H.M. Makonde, Molecular Characterization of low-density polyethene (LDPE) degrading bacteria and fungi from Dandora Dumpsite, Nairobi, Kenya, Int. J .Microbiol. 2018 (2018) 1-10. https://doi.org/10.1155/2018/4167845

[55] A. Belhaj, N. Desnoues, C. Elmerich, Alkane biodegradation in Pseudomonas aeruginosa strains isolated from a polluted zone: identification of alkB and alkB-related genes, Res. Microbiol. 153 (2002) 339–344. https://doi.org/10.1016/S0923-2508(02)01333-5

[56] K. Nakamiya, G. Sakasita, T. Ooi, S. Kinoshita, Enzymatic degradation of polystyrene by hydroquinone peroxidase of Azotobacter beijerinckii HM121, J. Ferment. Bioeng. 84 (1997) 480–482. https://doi.org/10.1016/S0922-338X(97)82013-2

[57] Y. Yang, J. Yang, W.M. Wu, J. Zhao, Y. Song, L. Gao, R. Yang, L. Jiang, Biodegradation and mineralization of polystyrene by plastic-eating mealworms: part 1. chemical and physical characterization and isotopic tests, Environ. Sci. Technol. 49 (2015) 12080–12086. https://doi.org/10.1021/acs.est.5b02661

[58] Y. Yang, J. Yang, W.-M. Wu, J. Zhao, Y. Song, L. Gao, R. Yang, L. Jiang, Biodegradation and mineralization of polystyrene by plastic-eating mealworms: part 2. role of gut microorganisms, Environ. Sci. Technol. 49 (2015) 12087–12093. https://doi.org/10.1021/acs.est.5b02663

[59] A.A. Shah, F. Hasan, A. Hameed, S. Ahmed, Biological degradation of plastics: A comprehensive review, Biotechnol. Adv. 26 (2008) 246–265. https://doi.org/10.1016/j.biotechadv.2007.12.005

[60] M.I. Ali, S. Ahmed, I. Javed, N. Ali, N. Atiq, A. Hameed, G. Robson, Biodegradation of starch blended polyvinyl chloride films by isolated Phanerochaete chrysosporium PV1, Int. J. Environ. Sci. Technol. 11 (2014) 339–348. https://doi.org/10.1007/s13762-013-0220-5

[61] M.I. Ali, S. Ahmed, G. Robson, I. Javed, N. Ali, N. Atiq, A. Hameed, Isolation and molecular characterization of polyvinyl chloride (PVC) plastic degrading fungal isolates, J. Basic Microbiol. 54 (2014) 18–27. https://doi.org/10.1002/jobm.201200496

[62] T. Sumathi, B. Viswanath, A. Sri Lakshmi, D.V.R. SaiGopal, Production of Laccase by Cochliobolus sp. isolated from plastic dumped soils and their ability to degrade low molecular weight PVC, Biochem Res Int. 2016 (2016). https://doi.org/10.1155/2016/9519527

[63] I.D. Prijambada, S. Negoro, T. Yomo, I. Urabe, Emergence of nylon oligomer degradation enzymes in Pseudomonas aeruginosa PAO through experimental evolution., Appl. Environ. Microbiol. 61 (1995) 2020–2022.

[64] T. Deguchi, M. Kakezawa, T. Nishida, Nylon biodegradation by lignin-degrading fungi, Appl. Environ. Microbiol. 63 (1997) 329–331.

[65] I. Jordanov, D.L. Stevens, A. Tarbuk, E. Magovac, S. Bischof, J.C. Grunlan, Enzymatic Modification of Polyamide for Improving the Conductivity of water-based multilayer nanocoatings, ACS Omega. 4 (2019) 12028–12035. https://doi.org/10.1021/acsomega.9b01052

[66] A. Kiumarsi, M. Parvinzadeh, Enzymatic hydrolysis of nylon 6 fiber using lipolytic enzyme, J. Appl. Polym. Sci. 116 (2010) 3140–3147. https://doi.org/10.1002/app.31756

Materials Research Forum LLC

https://doi.org/10.21741/9781644901335-5

Chapter 5

Plastic Biodegradation

Bhupender Singu[1], Karuna Nagula[2], Pravin D. Patil[3*], Manishkumar S. Tiwari[4]

[1]Food Engineering and Technology Department, Institute of Chemical Technology, Mumbai, 400019, Maharashtra, India

[2]Parul Institute of Pharmacy & Research, Parul University, Vadodara, 391760, Gujarat, India

[3]Department of Basic Science and Humanities, Mukesh Patel School of Technology Management and Engineering, SVKM's NMIMS University, Mumbai, 400056, Maharashtra, India

[4]Department of Chemical Engineering, Mukesh Patel School of Technology Management and Engineering, SVKM's NMIMS University, Mumbai, 400056, Maharashtra, India

* dr.pravinpatil.ict@gmail.com; pravin.patil@nmims.edu

Abstract

Plastics that are degraded by microbial or enzymatic activity are known as biodegradable plastics. Biodegradable plastics are an alternative to conventional plastics that are chemically synthesized and are responsible for causing environmental pollution due to unwanted accumulation occurring via disposal practices. There was a serious need to introduce biodegradable plastics in the market since the level of plastic pollution in the air, water, and soil has reached its threshold values. The non-biodegradable plastics are increasingly accumulating in the environment, which can be a threat to the planet in the coming future. This chapter provides detailed insight into biodegradables polymers, mostly aliphatic polyesters that are considered as a solution against synthetic plastic. It also gives brief information on the current scenario of plastic biodegradation, recent advancements, opportunities, and future challenges. Also, it comprises precise strategies currently used at a laboratory scale to enhance biodegradation of classical synthetic plastics (e.g., polyethylene, polystyrene, etc.). Moreover, the factors affecting the biodegradation process and the characterization techniques being employed to assess degradation extent are also discussed. The overall work focuses on thrust areas to be improved concerning environmental safety and sustainable vision.

Keywords

Bioremediation, Bio-Degradation, Plastics, Bio-Degradable Plastics, Bio-Based Polymers, Enzymes

Materials Research Forum LLC
https://doi.org/10.21741/9781644901335-5

Contents

1.   Introduction

The plastic industry is one of the leading sectors in the global market. The urban and rural population is highly dependent on plastic material for their day-to-day activities. The use of plastic has made life simple and more comfortable by providing convenience in almost every need. It is difficult to imagine modern life without the use of plastics as these materials are durable, waterproof, and do not corrode. Plastic is a cheaper and lightweight alternative to metal while being easily moldable and transportable makes it widely adoptable by industries. Despite having numerous applications, they raise considerable disposal concerns. Natural degradation of plastics may take hundreds of years, and it is not suitable for landfills either. Plastics cannot be incinerated as they are known to release toxic gases that pollute the environment and affect living beings. Problems associated with environmental pollution and waste material management have become more threatening with bio-resistant synthetic plastics in the past few decades. The focus of the research community has shifted towards avoiding further adverse effects on the global environment while initiatives have been taken by respective countries to use environmental-friendly materials that can replace conventional synthetic plastics. There are various technologies to breakdown the polymers used in the manufacturing of plastic products. Still, some have drawbacks such as the involvement of time-consuming processes, high process cost, and the release of toxic gases during degradation. The alternative solution to prevent pollution caused by the disposal of plastics is to recycle the used/waste plastic material or use biodegradable plastics or bio-based materials.

2.   Environmental impact of plastics

2.1   Ocean and land pollution

The most significant sources of plastic pollution are landfills. The plastic from landfills ultimately goes to the sea, causing contamination in the aquatic life of the ocean. The degradation of conventional plastics in sea environment is a prolonged process and takes centuries to occur, which leads to their accumulation in the oceans. The sea is the ultimate receptacle for all the waste produced on land (80% of the trash found at sea comes from the land). The plastic debris found on the surface of the water has predominant particles smaller than 5mm, commonly known as microplastics. It is estimated that the concentration of microplastics in the Mediterranean Sea will increase by 8% in the next 30 years [1]. The ocean gyres are continuously producing generations of microplastic via fragmentation of plastics. These microplastics get dispersed all over the oceans. These fragments are very stable and can sometimes persist for up to 1000 years in the marine environment [2]. The plastic present on the land in the form of landfills does not have the opportunity to form small particles or microplastics. The so-called "biodegradable" plastics are emerging in the global market that promises to reduce the impact of plastic waste at sea. However, the degradation process can lead to excess formation of bacterial biomass or mineralization, which in turn is causing microbial pollution in the affected environment.

2.2   Disruption of food chains due to the toxicity of plastic

The majority of plastics found in the contaminated region of sea and soil are non-degradable plastics such as polyethylene (PE), polypropylene (PP), and polyethylene terephthalate (PET). In the environment, pollution by plastics can have several consequences. Other than the visual pollution they cause, plastics affect marine organisms directly or indirectly at different levels of the food chain [3]. Chemically, plastics consist of identical sequences (or polymers) of carbonaceous molecules, mainly hydrocarbons, toxic to many organisms, capable of accumulating along food chains. Bioaccumulation of plastic in the environment is a significant issue concerning the non-degradability of plastic. In the accumulation zones, the concentration of microplastics observed (0.5 to 5 mm in size) is comparable to that of zooplankton (between 0.005 mm and more than 50 mm). The Mediterranean Sea, for example, has microplastic/zooplankton ratios between 1/10 to 1/2 [4]. The risk for zooplankton predators (i.e., fish) from ingesting microplastics is therefore considerable. The residence time of plastic in small pelagic fish has been estimated between 1 day to 1 year [5]. Fragments of ingested microplastics are found in animal droppings and can sink with corpses or even be transferred to predators and thus reach the upper echelons of the food chain [2]. Plastics are also observed as vectors of dispersion of toxic compounds, which can also accumulate in food chains. These compounds can be directly present in the composition of plastics or adsorb on their surface. In the first case, these are additives (phthalates, biphenyls, etc.) incorporated into certain plastics to increase their resistance. Various studies have shown that these compounds can be toxic to many animals and humans [6]. Other toxic compounds (hydrocarbons, pesticides, dichlorodiphenyltrichloroethane (DDT), polychlorinated biphenyl (PCB), etc. can be adsorbed on plastics, which is likely to increase their dispersion and persistence at sea, leading to accumulation in the highest trophic levels [7]. The disastrous effects of ingestion of plastic debris confused with prey are also well documented, with severe consequences on the digestive systems of animals such as fish, birds, sea turtles, and marine mammals, sometimes leading to their death [8]. This debris is also considered to be vectors for the dispersion of toxic algae [9] and pathogenic microorganisms [10]. Moreover, another significant risk could be the use of additives in the preparation of plastics that can be observed in the form of toxins such as bisphenol-A. It is a cancer-causing agent that can also impair the immune system while leading to early puberty and sometimes may trigger the development of obesity and diabetes in humans and animals [11].

3.   Types of bioplastics

Though plastic can be degraded in various ways (Fig.1), bio-degradation of plastic is the most efficient and environment-friendly approach. However, the extent of degradation of plastic in a biological environment is highly influenced by the type of plastic that is being degraded. Bioplastics can be categorized on the basis of their source of origin, the extent of degradability. There are mainly three categories in bioplastics; a) bio-based plastics, b) oxo-biodegradable plastics, and c) bio-degradable plastics.

3.1   Bio-based plastics (Hydro-biodegradable)

Bio-based polymers are plastic materials synthesized from renewable biomass sources such as sugarcane, sawdust, straw, corn starch, vegetable fats and oils, recycled food waste, etc. The raw material is mostly agricultural by-products and other recyclable materials; therefore, the cost of ingredients is comparably cheaper. These materials uptake water while the degradation process is mainly enzyme-mediated. As per the

Degradation of Plastics   Materials Research Forum LLC
Materials Research Foundations **99** (2021) 111-144   https://doi.org/10.21741/9781644901335-5

European standards of EN 13432 plastic, the material should be susceptible to degradation by microorganisms in its natural environment. Even if this type of plastics meets the standards, its degradation in the "natural" environment remains subject to controversy [12]. Bio-based polymers are sustainable, unlike conventional fossil resources such as petroleum oil and natural gas, which leave a high carbon footprint. Interestingly, the biobased plastics are known as carbon-neutral or carbon offset materials since they do not contribute CO_2 to the environment even after incineration. The raw material for bioplastics is generally derived from sugar derivatives, including cellulose, starch, lactic acid, and is usually combined with conventional plastics for recycling. However, bioplastics are not always biodegradable, and some may be more resistant than traditional plastics. Though the share of bioplastics in the global market is only 1.1 % as of 2019, it is rapidly increasing with the demand and awareness. Examples of these plastics are polyesters, polyamides, polysaccharides, etc.

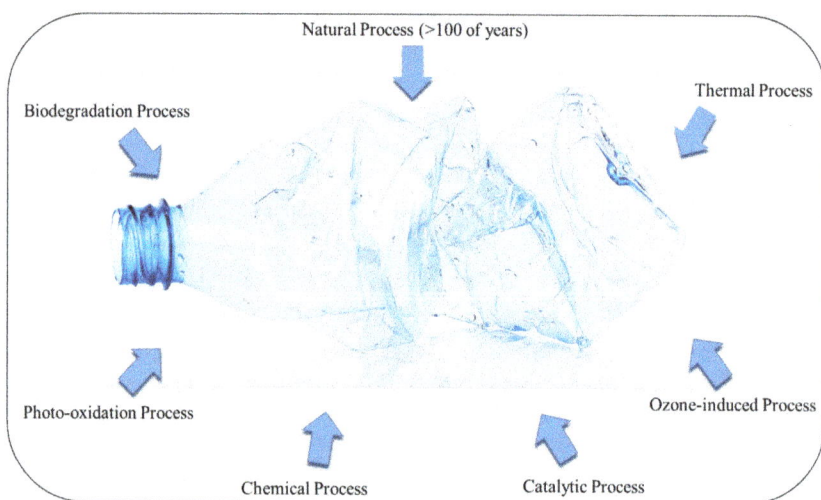

Fig. 1 Degradation of plastics using various processes.

3.2   Oxo-biodegradable plastics

Oxo-biodegradable plastics are a type of bioplastic that can be defragmented but are not depolymerized at a molecular level. They uptake oxygen due to the presence of catalytic enzymes. It also requires UV light for the process of biodegradation. However, O_2 and UV lights are not sufficiently found in landfills, making the process of degradation a bit difficult. Typically, during the process of degradation, they form oxidized fragments by oxidative degradation catalyzed by chemicals or additives present in it. These fragments are further amiable to microbes, which are then degraded by enzymes produced by the specific microbes [13]. This type of plastic may take up to 36 months for>60% degradation to occur. Oxo-biodegradable forms of plastic have the same primary components as conventional plastics (polyethylene, polypropylene, polystyrene, even production chains). Moreover, the addition of biodegradable stabilizers or additives possibly works as pro-oxidant and predicts the lifespan of products while facilitating biodegradation using microorganisms. There is a need to generate well-documented work on the abiotic degradation of these plastics so that the demonstration of their biodegradation remains a subject of ambiguity in the field. However, developments in additive formulations seem promising. To date, only one additive named "d2w" has obtained an eco-label (environmentally friendly products) according to ISO 14020: 2002 and 14024 standards: 2004 concerning the degradation of oxo-biodegradable plastics and is ready for commercial exploration. There is still a broad scope to develop more biodegradable additives, which will direct the degradation of oxo-plastics efficiently. Several forms of this type of plastic are polyolefins, polyvinyl alcohol, etc [14].

3.3   Biodegradable plastic

Biodegradable plastics are polymers made from synthetic, semisynthetic, natural polymers or a combination that can be degraded into water and carbon dioxide by the activity of living microorganisms. The American Society for Testing and Materials (ASTM) defines biodegradable plastic as "a degradable plastic in which the degradation is the result of the action of naturally occurring microorganisms such as bacteria, fungi, and algae" [12]. By another definition given by the European standard, EN 134322007 plastic shows biodegradability at a threshold of at least 90% degradation, in a maximum period of six months, under composting conditions (where microbiologically active environment provided with specific conditions of humidity and temperature). A plastic (or a polymer) is called degradable when it breaks down to smaller (monomeric) subunits and loses its original properties. Extracellular depolymerase enzymes of microorganisms are majorly responsible for the depolymerization of plastics. Aromatic polyeasterase is also one of the enzymes identified in a few bacteria that play an important role in degradation. The main difference between biodegradable polymers and bioplastics is that biodegradable polymers can be completely (100 %) degraded with the help of enzyme-producing microorganisms, which are mostly present in the soil. On the other hand, bioplastics may or may not be able to degrade by the microorganisms and sometimes may be found more resistant to degradation than conventional plastics. The technology used to prepare biodegradable polymers is mainly focused on maintaining their physical properties where it could replace conventional plastics while enhancing its biodegradability at the same time. A polymer is evaluated for its biodegradability based on its chemical structure along with physical properties such as solubility in water, crystallinity, glass transition temperature, melting point, storage modulus, hydrophobic, hydrophilic, etc. The lower rate of crystallization or crystallinity can also enhance its biodegradability. The complete decomposition of plastics can only be possible when a specific organism is identified and optimized for its growth at various parameters such as temperature, oxygen, pH, moisture content, nutrients, etc., to achieve the highest enzymatic activity that ultimately leads to the maximum extent of degradation. Since most aliphatic polyesters can be degraded with the help of microorganisms and their enzymes, biodegradable polymers are found to be a promising solution to tackle the plastic accumulation problem as they are environmentally friendly and known to increase the fertility of the soil.

3.4   Types of Biodegradable Plastics

Poly(ethylene adipate)

Polyethylene adipate (PEA) is an aliphatic polyester that is widely available in the market due to its low manufacturing cost as compared to other biodegradable plastics. PEA can be synthesized employing a polycondensation reaction between adipic acid and ethylene glycol. PEA is readily miscible with most other polymers except the low-density polyethylene (LDPE). Being an aliphatic polyester, PEA possesses poor mechanical properties. However, decreasing the crystallinity of the material can enhance the biodegradability and hydrophobicity of the final product [15]. Moreover, the low molecular weight of PEG as compared to other polymers further aids in enhancing its biodegradability. Penicillium species are found to produce lipase enzyme that breaks down the Polyethylene glycol (PEG) structure during degradation [16].

Polycaprolactoneor

Polycaprolactoneor (PCL) is a biodegradable aliphatic polyester that has broad range of applications from biomedical to the textile industry. PCL has a low melting point of around 60 °C and a glass transition temperature of about -60 °C, helping it to be readily miscible with other polymers and can be molded into various shapes. Its high biodegradability property and biocompatibility with the human body have led these biomaterials to be used in drug delivery systems, tissue engineering, prosthetics, and sutures [17]. These FDA (Food and Drug Administration) approved biomaterials are safe to use except having mild undesirable effects. PCL is often mixed with other polymers in

Materials Research Forum LLC

https://doi.org/10.21741/9781644901335-5

various concentrations to achieve desired characteristics in the final product. Moreover, the excellent physical properties of PCL enable the fabrication of a variety of structures and are used as a raw material in 3D printers.

Poly(beta-propiolactone)

Poly(beta-propiolactone), also known as PPL, is a polymer having two methylene units packed in between successive ester groups. It is a completely biocompatible and biodegradable polymer, having a similar molecular structure to those of Poly hydroxybutyrate (PHB) and PCL. Though the melting point is around 85 °C, this semi-crystalline aliphatic polyester is flexible at ambient temperature. PPL has a high molecular weight, can completely replace the conventional plastics that are used in the manufacturing of single-use cutlery, bottles, straws-cups, etc.

Polybutylene succinate

Polybutylene succinate (PBS) is a biodegradable plastic that can be easily decomposed into water and carbon dioxide using some microorganisms. The capacity to manufacture PBS is continuously being increased every year due to its high demand. PBS is a thermo-stable biodegradable aliphatic polyester polymer that contains polymerized units of butylene succinate. The synthesis of PBS is generally done by esterification of succinic acid with 1,4-butanediol. Later they are trans-esterified under vacuum using an appropriate catalyst to form a high molecular weight polymer. Since PBS possesses high heat resistance over other biodegradable polymers, these properties support the broad range of applications such as food and cosmetic packaging, fishing nets, disposable cutlery, and drug encapsulation systems used in the biomedical field. Further, they can also be used for the sustained release of pesticides and fertilizers [18].

Poly(ethylene succinate)

Polyethylene succinate (PES) is an aliphatic synthetic polyester having a melting point between 103 to 106 °C. The synthesis of PES is done either by polycondensation of succinic acid and ethylene glycol or by ring-opening polymerization of succinic anhydride with ethylene oxide using dicarboxylic acids. Similar to PBS, PES also widely used commercial polyesters with a reasonably high melting temperature and desired mechanical properties, which are now replacing the conventional non-biodegradable polymers such as polypropylene and low-density polyethylene. The lower crystallization rate and crystallinity in PES enable it to higher biodegradation rate as compared to PBS. The crystalline growth kinetics of PES was studied [19] and compared with the biodegradation process. Thermophilic *Bacillus* sp. TT96, *Paenibacillus* sp., and the fungal species of *Aspergillus clavatus* were found to be degrading PES films [20]. Among many others, *Bacillus pumilus* was found to have the highest degradation rate.

Polyhydroxybutyrate

Poly hydroxybutyrate (PHB) falls under the class of polyester that is biologically based, non-toxic, and biodegradable in nature. Biologically based means the raw materials required to produce PHB are found in nature, and *Bacillus* and *Streptomyces* microbes that are commonly found in the soil can easily breakdown these polymers into water and carbon dioxide. Typically, PHB is produced by the fermentation process, where the fully-grown microorganisms are nutritionally starved to produce bio-based polymers. This process is ecofriendly and scalable but demands high manufacturing costs when compared to conventional plastics. Therefore, scientists have come up with a transgenic plant containing genes responsible for PHB production. Now the production and the harvesting of PHB molecules are simpler and faster. The product Biopol is internal sutures (surgical stitches) manufactured for medical industries. The product TephaFLEX (used in the biomedical sector) is another type of product that can be produced using a fermentation process that is being employed by recombinant microbial strains [21].

Polylactic acid (PLA)

Poly lactic acid (PLA) is a highly thermal resistant aliphatic polyester that has derived from renewable resources like corn starch, sugar cane, cassava, etc. The raw material used for polymer preparation, such as lactic acid, can be produced by fermentative biotechnological processes using agricultural by-products and microorganisms. Its melting temperature is around 150 to 160 °C that enables its application in extended areas where other biodegradable polymers fail, making it the second-most consumed biodegradable polymer globally. PLA can replace other conventional plastics as it shows similar characteristics to polyethylene (PE), polypropylene (PP), and polystyrene (PS). The application of PLA is mostly in the manufacturing of single-use plastics, bottles and films, biomedical devices such as pins, screws, rods, etc. The melting properties of the PLA enable its application as a raw material for 3D printing as well.

4.   Biodegradation of plastic

The recalcitrant materials left in the environment act as contaminants; this is called bioaccumulation. For the growth and reproduction of microorganisms, several essential elements like carbon, nitrogen, sulfur, magnesium, etc., along with energy, are needed. Depending on the type of plastic, the suitable condition and the microorganism will help in the biodegradation. Biodegradation refers to the use of microbial functions for the conversion of organic substrate or polymer to small or simple molecular fragments, which can be further converted into water and carbon dioxides [2,22,23]. The biodegradation of polymer refers to a change in polymer structure via changing the polymer properties caused by the transformative action of a microbial enzyme, reduction in molecular weight; changes occurred on the surface, and enhanced mechanical strength [2]. Biodegradation of plastics can happen in two possible conditions; aerobic and anaerobic. In the first condition, oxygen acts as an electron receptor, while in anaerobic conditions, the other compounds such as sulfate or nitrate act as an electron receptor [24]. The oxidation reactions of organic compounds are highly exothermic (1 mole of oxygen, i.e., electron receptor has 435 kJ of energy) to give products; water and carbon dioxide. Microorganisms utilize the part of the available energy, and rest is lost as heat [25,26].

4.1   Aerobic biodegradation of plastic

Aerobic degradation refers to the deterioration of organic matter using oxygen and suitable microorganisms. CO_2 and H_2O, along with the cellular biomass of organisms, are produced during plastic degradation under aerobic conditions [2]. The aerobic degradation is more efficient in comparison to anaerobic digestion and provides more energy due to O_2 presence [2,27]. Aeration (either forced or natural) is used to provide enough oxygen and get rid of excess water present during the process. The aerobic degradation is fast and does not produce the pungent gases as in anaerobic digestion.

4.2   Anaerobic biodegradation of plastic

Anaerobic degradation of plastic refers to the breakdown of complex structures to simple forms such as CH_4, organic acids along with CO_2, and H_2O, without a supply of oxygen [28,29]. Under anaerobic conditions, the sulfate or nitrate ions act as a receptor and are less efficient. In the absence of oxygen, the amount of energy produced in the anaerobic condition is less as compared to aerobic digestion. Hence, anaerobic digestion is slow and requires more time for completing the degradation [2,30]. In the landfill waste, firstly, aerobic bacteria will be utilizing the oxygen present in the system to start the degradation, followed by the anaerobic digestion to get the complete biodegradation of waste. The anaerobic degradation will stop if there is an introduction of oxygen during the process, and hence the anaerobic condition has to be maintained.

5.   Stages of biodegradation of plastic

The two primary sites of plastic pollutions are landfills and oceans. The stages of biodegradation of plastic are majorly affected by these sites. Biodegradation is caused predominantly by the biological action of microbes, but in nature, it is a combination of abiotic and biotic activity. The abiotic activity includes the breakdown due to UV light irradiation and ozone, followed by the mechanical breakdown of plastic due to ocean gyres. The synergistic action of both leads to the efficient biodegradation of plastics.

Figure 2 depicts the summary of different stages of plastic biodegradation by microorganisms. A plastic will first undergo abiotic degradation (not biological), which includes physical and chemical degradation. Physical (wave, temperature, and UV) and chemical (oxidation or hydrolysis) degradation will contribute to the weakening of the polymer structures and reduce the plastic into smaller pieces [31]. Followed by the second step, i.e., biological degradation, which has four successive stages [24,32].

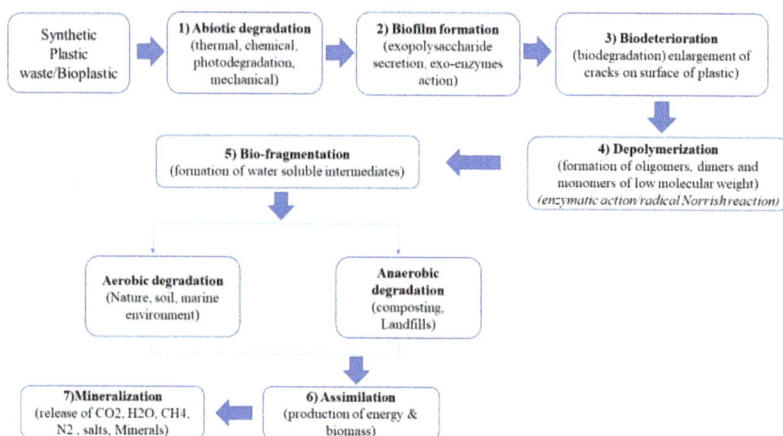

***Fig. 2** The summary of different stages of plastic biodegradation by microorganisms.*

5.1 Abiotic deterioration

Deterioration refers to the artificial degradation of various properties (chemical, physical, and mechanical) of plastic. These include thermal, chemical, mechanical, and photodegradation. Photodegradation occurs due to the sun-sensitive components that trigger the physical disintegration of plastic when exposed to sunlight [33]. The abiotic deterioration sometimes has a synergistic effect or helps to initiate the biodegradation process.

5.2 Bio-deterioration or biotic deterioration

Bio-deterioration refers to the deterioration of the chemical and physical properties of plastics due to the action of microorganisms, which will result in the superficial degradation of plastics [34,35]. The microorganisms forming biofilms have a major role in plastic biodegradation. They are highly effective in the first stage of biodegradation.

5.3   Biofilm formation

The first step of biodegradation of plastics is colonization or biofilm formation of bacteria or fungi cells onto the surface of the plastic. These biofilms of cells are attached to the surface of plastic due to the exopolysaccharide secreted by them. The degradation of the polymer takes place with the help of exo-enzymes secreted by the microbes [36]. The cells take up the polymer fragments for growth, in turn, secrete exopolysaccharides. Chemical degradation can also be orchestrated by the great diversity of species present in the biofilm, such as the production of acidic compounds by chemo-lithotrophic and chemoorganotrophic bacteria. The microbes exert mechanical, chemical, and enzymatic action on the plastics, leading to the changes in properties of the polymer. The overall process is dependent on various environmental factors such as humidity, atmospheric pollutants, and weather, especially in the case of degradation of plastic in the sea. The microbial biofilms give rise to different chemical and physical deterioration processes. The microbial biofilm formation secretes extracellular polymeric substances (EPS), which reinforce the adhesion to the surface of the plastic. This EPS helps the growth of microorganisms inside the pores, which results in cracking of pores, and hence physical properties deteriorate [37]. In the development of biofilm, the microorganism secretes the acid compounds. Chemo-lithotrophic bacteria release nitric acid, nitrous acid, or sulfuric acid. Similarly, they release other types of organic acids, such as glutaric, oxalic, gluconic, citric, fumaric, and glyoxylic acid. The release of acids results in a change in pH inside the pores and helps to degrade the structure of plastic, called chemical deterioration resulted from biofilm formation [10].

5.4   Depolymerization

The hydrophobic nature of the polymer, along with lengthy chain structure, makes microorganisms incapable of directly transporting them inside the cells [30]. The changes in the chemical and physical properties due to the formation of biofilms results in the breaking of plastic material in small molecule, i.e., dimers, oligomers, and monomers of low molecular weight. This will help to accommodate the small molecules inside the cells, which can be further converted to H_2O and CO_2 [26].

5.5   Bio-fragmentation

Fragmentation refers to the degradation of plastic in terms of modifying the mechanical, physical, and chemical properties of a material. Bio-fragmentation refers to the action of bacterial enzymes released outside cells to cleave plastic polymers into shorter sequences; oligomers and monomers [38]. Oxygenases, for example, make plastic polymers more water-soluble, making them easily degradable by bacteria. Lipases and esterases can individually attack carboxylic groups and endopeptidases amine groups. Different bacterial species can be employed in the process of bio-fragmentation [31].

5.6 Assimilation

It involves the transfer of plastic molecules of size <600Daltons (Da) into bacterial cells, followed by their transformation into cellular compounds and biomass [39]. The bio-deterioration and fragmentation of plastic materials result in the formation of small substrates, monomers. Its incorporation inside the microbial cells is called assimilation. During the process, microorganisms utilize the small molecules to obtain energy in the form of electrons along with essential elements for cell growth (carbon, nitrogen, oxygen, phosphorus, sulfur, and so forth) [2].

5.7 Mineralization

The process of biodegradation in which the plastic polymer is completely converted to carbon dioxide is called mineralization [26]. It corresponds to the complete degradation of plastic into oxidized molecules (CO_2, N_2, CH_4, and H_2O). However, the process in the natural environment is much more complex and involves many bacterial species.

6. Molecular mechanism of plastic biodegradation

Microorganisms or microbes consist of a broad range of living organisms that can live alone or in groups and are trillions in number, including bacteria, protozoa, fungi, and some algae. They can be heterotrophic or autotrophic, depending on their food production capacity. These microorganisms can actively participate in the biochemical transformation of several substrates. The conversion of organic chemicals into simple chemicals is termed biochemical transformation. Various bacteria, fungi, including yeast and molds, and few insects have been reported for plastic biodegradation, as depicted in Table 1 [40-51].

Materials Research Forum LLC
https://doi.org/10.21741/9781644901335-5

Table 1 Biodegradation of plastics using microorganisms [40-51]

S. No.	Organism	Plastic	Method	Results	References
	Bacteria				
1.	Heterotrophs (*Burkholderia* sp.), Nitrifying bacteria (*Nitrosomonas* sp. AL212, Nitrobacter winogradkyi), Type I methanotrophs (*Methylobactor* sp. and *Methylococcuscapsulatus*), Type II methanotrophs (Methylocystic sp., Methylocella sp.)	PE, PP, HDPE, LDPE, PS	Semi-aerobic landfill Bacterial degradation	Epoxides, alcohol detected after 3months, weight loss of 15-20% HDPE at 1mL/min, long degradation time, high aeration to increase the reaction rate.	[40]
2.	*Pseudomonas* sp. AKS2	Polyethylene succinate (PES)	Bioaugmentation in soil,	A higher number of PES-degrading organisms were found in the bio augmented microcosm.	[41]
3.	*Bacillus subtilis*	PE	UV irradiation and surfactant addition	9.26% degradation in 30 days	[42]
4.	*Ideonellasakaiensis*201-F6	PET Poly(ethylene terephthalate)	No pretreatment	Novel strain was obtained	[42]
5.	*Bacillus subtilis* V8, *Acinetobacter calcoaceticus*V4, *Pseudomonas putida* C 25, *Pseudomonas aeruginosa* V1	LDPE Low density polyethylene	UV light of 365nm wavelength, Nitric acid, 80°C Surfactant tween (0.05%v/v)	Degradation by each bacterium was at different rates. But enhancement in degradation has been obtained	[43]

6.	*Bacillus amyloliquefaciens*	Linear low-density polyethylene (LLDPE)	Gamma irradiation, at 90°C for 7 days	3.2% weight loss	[44]
7.	*Pseudozymaantarctica*	Polybutylene succinate-co-adipate (PBSA), polybutylene succinate (PBS)	5 weeks	Pre-treatment enhanced the biodegradation in soil	[45]
	Fungi				
8.	*Engyodontium album* MTP091 (F2), *Phanerochaetechrysosporium* NCIM 1170 (F1) and	ST-PP: Starch blended, MI-PP: metal ions blended polypropylene	UV irradiation for 10 days	79-57% weight loss	[46]
9.	*Trametes versicolor*	PVC	Liquid inoculation by mycelial fragments and subsequent fermentation	Degradation occurred in 21days	[47]
10.	*Pleurotusostreatus*	Biodegradable PE	120 days sunlight exposure, 30 days incubation by fungi	UV exposure before composting was found highly efficient.	[48]
	Insects				
11.	*Zophobasatratus*	Polystyrene	Containers stored in incubators maintained at 25 °C	In 16 days, 36.7% of ingested Styrofoam was mineralized into CO_2.	[49]
13.	*Tenebrio molitor*	Polystyrene and polyethylene	Containers were stored in incubators maintained at 25 °C and 70% humidity	Conversion up to $49.0 \pm 1.4\%$ of the ingested PE into a putative gas fraction (CO_2) was observed.	[50]

	Microalgae/Diatoms				
14.	*Phaeodactylum tricornutum*	PET, PET glycol	Active degradation at 21oC & 30oC, PETase secreting gene was inserted in microalgae and expressed	degradation products terephthalic acid were detected, 80-fold high activity on PET glycol than PET	[51]

Several studies have attempted to describe the physical, chemical, and biological stages involved in the decomposition of plastic [8]. Biological degradation is carried out by microorganisms, mainly bacteria [28]. The degradation capacity of different types of plastics by bacteria has been widely discussed in the literature, showing a wide variety of bacteria capable of degrading them [30,34,35]. Even though bacterial degradation of plastics is occurring in nature, the current environmental challenges have to be aimed at better understanding and characterizing the biodegradation of plastics by bacterial communities. Bacteria form the primary source of biodegradation, which occurs naturally at the sea coast and in the sea. The most abundant organisms in the oceans (~100 million bacteria and>500 species per liter of seawater) are invisible to the naked eye, have extremely varied metabolic capacities. In their natural environment, bacteria play the role of garbage collectors of the oceans (saprophytic organisms) since they re-mineralize half of the organic carbon that comes from waste in the food chain. Many bacteria are also specialized in the degradation of hydrocarbons, a major component of plastics, are known as hydrocarbonoclastic bacteria. Chemi-lithotrophic and -organotrophic bacteria are the major classes of bacteria responsible for plastic biodegradation in nature. There are few reports on the usage of fungal species for plastic biodegradation as well, including *Rhodococcus ruber, Phanerochaete chrysosporium, Pleurotus Ostreatus, etc.* The mechanism of degradation of insoluble synthetic hydrophobic polymers by microorganisms is yet not clearly explained in reports. Microorganisms follow a very complex mechanism in synthetic polymer degradation. Lipolytic enzymes such as lipase and esterase can hydrolyze not only fatty acid esters and triglycerides but also aliphatic polyesters. Laccases oxidize the polyethylene hydrocarbon chain, whereas alkane monooxygenases have been reported in the breakdown of polyethylene.

Table 2   *Factors affecting the biodegradability of various types of plastics [53-58]*

Sr. No.	Name of the Plastics	Melting Point (°C)	Glass Transition Point (°C)	Crystallinity (%)	Enzyme	Incubation Temperature (°C)	Microorganism	Reference
1	Poly-ethylene adipate (PEA)	55	-50	28.1	lipase	45	*Penicillium* sp.	[53]
2	Polycaprolactone (PCL)	60	−60	53.6	lipase, cutinase	45	*Streptomyces thermoviolaceus subsp.thermovio laceus 76T-2*	[54]
3	Polybutylene succinate (PBS)	114.1	−28.5	49.1	lipase B	37	*Candida antarctica*	[55]
4	Poly-ethylene succinate (PES)	103–106	-6.15	64	hydrolytic enzyme	30	*Bacillus and Paenibacillus*	[56]
5	Polyhydroxybutyrate (PHB)	175	2	60	PHA depolymerases	30	*AcidovoraxSp.*	[57]
6	Polylactic acid (PLA) or polyl actide	170	60	37	lipase, esterase, and alcalase	40	*Bacillus licheniformis*	[58]

7.   Factors affecting the biodegradation of plastic

The extent of biodegradation of bioplastic can be influenced by several factors such as the type and amount of microbial community present in the surrounding, temperature, moisture content, available gases to exchange, along the O_2 availability. Moreover, several factors such as the pH of the system, distribution of polymers at the molecular level, and the presence of chemical linkage influence the biodegradable properties of the material. The biodegradation of plastic in the soil is majorly altered by temperature, as detailed in a prior study [52]. The biofilm-forming microbial community can lead to a fast-paced degradation than others. Furthermore, the type of enzymes produced by the organisms also affects the biodegradation rates. The degradation rates differ according to the location where the biodegradation is taking place. In soil, degradation depends on whether the environment is supplied with dry or humid air and the type of site of disposal (soil or a landfill). Moreover, a composting environment is drastically influenced by the location of degradation (sewage, freshwater, or marine environment). The moist air or aquatic environment can commence the first step of the hydrolysis of polymer since radical formations accelerate these reactions. Table 2 [53-58] depicts the factors affecting the biodegradability of various types of plastics. These factors affecting the degradation process can further be classified as follows: a) environmental parameters (abiotic/biotic) and b) type of plastic (polymer properties).

7.1   Environmental Parameters

All polymer types are susceptible to biodegradation to some extent considering their multi-component system. The organic constituents (hardeners, resins, etc.) of their composition support the degradation. Polymer structure can comprise random copolymers while varying its structure. A branched network structure of the polymer can influence the cleavage of the polymer chains. An ample amount of qualitative/semi-quantitative information has been documented concerning the several factors that influence the overall degradation process of polymer in biological environments. Environmental parameters affecting the biodegradation process can further be classified as abiotic and biotic factors.

Abiotic factors

Abiotic (environmental) factors such as humidity (moisture), temperature, pH, sunlight, available gases to exchange along with O_2, and other stress conditions direct the abiotic degradation of the polymer. These factors further affect the growth of microbes and the expression of enzymes that are actively involved in the process [59]. Elevated moisture and temperature can enhance the hydrolysis process while increasing microbial activities that ultimately improve the depolymerization of polymers, making it easier for microbes to attack targeted sites [60].

Biotic factors

Several biotic factors that largely influence the biodegradation process are as follows; a) expression of enzyme, b) enzyme-substrate (E-S) complex formation, and c) hydrophobicity.

Expression of enzyme (extracellular)

The type of extracellular enzyme produced in the process shows its efficiency concerning its ability to attack available sites on that specific polymer and varies as the type of polymer changes. For instance, fungal species (*Aspergillus flavus* and *Aspergillus niger*) can express enzymes that efficiently attach 6-12 carbon di-acid-derived (aliphatic) polyesters when compared to other types of monomers [61].

Enzyme-substrate (E-S) complex formation

E-S complex formation can be described in two ways; lock-key or induced-fit model [62]. The activity of an enzyme is directly influenced by the geometry of both units involved in complex formation. Plastic, as a substrate, must possess a geometry that aligns with the enzyme in order to attain better enzymatic action that ultimately yields a better rate of degradation.

Hydrophobicity

Though the water solubility of the polymer does not assure its degradation, the strong hydrophobic nature of the polymer can make the degradation process difficult for microbial communities. Several enzyme-producing microbes may be able to degrade the plastic more efficiently when it has low hydrophobicity compared to others [63].

7.2   Type of plastic (Polymer properties)

Properties of polymer used in the synthesis of bioplastic also influence the overall degradation process. Several properties of polymers that affect the process are described below.

Conformational flexibility

The flexibility of polymers directs the rate of degradation by affecting the enzyme-substrate complex formation. Since most of the polymers are cross-linked and branched, they hinder the reach of microbes for possible attack sites. For instance, aliphatic polyester (highly flexible) is degrade when compared to PET (highly rigid) [64].

Molecular weight (MW)

Polymers with low molecular weights are easy to degrade for microbial communities during microbial transformation. Synthetic polymers (mol. wt. below 1000) are observed to be more favorable to the biodegradation process when compared to the polymers with higher molecular weight [53].

Crystallinity

Polymer with amorphous physical properties is easily susceptible to microbial degradation when compared to highly crystalline structures. An amorphous structure can accommodate more microbial interactions with polymer since they are loosely packed [65]. Moreover, the amorphous structure allows more permeability to O_2 that plays a crucial role in the formation of peroxy radicals that are actively involved in the dissociation of C–H bonds of the substrate (polymer) [66].

Melting temperature (T_m) and the size of the polymer

Polymers with lower T_m support the degradation process, whereas polymers with higher T_m are harder to depolymerize [53]. Similarly, the size of the polymer also affects the process by generating hindrance in mass transfer due to the lower surface area available in large-sized polymers. On the contrary, polymers with small sizes have a larger surface area, leading to higher rates of mass transfer that allow better interaction between microbes and polymers.

Plasticizer or additive

The type of plasticizer or additive used for the synthesis of plastic also partially influences the rate and extent of degradation. Chemically synthesized polymers are modified by blending additives during the synthesis. Co-monomers blended plastics are easy to degrade since it reduces the crystallinity of the material that favors microbial degradation by providing easy access to H_2O. Moreover, residues from catalytic processes, along with pigments/fillers, can create resistance to degradation. In contrast, the addition of metals in polyolefins can promote the pro-oxidants in the system leading to a better extent of degradation [67].

Mode of synthesis

The stability of the polymer is also influenced by the mode of synthesis employed. For instance, polystyrene (having peroxide residues) produced may favor a better photodegradation when synthesized in its free radical form compared to the photo-stable form. Similarly, polypropylene (PP) in copolymerized form is less susceptible to degradation than the PP synthesized employing a catalytic-polymerization (bulk) process [68].

8. Characterization/analytical estimation of plastic biodegradation efficiency

Biodegradability of a particular plastic can be assessed primarily by two parameters. The first standard is an assessment practice that precisely mimics the proposed environment that also defines a method to measure the extent of degradation. The second standard comprises a designed standard that designates a minimum value to set the biodegradation. Both standards are essential to establish the biodegradation performance of material adequately. For biodegradation, performance is checked by simply measuring the weight loss of plastic before and after biodegradation. There are some gravimetric methods that are used to measure the extent of biodegradation. The qualitative analysis of biodegradation can be done by sophisticated instruments such as scanning electron microscopy (SEM) and Fourier-transform infrared spectroscopy (FT-IR) that help analyze the changes in the functional group on the surface of the plastic material. Plastic waste has been disposed of in landfills leading to the formation of solid waste environments. Similar environments for plastic deposition include home/industrial compost, litter pits/landfills, anaerobic digestion systems, and ocean sites. The test methods and biodegradation performance standards established to date are only being explored for two types of disposal environments; industrial compost and marine environment. However, only test method standards are available for anaerobic-digestion and landfill environments. The extent of biodegradation means checking the performance; the standards for landfill, anaerobic-digestion systems, and home composting environments are not feasible at this moment. Therefore, the majority of plastics and bioplastics can claim coinciding performance standards of biodegradation with respect to industrial composting and marine environments, among others. To characterize and analyze the extent of degradation, several techniques can be employed in the process.

8.1 Dynamic Mechanical Analysis using universal testing systems

Dynamic Mechanical Analysis (DMR) includes the study of several mechanical properties of plastics such as tensile strength, polymer modulus, and elongation. The various physical properties relating to change in density, viscosity, glass transition temperature, amorphous and crystalline nature, and morphology are also considered to assess the information of degradation.

8.2 Scanning Electron Microscopy

The changes in the surface morphology of polymer can be studied with the help of Scanning Electron Microscopy (SEM). The method involves the sputter coating of the sample with gold or some metal ion (platinum) before analyzing. Researchers have reported the use of variable pressure SEM at suitable magnification as an alternative for direct analysis of samples [42].

8.3 Thermogravimetric analysis

Thermogravimetric analysis (TGA) records the change in the mass of samples with respect to heating. Dynamic thermogravimetry helps to study the temperature corresponding to the highest rate of decomposition along with the activation energy and decomposition reaction rate [69].

8.4   Fourier transform infrared spectroscopy

Fourier transform infrared spectroscopy (FTIR) helps to find the changes in the chemical structure of polymer due to the biodegradation [33,70,71]. It can provide information about reactions, hydrogen bonding, copolymer composition in solid and liquid form, end group detection, and chemical-physical structure changes that occurred during/after degradation.

8.5   Nuclear Magnetic Resonance Spectroscopy

The degradation reaction pathway is very useful information to understand the role of microbes and conditions in the degradation. Nuclear Magnetic Resonance Spectroscopy (NMR) is very useful to analyze this information along with the formed product composition using ^{1}H and ^{13}C [28,50].

8.6   Gel permeation chromatography

Gel permeation chromatography (GPC), also referred to as size exclusion chromatography (SEC), is very useful to analyze the distribution of molecular mass of polymer and alterations in the molecular weight of degraded polymer [72,73].

8.7   High-pressure liquid chromatography

High-pressure liquid chromatography (HPLC) is used for the separation of formed molecules after degradation and depends on the size of molecules [74]. The product formed after the metabolic degradation of xenobiotics can be analyzed using reverse-phase HPLC.

8.8   Gas chromatography-mass spectroscopy

Gas chromatography is also used to separate the formed products based on boiling point and chemical polarity. Coupling GC with MS (GC/MS) helps to identify and quantify the volatile components, residual monomers, trace contaminants, and for on-line or at-line monitoring of reaction.

8.9   Biological analysis

Various biochemical assays, like protein analysis, fluorescein diacetate analysis, and adenosine triphosphate assays, can be used to evaluate the metabolic activity of cells in the biofilm and culture [75–77].

8.10   Molecular techniques

Though depolymerization of the polymer during degradation can take a more prolonged time, genetic engineering can be efficiently employed once the pathway of degradation is established. Genetic engineering can be utilized to obtain several desirable features in the process, such as reduction of reaction time, enhancement of regulatory mechanisms, modification of enzymes involved in the process, and making plastic easily amiable to microbial degradation. DNA/Protein sequences can be employed to assess the results obtained from genetic modifications to understand the correlation of genetic alterations with the process of biodegradation.

9.   Value addition to biodegradation of plastics

The biodegradation process by microbial break down of solid matter can result in other essential substances. A nutrient-rich form of soil (Humus), CO_2, and methane gas are some of the crucial by-products of biodegradation that add value to the overall process. Methane, a potent greenhouse gas, can be a potential root of clean energy. CO_2, along with methane, can be employed to generate electricity using the internal combustion process. Methane harnessed during biodegradation limits its negative impact as a greenhouse gas since it is being used to generate a clean form of energy, electricity. Further, it also sidesteps the use of dwindling fossil sources for the generation of methane, which again adds up the value to the practice. Nevertheless, landfills can accommodate biodegradation of plastic waste while producing methane as a by-product that can be ultimately employed to obtain clean energy, makes the whole biodegradation process a crucial step towards a sustainable environment. Microbial communities can actively participate in the biodegradation process, where they break down the polymer by expressing different forms of the enzyme, reducing the time of decay from centuries to years (sometimes months, depending on the environment they are exposed to). Considering the amount of plastic disposed of in landfills every day, biodegradable additives (plasticizers) blended with biodegradable plastic can help smoothen the process of biodegradation while obtaining millions of megawatts of clean energy. Some of the new biopolymers can also be obtained from these environments, which can further add up to the more modern biodegradable plastics of microbial origin.

10.   Challenges in plastic biodegradation

The current challenge involved in this area are as follows: (1) To gain a better understanding of the mechanisms of biodegradation of plastics by natural communities, (2) Capturing a well-defined list of species that colonize plastics and capable of degrading them, and (3) Determining the molecular mechanisms involved in the degradation of different compositions of plastics. Currently, many researchers are contributing to the issue are attempting to answer these questions. A coupled DNA-SIP and high-speed pyrosequencing approach are already used to identify bacteria capable of degrading polycyclic aromatic hydrocarbons [78,79]. This approach is based on the isotopic labeling of plastics and the monitoring of their incorporation by bacteria to gain access to the functional community of "plasticlast bacteria," communities that are actively involved in the process of biodegradation.

Several attempts are being made in the field of research to develop techniques along with novel microbial communities that have the potential to alter the future of biodegradable plastic. A plastic manufacturer (Environmental Products, Inc.) has discovered a patented technology, TDPA™ (Totally Degradable Plastic Additive), that enables the degradation of common types of plastic under a monitored environment. The technology has been well-documented by researchers where it was found to be favorably working in the degradation of common plastics [80]. Thermal and photolytic pre-treatments can be applied to promote the abiotic degradation of low-density polyethylene (LDPE). The extent of biodegradation can be examined by monitoring the initial changes in molecular weight while analyzing other structural parameters as well. The degree of crystallinity, spectroscopic properties, and tensile strength are among several essential characteristics that can be assessed during the degradation process. Similarly, the biotic environment further helps in the overall degradation of polymers [80]. LDPE–TDPA™ oxidized samples showed 50-60 % of biodegradation during a specific duration (>18 months) while producing CO_2 in the process mediated by microbial communities of soil in a closed biometric flask. Research unveiled that the LDPE–TDPA™ formulations are found to effectively promote oxidation and biodegradation of PE in soil. Similarly, the discovery of an invertebrate species (insect larvae), which are reported with higher degradation rates, can reduce the waste polymer by ingestion followed by degradation in the gut, by expressing several enzymes is interesting and must be explored further. However, several areas are yet to be investigated concerning the cumulative time required for oxidation/biodegradation and the rate of degradation under varying environmental conditions.

Conclusions and future prospects

Various national and international research efforts have been encouraged in recent years, given the extent of pollution led by plastics in multiple environments. Understanding the mechanisms of biodegradation of plastics is still in its infancy. Though certain mechanisms have been observed in laboratory conditions, their study in a natural

Degradation of Plastics
Materials Research Foundations **99** (2021) 111-144

Materials Research Forum LLC
https://doi.org/10.21741/9781644901335-5

environment remains largely unexplored. The molecular mechanisms of bio-deterioration, bio-fragmentation, bio-assimilation, and bio-mineralization of plastics are yet to be fully understood. The diversity of microorganisms associated with these different stages of biodegradation is also in a primitive stage. Understanding these processes is very important in order to define the biodegradation rates of plastics and to better predict the future of so-called "biodegradable" plastics. The primary site of plastic pollution is the landfills, which ultimately push the waste into the sea. The sea is the ultimate receptacle for all the waste produced on land. Though spreading awareness can limit the waste to some extent, it cannot be wholly relied upon, and a significant step towards excluding plastic waste from the environment needs to be taken. Replacing conventional plastic with bio-based and biodegradable plastic seems to be the only option to tackle this global environmental crisis. Nevertheless, polymer recycling technology based on the conclusions is a potential opportunity that has still to be tapped. Finally, it can be concluded that for environmental safety, we have to restrict or reduce the use of non-degradable polymers while simultaneously encouraging the development of more efficient biodegradable plastics that can replace the traditional plastic to make the planet earth a sustainable biosphere, again.

References

[1]   L.C.M. Lebreton, S.D. Greer, J.C. Borrero, Numerical modelling of floating debris in the world's oceans, Mar. Pollut. Bull. 64 (2012) 653-661. https://doi.org/10.1016/j.marpolbul.2011.10.027

[2]   J.A. Glaser, Biological degradation of polymers in the environment, in: J.A. Glaser (Eds.), Plastics in the environment 2019. Intech Open, London, 2019, pp.1-23. https://doi.org/10.5772/intechopen.85124

[3]   S.L. Wright, R.C. Thompson, T.S. Galloway, The physical impacts of microplastics on marine organisms: a review, Environ. Pollut. 178 (2013) 483-492. https://doi.org/10.1016/j.envpol.2013.02.031

[4]   A. Collingnon, J.H. Hecq, F. Glagani, P. Voisin, F. Collard, A. Goffart, Neustonic microplastic and zooplankton in the north western Mediterranean sea, Mar. Pollut. Bull. 64 (2012) 861-864. https://doi.org/10.1016/j.marpolbul.2012.01.011

[5]   R.G. Asch, P. Davidson, Plastic ingestion by mesopelagic fishes in the North Pacific Subtropic Gyre, Mar. Ecol. Prog. 423 (2011) 173-180. https://doi.org/10.3354/meps09142

[6]   D. Lithner, A. Larsson, G. Dave, Environmental and health hazard ranking and assessment of plastic polymers based on chemical composition, Sci. Total Environ. 409 (2011) 3309-3324. https://doi.org/10.1016/j.scitotenv.2011.04.038

[7]   E.L. Teuten, J.M. Saquing, D.R.U. Knappe, M.A. Barlaz, S. Jonsson, A. Björn, S.J. Rowland, R.C. Thompson, T.S. Galloway, R. Yamashita, D. Ochi, Y. Watanuki, C. Moore, P.H. Viet, T.S. Tana, M. Prudente, R. Boonyatumanond, M.P. Zakaria, K. Akkhavong, Y. Ogata, H. Hirai, S. Iwasa, K. Mizukawa, Y. Hagino, A. Imamura, M. Saha, H. Takada, Transport and release of chemicals from plastics to the environment and to wildlife, Philos. Trans. R. Soc. B Biol. Sci. 364 (2009) 2027–2045. https://doi.org/10.1098/rstb.2008.0284

[8]   A.L. Andrady, Microplastics in the marine environment, Mar. Pollut. Bull. 62 (2011) 1596–1605. https://doi.org/10.1016/j.marpolbul.2011.05.030

[9]   G. Caruso, Microplastics as vectors of contaminants, Mar. Pollut. Bull. 146 (2019) 921–924. https://doi.org/10.1016/j.marpolbul.2019.07.052

[10]   E.R. Zettler, T.J. Mincer, L.A. Amaral-Zettler, Life in the "plastisphere": microbial communities on plastic marine debris, Environ. Sci. Technol. 47 (2013) 7137–7146. https://doi.org/10.1021/es401288x

[11]   J.N. Hahladakis, C.A. Velis, R. Weber, E. Iacovidou, P. Purnell, An overview of chemical additives present in plastics: migration, release, fate and environmental impact during their use, disposal and recycling, J. Hazard. Mater. 344 (2018) 179–199. https://doi.org/10.1016/j.jhazmat.2017.10.014

[12]   J.P. Harrison, C. Boardman, O. Callaghan, A. Delort, J. Song, J.P. Harrison, Biodegradability standards for carrier bags and plastic films in aquatic environments: a critical review, R. Soc. Open Sci. 65 (2018) 1-18. https://doi.org/10.1098/rsos.171792

[13]   S.M. Al-Salem, A.Y. Al-Nasser, M.H. Behbehani, H.H. Sultan, H.J. Karam, M.H. Al-Wadi, A.T. Al-Dhafeeri, Z. Rasheed, M. Al-Foudaree, Thermal response and degressive reaction study of oxo-biodegradable plastic products exposed to various degradation media, Int. J. Polym. Sci. (2019). https://doi.org/10.1155/2019/9612813

[14]   J.M.R. da Luz, S.A. Paes, M.D. Nunes, M. de C.S. da Silva, M.C.M. Kasuya, Degradation of oxo-biodegradable plastic by *Pleurotus ostreatus*, PLoS One. 8 (2013) e69386. https://doi.org/10.1371/journal.pone.0069386

[15]   R. Pantani, A. Sorrentino, Influence of crystallinity on the biodegradation rate of injection-moulded poly(lactic acid) samples in controlled composting conditions,

Materials Research Foundations **99** (2021) 111-144   https://doi.org/10.21741/9781644901335-5

Polym. Degrad. Stab. 98 (2013) 1089–1096.
https://doi.org/10.1016/j.polymdegradstab.2013.01.005

[16]   V.M. Pathak, Navneet, Review on the current status of polymer degradation: a microbial approach, Bioresour. Bioprocess. 4 (2017) 1-31.
https://doi.org/10.1186/s40643-017-0145-9

[17]   B. Azimi, P. Nourpanah, M. Rabiee, S. Arbab, Poly (ε-caprolactone) fiber: an overview, J. Eng. Fiber. Fabr. 9 (2014) 74–90.
https://doi.org/10.1177/155892501400900309

[18]   E. Rudnik, Compostable polymer materials – definitions, structures and methods of preparation, in E. Rudnik (Eds.) Compostable Polymer Materials (Second Edition), Elsevier, Netherlands, 2019 pp. 11-48. https://doi.org/10.1016/B978-0-08-099438-3.00002-1

[19]   Z. Qiu, S. Fujinami, M. Komura, K. Nakajima, T. Ikehara, T. Nishi, Non-isothermal crystallization kinetics of poly(butylene succinate) and poly(ethylene succinate), Polym. J. 36 (2004) 642–646. https://doi.org/10.1295/polymj.36.642

[20]   Y. Tokiwa, B.P. Calabia, C.U. Ugwu, S. Aiba, Biodegradability of plastics, Int. J. Mol. Sci. 10 (2009) 3722–3742. https://doi.org/10.3390/ijms10093722

[21]   L. W. McKeen, Introduction to the physical, mechanical, and thermal properties of plastics and elastomers, L. W. McKeen (Eds.) in: Plastics Design Library, The Effect of UV Light and Weather on Plastics and Elastomers (Fourth Edition), William Andrew Publishing, Norwich, 2019, pp 49-76. https://doi.org/10.1016/B978-0-12-816457-0.00003-4

[22]   T. Ahmed, M. Shahid, F. Azeem, I. Rasul, A.A. Shah, M. Noman, A. Hameed, N. Manzoor, I. Manzoor, S. Muhammad, Biodegradation of plastics: current scenario and future prospects for environmental safety, Environ. Sci. Pollut. Res. 25 (2018) 7287–7298. https://doi.org/10.1007/s11356-018-1234-9

[23]   Y. Zheng, E.K. Yanful, A.S. Bassi, A review of plastic waste biodegradation, Crit. Rev. Biotechnol. 25 (2005) 243–250. https://doi.org/10.1080/07388550500346359

[24]   N. Lucas, C. Bienaime, C. Belloy, M. Queneudec, F. Silvestre, J. E. Nava-Saucedo, Polymer biodegradation: mechanisms and estimation techniques – a review, Chemosphere. 73 (2008) 429–442. https://doi.org/10.1016/j.chemosphere.2008.06.064

[25]   N. Wierckx, T. Narancic, C. Eberlein, R. Wei, O. Drzyzga, A. Magnin, H. Ballerstedt, S.T. Kenny, E. Pollet, L. Avérous, K.E.O'Connor, W. Zimmermann, H.J. Heipieper, A. Prieto, J. Jiménez, L.M. Blank, Plastic biodegradation: Challenges and

Materials Research Forum LLC

https://doi.org/10.21741/9781644901335-5

Opportunities, In: Steffan R. (eds) Consequences Microbial Interactions with Hydrocarbons Oils, Lipids Biodegradation and Bioremediation, Springer International Publishing, Cham, 2018: pp. 1–29. https://doi.org/10.1007/978-3-319-44535-9_23-1

[26]   M. Shimao, Biodegradation of plastics, Curr. Opin. Biotechnol. 12 (2001) 242–247. https://doi.org/10.1016/S0958-1669(00)00206-8

[27]   N.S. Panikov, Microbial growth kinetics, Spinger Netherlands 1995, pp-378.

[28]   A.A. Shah, F. Hasan, A. Hameed, S. Ahmed, Biological degradation of plastics: A comprehensive review, Biotechnol. Adv. 26 (2008) 246–265. https://doi.org/10.1016/j.biotechadv.2007.12.005

[29]   M.A. Barlaz, R.K. Ham, D.M. Schaefer, Mass-Balance analysis of anaerobically decomposed refuse, J. Environ. Eng. 115 (1989) 1088–1102. https://doi.org/10.1061/(ASCE)0733-9372(1989)115:6(1088)

[30]   A. Sheel, D. Pant, Microbial depolymerization, in: Waste Bioremediation. Energy, Environment, and Sustainability. Springer, Singapore (2018) pp 61-103. https://doi.org/10.1007/978-981-10-7413-4_4

[31]   S.K. Ghosh, S. Pal, S. Ray, Study of microbes having potentiality for biodegradation of plastics, Environ. Sci. Pollut. Res. 20 (2013) 4339–4355. https://doi.org/10.1007/s11356-013-1706-x

[32]   R. Devi, V. Kannan, K. Natarajan, D. Nivas, K. Kannan, S. Chandru, A. Antony, The role of microbes in plastic degradation, in: R. Chandra (Eds.), Environ. Waste Management, 1st edition Taylor & Francis Group, LLC, Milton, 2015, pp 341-370. https://doi.org/10.1201/b19243-13

[33]   B. Singh, N. Sharma, Mechanistic implications of plastic degradation, Polym. Degrad. Stab. 93 (2008) 561–584. https://doi.org/10.1016/j.polymdegradstab.2007.11.008

[34]   G. O'Toole, H.B. Kaplan, R. Kolter, Biofilm formation as microbial development, Annu. Rev. Microbiol. 54 (2000) 49–79. doi.org/10.1146/annurev.micro.54.1.49

[35]   J.D. Gu, Microbial colonization of polymeric materials for space applications and mechanisms of biodeterioration: A review, Int. Biodeterior. Biodegradation. 59 (2007) 170–179. https://doi.org/10.1016/j.ibiod.2006.08.010

[36]   H.C. Flemming, Relevance of biofilms for the biodeterioration of surfaces of polymeric materials, Polym. Degrad. Stab. 59 (1998) 309–315. https://doi.org/10.1016/S0141-3910(97)00189-4

[37]   S. Bonhomme, A. Cuer, A.M. Delort, J. Lemaire, M. Sancelme, G. Scott, Environmental biodegradation of polyethylene, Polym. Degrad. Stab. 81 (2003) 441–452. https://doi.org/10.1016/S0141-3910(03)00129-0

[38]   A. Lugauskas, L. Levinskait, D. Pečiulyt, Micromycetes as deterioration agents of polymeric materials, Int. Biodeterior. Biodegradation. 52 (2003) 233–242. https://doi.org/10.1016/S0964-8305(03)00110-0

[39]   W. Guo, J. Tao, C. Yang, C. Song, W. Geng, Q. Li, Y. Wang, M. Kong, S. Wang, Introduction of environmentally degradable parameters to evaluate the biodegradability of biodegradable polymers, PLoS One. 7 (2012) e38341. https://doi.org/10.1371/journal.pone.0038341

[40]   S. Muenmee, W. Chiemchaisri, C. Chiemchaisri, Enhancement of biodegradation of plastic wastes via methane oxidation in semi-aerobic landfill, Int. Biodeterior. Biodegradation. 113 (2016) 244–255. https://doi.org/10.1016/j.ibiod.2016.03.016

[41]   P. Tribedi, A.K. Sil, Bioaugmentation of polyethylene succinate-contaminated soil with *Pseudomonas* sp. AKS2 results in increased microbial activity and better polymer degradation, Environ. Sci. Pollut. Res. 20 (2013) 1318–1326. https://doi.org/10.1007/s11356-012-1080-0

[42]   P.P. Vimala, L. Mathew, Biodegradation of Polyethylene Using *Bacillus Subtilis*, Procedia Technol. 24 (2016) 232–239. https://doi.org/10.1016/j.protcy.2016.05.031

[43]   V.M. Pathak, N. Kumar, Dataset on the impact of UV, nitric acid and surfactant treatments on low-density polyethylene biodegradation, Data Br. 14 (2017) 393–411. https://doi.org/10.1016/j.dib.2017.07.073

[44]   Č. Novotný, K. Malachová, G. Adamus, M. Kwiecień, N. Lotti, M. Soccio, V. Verney, F. Fava, Deterioration of irradiation/high-temperature pretreated, linear low-density polyethylene (LLDPE) by *Bacillus amyloliquefaciens*, Int. Biodeterior. Biodegrad. 132 (2018) 259–267. https://doi.org/10.1016/j.ibiod.2018.04.014

[45]   Y. Sameshima-Yamashita, H. Ueda, M. Koitabashi, H. Kitamoto, Pre-treatment with an esterase from the yeast *Pseudozyma antarctica* accelerates biodegradation of plastic mulch film in soil under laboratory conditions, J. Biosci. Bioeng. 127 (2019) 93–98. https://doi.org/10.1016/j.jbiosc.2018.06.011

[46]   D. Jeyakumar, J. Chirsteen, M. Doble, Synergistic effects of pre-treatment and blending on fungi mediated biodegradation of polypropylenes, Bioresour. Technol. 148 (2013) 78–85. https://doi.org/10.1016/j.biortech.2013.08.074

[47]   D. Dussault, B.F. Mayer, A. Jaouich, A. Karam, Biodegradation of a synthetic textile containing pvc, Int. J. Adv. Sci. Eng. Tech. (2017) 74–77.

[48]   J.M.R. da Luz, S.A. Paes, K.V.G. Ribeiro, I.R. Mendes, M.C.M. Kasuya, Degradation of green polyethylene by *Pleurotus ostreatus,* PLoS One. 10 (2015) e0126047. https://doi.org/10.1371/journal.pone.0126047

[49]   Y. Yang, J. Wang, M. Xia, Biodegradation and mineralization of polystyrene by plastic-eating superworms *Zophobas atratus*, Sci. Total Environ. 708 (2020) 135233. https://doi.org/10.1016/j.scitotenv.2019.135233

[50]   A.M. Brandon, S.H. Gao, R. Tian, D. Ning, S.S. Yang, J. Zhou, W.M. Wu, C.S. Criddle, Biodegradation of polyethylene and plastic mixtures in mealworms (larvae of *Tenebrio molitor*) and effects on the gut microbiome, Environ. Sci. Technol. 52 (2018) 6526–6533. https://doi.org/10.1021/acs.est.8b02301

[51]   D. Moog, J. Schmitt, J. Senger, J. Zarzycki, K.H. Rexer, U. Linne, T. Erb, U.G. Maier, Using a marine microalga as a chassis for polyethylene terephthalate (PET) degradation. Microb Cell Fact. 18 (2019) 171. https://doi.org/10.1186/s12934-019-1220-z

[52]   A. Pischedda, M. Tosin, F. Degli-Innocenti, Biodegradation of plastics in soil: the effect of temperature, Polym. Degrad. Stab. 170 (2019) 109017. https://doi.org/10.1016/j.polymdegradstab.2019.109017

[53]   J. Suzuki, K. Hukushima, S. Suzuki, Effect of ozone treatment upon biodegradability of water-soluble polymers, Environ. Sci. Technol. 12 (1978) 1180–1183. https://doi.org/10.1021/es60146a002

[54]   T.K. Chua, M. Tseng, M.K. Yang, Degradation of Poly(ε-caprolactone) by thermophilic *Streptomyces thermoviolaceus* sub sp. thermoviolaceus 76T-2, AMB Express. 3 (2013) 8. https://doi.org/10.1186/2191-0855-3-8

[55]   A. Wcisłek, A.S. Olalla, A. McClain, A. Piegat, P. Sobolewski, J. Puskas, M.E. Fray, Enzymatic Degradation of Poly(butylene succinate) copolyesters synthesized with the use of *Candida antarctica* lipase B, Polymers (Basel). 10 (2018) 688. https://doi.org/10.3390/polym10060688

[56]   Y. Tezuka, N. Ishii, K.I. Kasuya, H. Mitomo, Degradation of poly(ethylene succinate) by mesophilic bacteria, Polym. Degrad. Stab. 84 (2004) 115–121. https://doi.org/10.1016/j.polymdegradstab.2003.09.018

[57]   T. Kobayashi, A. Sugiyama, Y. Kawase, T. Saito, J. Mergaert, J. Swings, Biochemical and genetic characterization of an extracellular poly(3-hydroxybutyrate)

depolymerase from *Acidovorax* sp. strain TP4, J. Environ. Polym. Degrad. 7 (1999) 9–18. https://doi.org/10.1023/A:1021885901119

[58]   S.H. Lee, I.Y. Kim, W.S. Song, Biodegradation of polylactic acid (PLA) fibers using different enzymes, Macromol. Res. 22 (2014) 657–663. https://doi.org/10.1007/s13233-014-2107-9

[59]   J.G. Gu, J.D. Gu, Methods currently used in testing microbiological degradation and deterioration of a wide range of polymeric materials with various degree of degradability: A review, J. Polym. Environ. 13 (2005) 65–74. https://doi.org/10.1007/s10924-004-1230-7

[60]   K.L.G. Ho, A.L. Pometto, A. Gadea-Rivas, J.A. Briceño, A. Rojas, Degradation of polylactic acid (PLA) plastic in Costa Rican soil and Iowa State University compost rows, J. Environ. Polym. Degrad. 7 (1999) 173–177. https://doi.org/10.1023/A:1022874530586

[61]   M.K. Sangale, A review on biodegradation of polythene: the microbial approach, J. Bioremediation Biodegrad. 03 (2012) 1-9. https://doi.org/10.4172/2155-6199.1000164

[62]   J.G. Voet, Mechanism of enzyme action, in: D. Voet, J.G. Voet, C.W. Pratt (Eds.), Fundamentals of Biochemistry: Life At The Molecular Level, John Wiley and Sons, New Jersey, 2016, pp-322-345.

[63]   T. Nakajima-Kambe, Y. Shigeno-Akutsu, N. Nomura, F. Onuma, T. Nakahara, Microbial degradation of polyurethane, polyester polyurethanes and polyether polyurethanes, Appl. Microbiol. Biotechnol. 51 (1999) 134–140. https://doi.org/10.1007/s002530051373

[64]   J. D. GU, Microbial degradation of materials: general processes (Eds.) R. Winston, Uhlig Corros. Handb. John Wiley and Sons, New Jersey, 2000 pp 349–365.

[65]   H. Tsuji, S. Miyauchi, Poly(l-lactide): VI Effects of crystallinity on enzymatic hydrolysis of poly(l-lactide) without free amorphous region, Polym. Degrad. Stab. 71 (2001) 415–424. https://doi.org/10.1016/S0141-3910(00)00191-9

[66]   H. Morawetz, Macromolecules, an introduction to polymer science, J. Polym. Sci. Polym. Lett. Ed. 18 (1980) 153–154. https://doi.org/10.1002/pol.1980.130180225

[67]   Y. Orhan, J. Hrenović, H. Büyükgüngör, Biodegradation of plastic compost bags under controlled soil conditions, Acta Chim. Slov. 51 (2004) 579-588.

[68]   L. Tang, Q. Wu, B. Qu, The effects of chemical structure and synthesis method on photodegradation of polypropylene, J. Appl. Polym. Sci. 95 (2005) 270–279. https://doi.org/10.1002/app.21272

[69]   F. Carrasco, P. Pagès, Thermogravimetric analysis of polystyrene: influence of sample weight and heating rate on thermal and kinetic parameters, J. Appl. Polym. Sci. 61 (1996) 187–197. https://doi.org/10.1002/(SICI)1097-4628(19960705)61:1<187::AID-APP20>3.0.CO;2-3

[70]   I.S. Elashmawi, N.A. Hakeem, E.M. Abdelrazek, Spectroscopic and thermal studies of PS/PVAc blends, Phys. B Condens. Matter. 403 (2008) 3547–3552. https://doi.org/10.1016/j.physb.2008.05.024

[71]   A. Mohamed, S.H. Gordon, G. Biresaw, Polycaprolactone/polystyrene bio-blends characterized by thermogravimetry, modulated differential scanning calorimetry and infrared photoacoustic spectroscopy, Polym. Degrad. Stab. 92 (2007) 1177–1185. https://doi.org/10.1016/j.polymdegradstab.2007.04.012

[72]   G. Kale, R. Auras, S.P. Singh, Degradation of commercial biodegradable packages under real composting and ambient exposure conditions, J. Polym. Environ. 14 (2006) 317–334. https://doi.org/10.1007/s10924-006-0015-6

[73]   X. Peng, J. Shen, Preparation and biodegradability of polystyrene having pyridinium group in the main chain, Eur. Polym. J. 35 (1999) 1599–1605. https://doi.org/10.1016/S0014-3057(98)00253-5

[74]   F. Beltrametti, A.M. Marconi, G. Bestetti, C. Colombo, E. Galli, M. Ruzzi, E. Zennaro, Sequencing and functional analysis of styrene catabolism genes from *Pseudomonas fluorescens* ST, Appl. Environ. Microbiol. 63 (1997) 2232–2239. https://doi.org/10.1128/AEM.63.6.2232-2239.1997

[75]   M. Koutny, J. Lemaire, A.-M. Delort, Biodegradation of polyethylene films with pro-oxidant additives, Chemosphere. 64 (2006) 1243–1252. https://doi.org/10.1016/j.chemosphere.2005.12.060

[76]   M. Koutny, M. Sancelme, C. Dabin, N. Pichon, J. Lemaire, M. Koutny, M. Sancelme, C. Dabin, N. Pichon, A.M. Delort, Acquired biodegradability of polyethylenes containing pro-oxidant additives to cite this version: HAL Id : hal-00021890, Polym. Degrad. Stabilty. 91 (2006) 1495–1503. https://doi.org/10.1016/j.polymdegradstab.2005.10.007

[77]   I.G. Orr, Y. Hadar, A. Sivan, Colonization, biofilm formation and biodegradation of polyethylene by a strain of *Rhodococcus ruber*, Appl. Microbiol. Biotechnol. 65 (2004) 97-104. https://doi.org/10.1007/s00253-004-1584-8

[78]   T. Gutierrez, J.F. Biddle, A. Teske, M.D. Aitken, Cultivation-dependent and cultivation-independent characterization of hydrocarbon-degrading bacteria in Guaymas Basin sediments, Front. Microbiol. 6 (2015) 1-12. https://doi.org/10.3389/fmicb.2015.00695

[79]   D.K. Allen, B.S. Evans, I.G.L. Libourel, Analysis of isotopic labeling in peptide fragments by tandem mass spectrometry, PLoS One. 9 (2014) e91537. https://doi.org/10.1371/journal.pone.0091537

[80]   M. Vaverková, F. Toman, D. Adamcová, J. Kotovicová, Study of the biodegrability of degradable/biodegradable plastic material in a controlled composting environment, Ecol. Chem. Eng. S. (2012). https://doi.org/10.2478/v10216-011-0025-8.

Degradation of Plastics
Materials Research Foundations **99** (2021) 145-162

Materials Research Forum LLC
https://doi.org/10.21741/9781644901335-6

Chapter 6

Recovery of Biodegradable Bioplastics from Different Activated Sludge Processes during Wastewater Treatment

S. Ghosh[1]*, S. Chakraborty[1,2]

[1]Centre for the Environment, Indian Institute of Technology Guwahati, Assam 781039, India

[2]Department of Civil Engineering, Indian Institute of Technology Guwahati, Assam 781039, India

*g.sayanti@iitg.ac.in

Abstract

To overcome the environmental hazards of petroleum based plastics, synthesis and use of microbial bioplastics became popular. Polyhydroxyalkanoates (PHAs) are biodegradable biopolymers having plastic like properties mainly used in tissue engineering and packaging. Bacteria can produce bioplastics in carbon abundance. Activated sludge process is a simultaneous process for treating wastewater and producing PHAs. Wastewaters are treated by using mixed sludge, aerobic granular sludge and chemically treated sludge which provided more than 40% PHA yield. This chapter describes the PHA structure, synthesis pathways, types of wastewaters and activated sludge processes used with reactor parameters and environmental factors effecting PHA productions.

Keywords

Polyhydroxyalkanoates, Activated Sludge, Wastewater, Pathways, Synthesis Conditions, Extraction, Quantification

Contents

1.   Introduction

The ever-expanding production and extensive use of petroleum based plastics are one of the major environmental concerns of modern civilization. Starting from packaging to utensil, transport, furniture, pipelines and in different industrial sectors plastic usage has become an integral part of our daily life. Polyamides, polyethylene, polyester, polypropylene, polyvinyl chloride (PVC) are the most commonly used plastic polymers nowadays. However, non-biodegradability, slow environmental accumulation and carbon dioxide (CO_2) emission are the major threats of plastic pollution [1]. Generation of approximately 34 million tons of plastic wastes has been reported throughout the globe and about 93% of them are disposed in oceans and utilized in landfill purposes [2].

Synthesis of bio-based biodegradable bioplastics is the environmentally viable alternative to fossil fuel based plastics. First bioplastic was produced and commercialized by Imperial Chemical Industries in 1980s under the trade name 'Biopol'. Polyhydroxyalkanoates (PHA) are polyesters with thermoplastic properties largely known as biodegradable biopolymer. Among all the PHAs, polyhydroxybutyrate (PHB) and its copolymer polyhydroxyvalerate (PHV) combine together to form a biocompatible polymer (poly(3-hydroxybutyrate-*co*-3-hydroxyvalerate)) known as PHBV[3]. Microbes are capable of producing PHA in abundance of external carbon source (feast phase) and store them for future use in absence of food (famine phase) and further convert it into energy for metabolic activities [4]. According to the literature, *Ralstonia eutropha* mediated and volatile fatty acid (VFA) enhanced pathways are the main proposed pathways for bioplastic production. Halophilic bacteria *Haloferax mediterranei, Ralstonia eutropha,* and genetically engineered recombinant *Bacillus subtilis* str. pBE2C1 and *Bacillus subtilis* str. pBE2C1AB are largely reported PHA producing strains [5].Genetically modified bacteria were reported to accumulate more than 90% PHA of their cell dry weight by utilizing sugar substrate [6]. But to avoid the use of expensive pure culture bacteria and sugar substrates, activated sludge and wastewater can be used as a simple, renewable and low cost method of bioplastics production. Pie charts of Figure 1 [5] are describing the comparisons of global bioplastic market and traditional plastic markets.

The latest research trend in wastewater bio-refineries (WWBR) is driving the researchers' attention towards resource recovery during wastewater treatment. Wastewaters are considered of having more than 50% of the lost resources [6]. Hence, complex wastewater can be used as C, N, P (nutrient) enriched renewable feedstock for bacterial growth and precursor of PHA synthesis. Activated sludge process is a well known biological wastewater treatment used in both municipal and industrial wastewater treatment. This process involves the use of large amount of bio-sludge which can simultaneously treat wastewater and can recover renewable biomaterials.

Activated sludge based PHA production has been extensively studied mainly in palm oil mill effluent (POME), paper and pulp mill wastewater, dairy wastewaters, molasses spent wash and olive oil mill wastewater containing recalcitrant pollutants [4, 7-10]. Variation

Materials Research Forum LLC

https://doi.org/10.21741/9781644901335-6

in reactor parameters and conditions has significant impact in PHA yield. For microbial mixed cultures, aero dynamic feeding (ADF) strategy and regulation in feast famine regime and application of selection pressure enhanced PHA yield. However, sludge modification was proved to be fruitful in enhancing the PHA production. Several scientists conducted PHA extraction from different activated sludge like secondary municipal wastewater sludge, dissolved oxygen (DO) controlled municipal sludge, chemically treated and aerobic granular sludge [4, 11, 12].

Hence activated sludge process is an eco-friendly and economic process for environmental remediation as well as biodegradable bioplastic production leading to circular economy.

GLOBAL PLASTICS MARKET

- Bioplastic market expected to grow at **30% CAGR 2013-2030**

- Traditional plastics expected to grow 3% annually

40% market share

$324B

Bioplastics:
<1% market share
$3.75B

4% market share
$21B

$455B
2013

$540B
2019

$803B
2030

Bioplastics   Oil-based plastics

Source: Grand View Research 2014, European Bioplastics 2013, BCC Research 2014, Nexant Inc. 2012

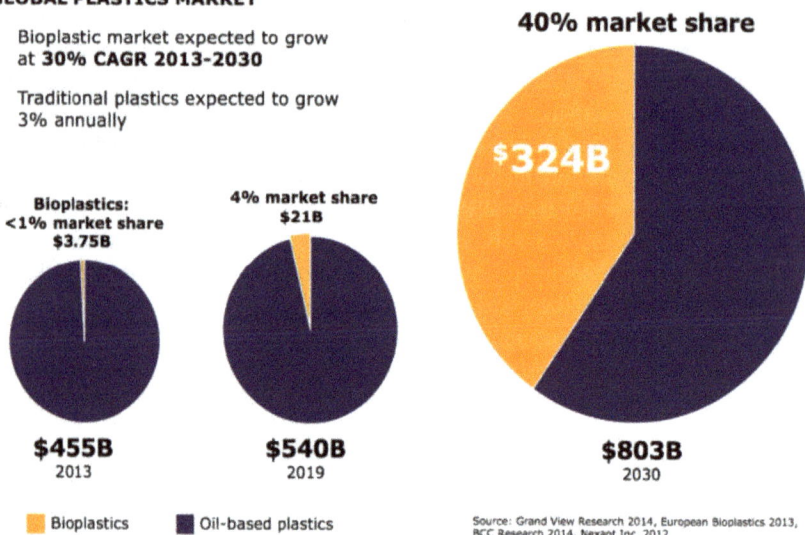

Figure 1 The pie charts are depicting the global scenario for oil-based traditional plastics and bio-based bioplastics which is indicating that bioplastics hold 40% market share [5].

2. Chemistry and application of bioplastics

Degradation of Plastics   Materials Research Forum LLC
Materials Research Foundations **99** (2021) 145-162   https://doi.org/10.21741/9781644901335-6

Polyhydroxyalkanoates are environment friendly, non-toxic biocompatible polyesters which are almost biodegradable in nature. Bacterial PHA can be produced by bacteria or haloarchaea in absence of carbon and nutrient sources. They are thermoplastics having very high melting point and degradation temperature of 179 °C [3]. PHAs are ultraviolet (UV) stable bioplastics soluble in chloroform. Their strength and flexibility can be modified by varying valerate percentages [5].

The structure and length of the PHA polymer chain depends on the type of microorganism and carbon source used for the synthesis, where polyesters are connected by ester bonds. PHAs generally carry one pendant alkyl group (R) in it where the R can vary from methyl (C_1) to tridecyl (C_{13}) [13]. However, PHAs contain a minimum of 100 and maximum of 35000 hydroxyacid (HA) monomers in it. The HA monomers generally prevail in three forms: 3, 4 and 5 hydroxyalkanoates [14]. They can be further divided into three classes on the basis of the length of their side chain. They are

- Short chain length PHAs (scl-PHAs) containing short chains having 3 to 5 carbons. They have rigid crystalline structure,

- Medium chain length PHAs (mcl-PHAs) possess medium length chains containing minimum 6 and maximum 14 carbons having a soft and sticky characteristics and

- Large-chain-length PHAs (lcl-PHAs) are having 15 or more carbons [15].

PHAs are mostly used in packaging and coatings. Due to the unique biocompatibility properties, PHAs are also largely being used in pharmaceutical and medical usage like wound bandage, biomedical devices like bone scaffolds (used in tissue engineering implants like heart valves), orthopedic pins, heart stents, nerve guides etc. Moreover, there are also several other applications of bioplastics like in making electronic gadgets (cell phone), encapsulation, crop protection by plastic films, diaper making (personal hygiene) and in formulating adhesives [16].

3. PHA synthesis pathway

The most well-known PHA production pathway is given in Figure 2 [5]. *Ralstonia eutropha* mediated PHA synthesis pathway is used in industrial scale production [17]. Generally, scl-PHAs are produced in this process. The biosynthesis is initiated with the conversion of sugar substrate or organic acids to acetyl coenzyme-A (acetyl-CoA) via glycolysis and beta-oxidation. β-ketotiolase (PhaA) enzyme helps in the condensation of two molecules of acetyl-CoA to produce acetoacetyl coenzyme-A. It is further converted to (R)-3-hydroxybutyril-CoA by acetoacetyl-CoA reductase (PhaB) by reducing nicotinamide adenine dinucleotide phosphate (NADPH) to $NADP^+$. Finally, 3-

Degradation of Plastics

Materials Research Foundations **99** (2021) 145-162

Materials Research Forum LLC

https://doi.org/10.21741/9781644901335-6

hydroxybutyril-CoA monomer is polymerized into P(3HB) by PHA synthase enzyme (PhaC).

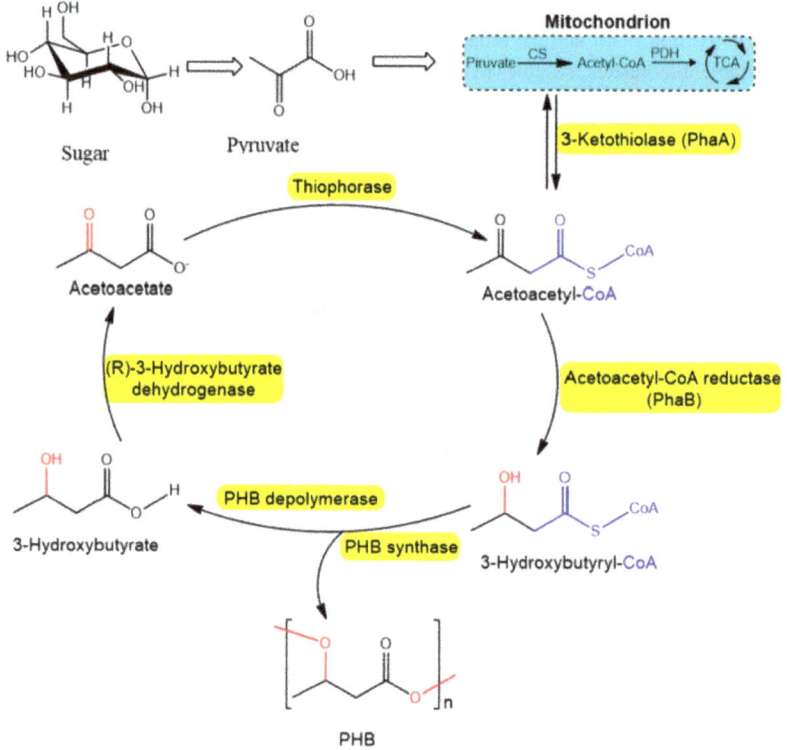

Figure 2 The first discovered pathway for PHA biosynthesis [5].

Mixed microbial culture (MMC) mediated PHA biosynthesis is generally described in 3 following steps [18].

- In *Step 1* organic sugar substrate is converted into VFA enriched fermented product.
- In *Step 2* sequencing batch reactor (SBR) is operated with optimized operating conditions and microenvironment to select the PHA accumulating microorganisms among the mixed culture.

- *Step 3* is the final stage where the biomass is cultivated with VFA feeding which resulted into PHA production under optimal conditions. The biosynthesized PHA is further harvested by downstream processing of extraction.

4.   Activated sludge process for PHA production

The microbial diversity of activated sludge made that process suitable for bioplastic synthesis. Modified sludge processes are proved to be more efficient in bioplastic yields. Mixed, aerobic granular and chemically treated sludge processes are described below.

4.1   Mixed activated sludge process

Wen and co-workers [19] used bulking sludge for PHA production. They operated two SBRs , one with aerobic dynamic feeding (ADF) model (R2) and another one by anaerobic phase followed by an aerobic phase (R1) in order to find out mixed cultures with high PHA accumulating capacities. Due to the incorporation of anaerobic phase before aerobic phase, in R1 sludge settleability was improved but the simultaneous increase in feast/famine ratio was responsible for less PHB accumulation in R1 than R2. Maximum PHB yield was 53% of the total suspended solids which was achieved at 10.2 h of a batch study under the starvation of nitrogen. The study indicated that filamentous bacterial sludge can also accumulate PHB while compared to the well settled sludge. The low oxygen requirement of the filamentous sludge helped to save energy required for PHA producing bacterial enrichment.

PHA production was obtained in a three-step process utilizing food waste as a renewable source [20]. In step I, bio-hydrogen was produced in acidogenic fermentation pathway. The VFA enriched effluent produced from bio-hydrogen reactor was utilized for PHA production. PHA production occurred in two stages. Step II was conducted for microbial enrichment and finally step III was for PHA production. PHA producing bacteria were cultivated inside a SBR , which was operated in two different cycles of CL-24 and CL-12. Maximum PHA recovery was achieved in CL-12 cycle in both Step II (PHA yield: 16.3% of dry cell weight with 84% VFA removal) and Step III (PHA yield: 23.7%; 88% VFA removal). The synthesized PHA obtained was basically in a copolymer form of [P(3HB-co-3HV)] of PHB and PHV. Hence this study established an innovative approach of using wastewater in simultaneous waste remediation and bioplastic production by using mixed culture sludge.

In another literature PHA extraction was reported by using municipal secondary wastewater sludge by improving different parameters like temperature, operation time and solid biomass concentration [21]. After parameter optimization, maximum PHA production was 0.605 g which was higher than the un-optimized operating conditions. The produced PHA was mostly dominated by 58% mcl PHA which was produced in comparatively low cost than the conventional bioplastics. Again, this study focused on the probable low cost biomaterial production by using waste resource with simultaneous waste minimization.

Different PHA copolymers were extracted from municipal activated sludge cultivated in fermented waste feedstock [11]. In biological wastewater treatment, nitrification occurs first where ammonia is converted to nitrate by ammonia and nitrite oxidizing bacteria (AOBs and NOBs) and finally it is removed as nitrogen via denitrification. Activated sludge can accumulate PHA along with nitrogen removal. Dissolved oxygen (DO) had a significant role in regulating PHA production and nitrification of the activated sludge process. The experiment finally suggested a PHA recovery method by utilizing nitrifying activated sludge where DO level should be maintained while (1) NOB growth and activity are mitigated, (2) nitrogen removal is observed, (3) alkalinity is controlled by denitrification, and (4) aeration is reduced.

4.2   Aerobic granular sludge process

Aerobic granulation technology is one of the most popular biological wastewater treatment strategies in current research trends for sustainable environment. Presence of microbial diversity, formation of even shaped compact sludge granules with high settling velocity and recalcitrant pollutant removal efficiency made this process significant for the researchers. Different types of domestic as well as industrial wastewaters have been successfully treated by aerobic granular sludge so far. However, many literatures have also reported the PHA accumulation by aerobic granules while treating the wastewater in aerobic granular reactors (AGR). Table 1 [4,7, 22-24] provides the literature details of aerobic granulation mediated PHA production while treating different wastewater with operating conditions, granule characteristics and PHA yield.

In Figure 3 [7] the PHA accumulation inside aerobic sludge granules is visible in different famine-period aeration rate of 2 L/min, 1 L/min, 0.5 L/min and 0 L/min, respectively.

Table 1 Aerobic granulation in different wastewater treatment with different operating conditions, granule characteristics and PHA yield [4,7, 22-24]

Wastewater	Operating condition	Granule characteristics	Pollutant removal efficiency	PHA yield	Ref.
Acetate wastewater (synthetic)	AGR volume: 2 L, HRT: 12 h, cycle time: 6 h, settling time: 2 min, Influent COD: 500-1000 mg/L, ammonia nitrogen: 25-12.5 mg/L, COD/N: 20-90	MLSS: 4-8 g/L, MLVSS: 4.5-10 g/L	COD: 80-99%	44%	[22]
Acetate wastewater (synthetic)	AGR volume: 2 L, HRT: 12 h, SRT: 10 days	MLVSS: 3-9 g/L, SVI: <40 mL/g, EPS: 6-30 mg/g VSS	COD: 90%	40 ± 4.6%.	[23]
Palm oil mill effluent (POME)	Cycle time: 6 h, settling time: 1.5 min, VER: 25%, H/D: 10, air flow rate: 3 L/min, COD: 51,000 mg/L	EPS: 40–45 µg/mL	COD: 90%, propionic and butyric acids: 100%	0.6833 mg PHA/mg biomass.	[4]
Palm oil mill effluent (POME)	AGR volume: 2L, VER: 25%, cycle time: 24 h, air flow rate: 3 L/min, OLR: 0.91, 1.82, 2.73 and 3.64 kg COD/m3 day.	Granule size: 0.35-2 mm	COD: 90%	0.66 to 0.87 g PHA/g CDW	[24]
Palm oil mill effluent (POME)	AGR volume: 2 L, cycle time: 6 h, air flow rate: 3 L/min	Granule size: 1.03 mm, SVI: 20-40 mL/g	COD: 90%	0.56 g PHA/g CDW	[7]

HRT: Hydraulic retention time, MLSS: Mixed liquor suspended solids, MLVSS: Mixed liquor volatile suspended solids, VSS: Volatile suspended solids, COD: Chemical oxygen demand, EPS: Extracellular polymerix substances, H/D: Height/diameter ratio, SRT: Sludge retention time, VER: Volume exchange ratio, SVI: Sludge volume index, CDW: Cell dry weight.

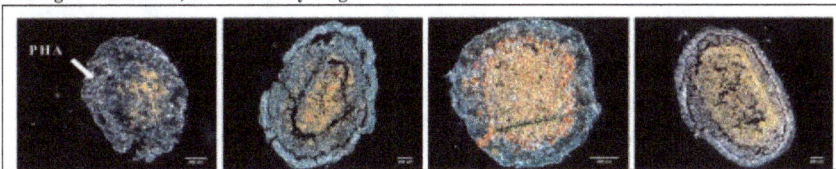

Figure 3 PHA accumulation inside aerobic sludge granules is visible in famine-period aeration rate of (a) 2 L/min, (b) 1 L/min, (c) 0.5 L/min and (d) 0 L/min [7].

Materials Research Forum LLC
https://doi.org/10.21741/9781644901335-6

4.3 Chemically modified sludge process

Currently, chemically enhanced primary sedimentation (CEPS) technique is extensively being used in Hong Kong as an effective sewage treatment process. The CEPS sludge consists of various organic compounds which can recover VFAs which can be further utilized for PHA production. Xu and co-workers [12] observed that addition of CaO_2 enhanced the disintegration of CEPS sludge and rapid VFA production occurred. The maximum VFAs yield was 455.8 mg COD/g VSS and the microbes requiring hydrolysis and acidogenesis were enriched with the VFA. CaO_2 treatment had no influence on the release of ammonia nitrogen (NH_4^+-N), but it proved to be significant in PHA biosynthesis by sludge microbes. The VFA enriched feedstock proved to be a suitable substrate for PHA production. After the cultivation phase in VFA enriched substrate the CEPS sludge achieved almost 22.3% PHA accumulation.

5. Types of wastewaters in bioplastic production

5.1 Palm oil mill effluent (POME)

Palm oil mill wastewaters are enriched with VFAs. Aerobic granulation mediated PHAs were generally produced by using POME where more than 90% influent COD was removed by complete removal of VFAs [4]. Maximum 0.66 to 0.87 g PHA/g CDW PHA accumulation was reported by Vjayan and Vadivelu [7].

5.2 Paper pulp mill wastewater

PHA production was also carried out by using paper mill wastewater [8]. The biosynthesis process was carried out into the following steps. (1) A single batch process was required for the acidogenic fermentation of the paper mill wastewater, (2) PHA producing bacterial enrichment in batch mode operation in a feast-famine regime, (3) maximizing the PHA yield by PHA accumulating strain enrichment in batch operation, and finally (4) the selective pressure was also required for microbial enrichment. However, the study resulted into 77% yield of PHA of cell dry weight within 5 h of batch study after microbial enrichment.

5.3 Dairy wastewater

Dairy wastewater activated sludge was used as the inoculums to produce PHA by using milk whey [9]. The biosynthesis was carried out into several steps by pre-treating the sludge to reduce protein concentration, optimizing a suitable carbon/nitrogen (C/N) ratio for maximum PHA yield and then pH correction for synthesis. The study showed that at a

C/N=50 about 13.82% PHA yield was possible without pH correction. Another advantage of using dairy wastewater was no requirement of aseptic condition.

5.4   Molasses spentwash

A potential low cost feedstock molasses spent wash was utilized by waste activated sludge bacteria to produce biodegradable biopolymer polyhydroxybutyrate (PHB) [25]. The synthesized PHB was visible inside sludge biomass under fluorescence microscopy. PHB was further characterized by fourier transform-infra-red spectroscopy (FT-IR) and ^{13}C nuclear magnetic resonance (NMR). At C:N=28 condition, maximum 52% COD removal was achieved with maximum PHB accumulation of 28%.

5.5   Olive oil mill wastewater

Olive oil mill wastewater was used as a no-cost substrate for PHA accumulation in a mixed microbial culture [10]. The process was carried out in multistages with phenol removal and recovery. Initially phenol was removed from the wastewater and selection of the PHA-producing bacteria was carried out inside an SBR. At optimum influent organic loading rate (OLR) of 4.70 g COD/L.d maximum PHA storage and yield was observed. Mass balance indicated that about 85% of the influent COD was converted into 10% PHA (volumetric productivity). Maximum PHA yield was found as 1.50 g PHA/L.d.

6.   Conditions for PHA production

6.1   Aerodynamic feeding (ADF)

Aerodynamic feeding (ADF) is a very popular feeding strategy for PHA production from mixed sludge. ADF helps in establishing a strong feast-famine regime in a SBR cycle which help the microorganisms to accumulate PHA in abundance of carbon source (feast phase).

Amulya and coworkers [20] employed ADF in a three-step process for PHA production from microbial mixed culture (MMC). In first step, bio-hydrogen was produced in acidogenic fermentation. In second step, VFA enriched substrate enriched the PHA producing microorganisms and in 3rd step finally PHA synthesis was obtained. The final product was a copolymer of [P(3HB-co-3HV)] of PHB and PHV.

Gobi and Vadivelu [4] were able to produce maximum 0.6833 mg PHA/mg biomass from aerobic granular sludge while treating palm oil mill effluent (POME).

6.2   Feast-famine regime

According to literature, during PHA production from activated sludge, high oxygen consumption was required to oxidize the external carbon source in feast phase. But in famine phase less oxygen is required for the oxidation of stored PHA and to convert it into metabolic energies. However, the reduction in oxygen consumption contributes to energy and cost reduction in commercial PHA production. Hence feast-famine regime of a rector cycle plays an important role in PHA accumulation by mixed microbial culture.

By regulating feast-famine regime and uncoupled carbon and nitrogen, Oliveira and coworkers [26] were able to produce 6.09 gPHA/L of PHA from mixed microbial culture by utilizing cheap cost industrial and agricultural waste feedstock.

6.3   pH

Impacts of microenvironment and pH values were also evaluated in PHA production from wastewater [27]. The study was conducted by varying the SBR microenvironment from aerobic to microaerophilic and also by regulating the reactor pH as 6, 7 and 8. The study showed that neutral pH with microaerophilic rector environment was more favourable for PHA production than other conditions.

6.4   Carbon/nitrogen ratio

High COD loading and low ammonia concentration was proved to be very effective in high PHA yield from activated sludge. Hence, selection of an optimum C/N ratio was important to enhance PHA production.

Fang et al. [22] observed that at 750 mg/L of COD and 8.5 mg/L of influent ammonia concentration maximum 44% PHB production was obtained from aerobic granules while treating synthetic acetate wastewater. The study also revealed that increasing ammonia concentration enhanced granule biomass concentration and settling velocity but reduced the PHB accumulation. So, it was suggested that by regulating influent ammonia concentration we can obtain maximum PHB yield with cultivating highly settleable compact sludge granules.

In another study, dairy wastewater activated sludge was used as the inoculum to produce PHA by using milk whey [9]. Finally, the study showed that at C/N=50 about 13.82% PHA yield was possible without pH correction.

6.5   Selective pressure

Chen et al. [28] proposed a novel approach of aerobic dynamic discharge (ADD) to provide physical selective pressure to enhance PHA production in PHA accumulating bacteria in a system having MMCs. The study provided a comparative analysis of ADD coupled with SBR operation with conventional aero dynamic feeding (ADF) method. Microbes treated with ADD mode resulted into maximum PHA yield of 0.72±0.07 Carbon mol PHA/Carbon mol acetate and maximum PHA content of 74.16±0.03% of biomass weight after 30 days of the ADD operation, which was higher than MMC operated with ADF mode. The study provided a new ADD strategy for high PHA accumulation possibility in SBR mediated MMC system.

7. PHA extraction process

PHA extraction methods can be classified into 2 categories.

- Chemical digestion with solvent extraction and
- Mechanical disruption

7.1 Chemical digestion with solvent extraction

Sodium hypochlorite, acetone, chloroform and sodium hydroxide are commonly used chemicals for PHA extraction. Figure 4 [29] provides a flowchart for aerobic granular biomass digestion and PHA recovery from granules by using four chemicals acetone, NaOCl, NaCl and NaOH.

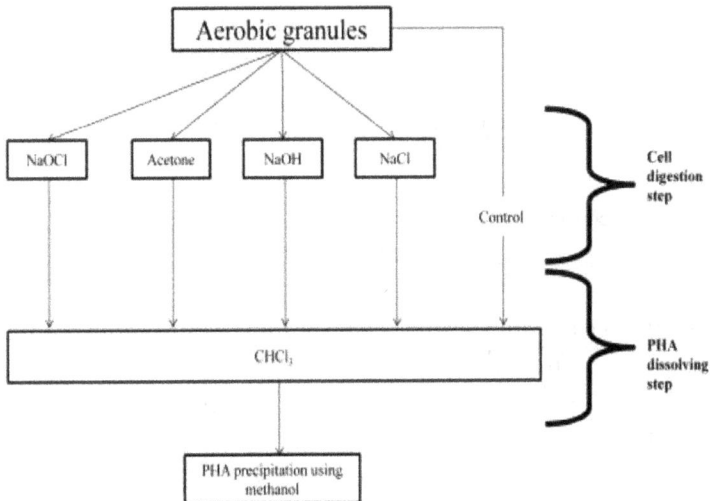

Figure 4 This flowchart shows aerobic granular biomass digestion and PHA recovery from granules by using four chemicals acetone, NaOCl, NaCl and NaOH [29].

7.1.1 Sodium hypochlorite extraction

About 12.5 mL sodium hypochlorite and 12.5 mL of chloroform is required for PHA extraction from 1 g of lyophilized biomass. After vortexing the mixture is incubated at 37° C temperature for 90 min and then after centrifugation phase separation is observed. The top layer having sodium hypochlorite is discarded and the chloroform layer containing PHA is collected for recovery [30].

7.1.2 Acetone extraction

For 1 g of lyophilized biomass, about 12.5 mL acetone is required to remove cell impurities. The mixture is kept at 65° C temperature and then the acetone layer is discarded and the biomass is subjected to PHA dissolving step [31].

7.1.3 Sodium hydroxide extraction

About 1 g of lyophilized biomass is digested by using 0.05 M NaOH. Finally, the biomass is immersed in 6.84 N concentrated sodium chloride solution inside a shaker incubator for 24 h. The sodium chloride layer is discarded and the biomass is subjected to PHA recovery [32].

7.2 Mechanical disruption

Mechanical disruption is generally carried out by providing high pressure bead milling or homogenization without using any chemicals. However, there are several drawbacks of this method [33]. They are:

- Cell damage due to mechanical pressure
- Long process time isrequired
- More power consumption
- Difficult to scale up the process.

8. Quantification method

After the extraction of PHA it is dried up and the resultant weight is recorded. The PHA yield is calculated by comparing with total cell dry weight [4]. The PHA yield is expressed by Eq. 1

$$Y_{PHA/CDW} = \frac{\text{Total amount of PHA,mg}}{\text{Cell dry weight,mg}} \tag{1}$$

Materials Research Forum LLC

https://doi.org/10.21741/9781644901335-6

In many literature PHA yield was also expressed with respect to COD consumption [4]. It is calculated according to Eq. 2.

$$Y_{PHA/COD} = \frac{Total\ amount\ of\ PHA, mg}{Total\ amount\ of\ COD\ converted, mg/L} \tag{2}$$

Conclusion and future scopes

In modern scenario, environmental viability and cost effectiveness of bioplastics are increasing its market potential. Like agro, plant and algal-based bioplastics, bacterial bioplastics are also being studied throughout the world. Activated sludge process provides a simultaneous flow of waste management and biomaterial recovery. Activated sludge process involves the use of cheap or no-cost wastewater and waste sludge bacteria which can remove a big part of the production cost. Extensive studies have been conducted for PHA synthesis in activated sludge process by utilizing POME, dairy wastewater and paper mill wastewater due to presence of VFA which can be used to enrich PHA accumulating bacteria. Many researchers were able to enhance PHA synthesis by regulating different feeding strategies like ADF, ADD, by controlling feast-famine regime, microenvironment and pH of the reactor and also by varying the C/N ratio and DO concentration.

However, the major drawback of this activated sludge mediated PHA production is the lack of purity of the product. Hence, more selective sludge bacteria and modified extraction techniques should be implemented to ensure purity of the product. In PHA extraction, use of large amount of organic solvents is also not very economic and environmentally viable. More studies on recycling and reuse of solvents and investigation of low cost extraction process are required. Further studies should explore the probable utilization of innovative cheap cost feedstocks to ensure more economic PHA production. Moreover, for successful PHA production, scale-up from lab scale to pilot and industrial scale is required to establish activated sludge based bioplastic production market.

References

[1] S.M. Emadian, T.T. Onay, B. Demirel, Biodegradation of bioplastics in natural environments, Waste Manage. 59 (2017) 526-536. https://doi.org/10.1016/j.wasman.2016.10.006

[2] S. Pathak, C. Sneha, B.B. Mathew, Bioplastics: its timeline based scenario & challenges, J. of Polym. and Biopolym. Phy. Chem. 2 (2014) 84-90. https://doi.org/10.12691/jpbpc-2-4-5

[3] H. Li, J. Zhang, L. Shen, Z. Chen, Y. Zhang, C. Zhang, Q. Li, Y. Wang, Production of polyhydroxyalkanoates by activated sludge: Correlation with extracellular polymeric

substances and characteristics of activated sludge, Chem. Eng. J. 361 (2019) 219-226. https://doi.org/10.1016/j.cej.2018.12.066

[4] K. Gobi, V. Vadivelu, Aerobic dynamic feeding as a strategy for in situ accumulation of polyhydroxyalkanoate in aerobic granules, Bioresour. Technol. 161 (2014) 441-445. https://doi.org/10.1016/j.biortech.2014.03.104

[5] R. Tarrahi, Z. Fathi, M.Ö. Seydibeyoğlu, E. Doustkhah, A. Khataee, Polyhydroxyalkanoates (PHA): From production to nanoarchitecture, Int. J. Biol. Macromol.(2019). https://doi.org/10.1016/j.ijbiomac.2019.12.181

[6] G. Mannina, D. Presti, G. Montiel-Jarillo, J. Carrera, M.E. Suárez-Ojeda, Recovery of Polyhydroxyalkanoates (PHAs) from wastewater: A review, Bioresour. Technol. (2019) 122478. https://doi.org/10.1016/j.biortech.2019.122478

[7] T. Vjayan, V. Vadivelu, Effect of famine-phase reduced aeration on polyhydroxyalkanoate accumulation in aerobic granules, Bioresour. Technol. 245 (2017) 970-976. https://doi.org/10.1016/j.biortech.2017.09.038

[8] Y. Jiang, L. Marang, J. Tamis, M.C. van Loosdrecht, H. Dijkman, R. Kleerebezem, Waste to resource: converting paper mill wastewater to bioplastic, Water Res. 46 (2012) 5517-5530. https://doi.org/10.1016/j.watres.2012.07.028

[9] F. Bosco, F. Chiampo, Production of polyhydroxyalcanoates (PHAs) using milk whey and dairy wastewater activated sludge: production of bioplastics using dairy residues, J. Biosci. Bioeng.109 (2010) 418-421. https://doi.org/10.1016/j.jbiosc.2009.10.012

[10] S. Campanari, F.A. e Silva, L. Bertin, M. Villano, M. Majone, Effect of the organic loading rate on the production of polyhydroxyalkanoates in a multi-stage process aimed at the valorization of olive oil mill wastewater, Int. J. Biol. Macromol.71 (2014) 34-41. https://doi.org/10.1016/j.ijbiomac.2014.06.006

[11] X. Wang, S. Bengtsson, A. Oehmen, G. Carvalho, A. Werker, M.A. Reis, Application of dissolved oxygen (DO) level control for polyhydroxyalkanoate (PHA) accumulation with concurrent nitrification in surplus municipal activated sludge, New Biotechnol. 50 (2019) 37-43. https://doi.org/10.1016/j.nbt.2019.01.003

[12] J. Xu, X. Li, L. Gan, X. Li, Fermentation liquor of CaO2 treated chemically enhanced primary sedimentation (CEPS) sludge for bioplastic biosynthesis, Sci. Total Environ.644 (2018) 547-555. https://doi.org/10.1016/j.scitotenv.2018.06.392

[13] Q. Chen, S. Liang, G.A. Thouas, Elastomeric biomaterials for tissue engineering, Prog. Polym. Sci.38 (2013) 584-671. https://doi.org/10.1016/j.progpolymsci.2012.05.003

[14] P.B. Albuquerque, C.B. Malafaia, Perspectives on the production, structural characteristics and potential applications of bioplastics derived from

polyhydroxyalkanoates, Int. J. Biol. Macromol.107 (2018) 615-625.
https://doi.org/10.1016/j.ijbiomac.2017.09.026

[15] M. Zinn, B. Witholt, T. Egli, Occurrence, synthesis and medical application of bacterial polyhydroxyalkanoate, Adv. Drug Deliv. Rev. 53 (2001) 5-21.
https://doi.org/10.1016/S0169-409X(01)00218-6

[16] F. Masood, T. Yasin, A. Hameed, Polyhydroxyalkanoates–what are the uses? Current challenges and perspectives, Crit. Rev. Biotechnol. 35 (2015) 514-521.
https://doi.org/10.3109/07388551.2014.913548

[17] K. Sudesh, H. Abe, Y. Doi, Synthesis, structure and properties of polyhydroxyalkanoates: biological polyesters, Prog. Polym. Sci.25 (2000) 1503-1555.
https://doi.org/10.1016/S0079-6700(00)00035-6

[18] C. Kourmentza, J. Plácido, N. Venetsaneas, A. Burniol-Figols, C. Varrone, H.N. Gavala, M.A. Reis, Recent advances and challenges towards sustainable polyhydroxyalkanoate (PHA) production, Bioengineering 4 (2017) 55.
https://doi.org/10.3390/bioengineering4020055

[19] Q. Wen, Z. Chen, C. Wang, N. Ren, Bulking sludge for PHA production: Energy saving and comparative storage capacity with well-settled sludge, J Environ Sci. 24 (2012) 1744-1752. https://doi.org/10.1016/S1001-0742(11)61005-X

[20] K. Amulya, S. Jukuri, S.V. Mohan, Sustainable multistage process for enhanced productivity of bioplastics from waste remediation through aerobic dynamic feeding strategy: Process integration for up-scaling, Bioresour. Technol. 188 (2015) 231-239.
https://doi.org/10.1016/j.biortech.2015.01.070

[21] M. Kumar, P. Ghosh, K. Khosla, I.S. Thakur, Recovery of polyhydroxyalkanoates from municipal secondary wastewater sludge, Bioresour. Technol. 255 (2018) 111-115. https://doi.org/10.1016/j.biortech.2018.01.031

[22] F. Fang, X.-W. Liu, J. Xu, H.-Q. Yu, Y.-M. Li, Formation of aerobic granules and their PHB production at various substrate and ammonium concentrations, Bioresour. Technol. 100 (2009) 59-63. https://doi.org/10.1016/j.biortech.2008.06.016

[23] J. Wang, W.-W. Li, Z.-B. Yue, H.-Q. Yu, Cultivation of aerobic granules for polyhydroxybutyrate production from wastewater, Bioresour. Technol. 159 (2014) 442-445. https://doi.org/10.1016/j.biortech.2014.03.029

[24] K. Gobi, V. Vadivelu, Dynamics of polyhydroxyalkanoate accumulation in aerobic granules during the growth–disintegration cycle, Bioresour. Technol. 196 (2015) 731-735. https://doi.org/10.1016/j.biortech.2015.07.083

[25] A.A. Khardenavis, A.N. Vaidya, M.S. Kumar, T. Chakrabarti, Utilization of molasses spentwash for production of bioplastics by waste activated sludge, Waste Manage. 29 (2009) 2558-2565. https://doi.org/10.1016/j.wasman.2009.04.008

[26] C.S. Oliveira, C.E. Silva, G. Carvalho, M.A. Reis, Strategies for efficiently selecting PHA producing mixed microbial cultures using complex feedstocks: Feast and famine regime and uncoupled carbon and nitrogen availabilities, New Biotechnol. 37 (2017) 69-79. https://doi.org/10.1016/j.nbt.2016.10.008

[27] K. Amulya, M.V. Reddy, M. Rohit, S.V. Mohan, Wastewater as renewable feedstock for bioplastics production: understanding the role of reactor microenvironment and system pH, J. Clean. Prod. 112 (2016) 4618-4627. https://doi.org/10.1016/j.jclepro.2015.08.009

[28] Z. Chen, Z. Guo, Q. Wen, L. Huang, R. Bakke, M. Du, A new method for polyhydroxyalkanoate (PHA) accumulating bacteria selection under physical selective pressure, Int. J. Biol. Macromol.72 (2015) 1329-1334. https://doi.org/10.1016/j.ijbiomac.2014.10.027

[29] K. Gobi, V. Vadivelu, Polyhydroxyalkanoate recovery and effect of in situ extracellular polymeric substances removal from aerobic granules, Bioresour. Technol. 189 (2015) 169-176. https://doi.org/10.1016/j.biortech.2015.04.023.

[30] S.K. Hahn, Y.K. Chang, B.S. Kim, H.N. Chang, Optimization of microbial poly (3-hydroxybutyrate) recover using dispersions of sodium hypochlorite solution and chloroform, Biotechnol. Bioeng. 44 (1994) 256-261. https://doi.org/10.1002/bit.260440215

[31] R.F. Gamal, H.M. Abdelhady, T.A. Khodair, T.S. El-Tayeb, E.A. Hassan, K.A. Aboutaleb, Semi-scale production of PHAs from waste frying oil by *Pseudomonas fluorescens* S48, Braz. J. Microbiol.44 (2013) 539-549. https://doi.org/10.1590/S1517-83822013000200034

[32] C. Estrela, C.R. Estrela, E.L. Barbin, J.C.E. Spanó, M.A. Marchesan, J.D. Pécora, Mechanism of action of sodium hypochlorite, Braz. Dent. J.13 (2002) 113-117. https://doi.org/10.1590/S0103-64402002000200007

[33] M.H. Madkour, D. Heinrich, M.A. Alghamdi, I.I. Shabbaj, A. Steinbüchel, PHA recovery from biomass, Biomacromolecules 14 (2013) 2963-2972. https://doi.org/10.1021/bm4010244

Degradation of Plastics
Materials Research Foundations **99** (2021) 163-178

Materials Research Forum LLC
https://doi.org/10.21741/9781644901335-7

Chapter 7

Photocatalytic Degradation of Plastic

Milan Malhotra[1]*, Lakshmi Pisharody1, Ansaf V. Karim[1], Sukanya Krishnan[1]

[1]Indian Institute of Technology Bombay (IIT-B), Powai, Mumbai Maharashtra -400076, India

*milanmalhotraiitb@gmail.com

Abstract

Due to rapid growth and modernization, the consumption of plastic has increased rapidly. However, due to the non-biodegradable nature of plastics, its management and disposal have become an environmental concern. The majority of plastics end up in landfill sites or oceans through rivers which is a threat to the marine ecosystem. Plastic can remain in the environment for thousands of years furthermore due to physical degradation plastics are converted into microplastics. Current techniques of recycling plastic require a significant amount of segregation which is not feasible due to economic constraints. Photocatalyst enhances the rate of degradation using light as a source of energy hence making the process economically feasible. This chapter provides a comprehensive review focusing on plastics, its pollution and type of polymers. Further, the chapter also reviews the various research conducted for the photocatalytic degradation of plastics.

Keywords

Plastics, Photocatalysis, Solid Phase Photocatalysis, Degradation, Waste Management

Contents

Materials Research Forum LLC
https://doi.org/10.21741/9781644901335-7

1. Introduction

Plastics are synthetic or semi-synthetic organic compounds containing long carbon chains as their repeated units [1]. These long-chain polymers are well-packed, possess excellent strength, durability, and higher molecular mass. The term plastic was originally derived from the Greek word 'plastikos' which means fit for moulding. They are classified based on their chemical structure: (i) thermosetting which degrades and turn into other substance after moulding and (ii) thermoplastics which can be reheated and moulded back to the original shape. These materials are lightweight, durable, and inert which makes them suitable for the manufacture of a very wide range of products [2]. Due to their higher chemical stability and non-biodegradability, they occupy a specific position in the environmental spectrum generating a huge amount of wastes. The current global production and consumption of plastic products increasing unabatedly high leading to unacceptable pollution [3]. The disposal of plastics has been recognized as a worldwide environmental problem. Due to its inert nature, most of the plastic materials end up in the marine environment or left behind in baches, municipal drainage systems, etc. [2]. It is generally accepted that material recovery and recycling is not a long solution for plastic pollution [4]. Various techniques for managing plastic waste have been developed over the years, such as, chemical, thermal, ultrasonic, etc. However, no such degradation method has been developed which single-handedly can handle this increasing plastic pollution. Photocatalytic degradation using a semiconductor catalyst can be efficient for the degradation of plastics from the environment. The method does not generate any toxic dioxins, inexpensive and the degradation process occurs at low temperatures [5,6]. The properties and applications of different types of plastics are shown in Table 1 [1,7,8]. This chapter gives a comprehensive review focusing on plastics, the type of plastics, and the problems associated with plastics. Further, the chapter also reviews the various research conducted in the field for the degradation of plastics emphasising more on the heterogeneous photocatalytic degradation of plastics.

Table 1   Properties of different type of plastics [1,7,8]

Polymer	Density (g/cm³)	Properties	Applications
Polypropylene (PP)	0.90-0.91	<ul><li>High melting point</li><li>chemical resistance</li><li>translucent</li></ul>	Bottles, food containers, toys
Low-density polyethylene (LDPE)	0.92-0.93	<ul><li>Low melting point</li><li>Tough and flexible</li></ul>	Packaging film, agricultural mulch, grocery bags
Polyvinyl chloride (PVC)	1.3-1.6	<ul><li>Hard</li><li>chemical resistance</li><li>Long term stability</li></ul>	Pipe, conduit, home siding, window frames
High-Density Polyethylene (HDPE)	0.93-0.97	<ul><li>Chemical resistance</li><li>Hard to semi-flexible</li><li>Soft waxy surface</li></ul>	Milk bottles, wire, and cable insulation, toys
Polyethylene terephthalate (PET)	1.3-1.4	<ul><li>High heat resistance</li><li>Tough</li></ul>	Transparent bottles, recording tape
Polystyrene (PS)	1.0-1.1	<ul><li>Rigid or foamed</li><li>Hard</li><li>Brittle</li></ul>	Eating utensils, foamed food containers

2.   History of plastic

Plastics the terminology which initially meant "pliable and easily shaped" is now analogous to synthetic polymers. Plastics have an indispensable role in daily life human activities and chores even though the large-scale production of these synthetic polymers dates only a few decades ago, approximately only 1950. The first synthetic polymer was developed by John Wesley Hyatt in 1869 [7]. The motivation for the invention was to develop an alternative for natural ivory which was then obtained by slaughtering wild elephants. Hyatt treated cellulose obtained from the cotton fibre with camphor which could then be crafted into various shapes. Later, new age synthetic plastic, such as Bakelite, was developed in 1907 and the widespread use of plastic outside the military sector was only after World War II [8]. The advantages of plastic were well known, they are inexpensive, durable, light weight as well as strong, with high thermal and insulation properties. Also, they are incredibly versatile thereby facilitating the production of a vast range of products. The discovery of plastic was then referred to as a saviour of the world from the destructive actions of humans. Thus, material wealth was widespread and accessible, and this was just the start of the plastic revolution.

World war further enhanced the necessity of plastic production due to an increase in demand and, there was a 300% surge in plastic production during that tenure in the United States of America. Materials such as nylon and plexiglass were synthesized during that tenure. Nylon, a synthetic cloth replaced the natural cloth which was used in parachutes, rope, and body armor, while, plexiglass was used in place of glass in aircraft windows. Thus, slowly plastic usage expanded and the expansion was not found to subside even after World War II [9]. Author Susan Freinkel thus correctly predicted, "In product after product, market after market, plastics challenged traditional materials and won, taking the place of steel in cars, paper, and glass in packaging, and wood in furniture."

However, the untarnished positive aspects of plastic did not last long when awareness about environmental pollution spread. Where the long persistence of plastics became a trouble to environmental activists. The accumulation of plastics in oceans was observed in the 1960s, a decade after the rapid expansion of the plastic industry in the USA [10].

3.   Types of plastic

The structure of plastic is typically polymers derived from a macromolecule chain, synthesized from various monomers through specific chemical reaction. Polyaddition or polycondensation are typical reactions for assembling chains of these monomeric units to form a plastic. Polyaddition may occur in a stepwise or continuous way, while polycondensation occurs in the stepwise mode with liberation of water or ammonia molecule (Fig. 1) [9].

Based on the macromolecular structure plastics are categorised into four major classes namely, thermoplastics, thermosets, elastomers, and polymer compounds [9]. Thermoplastics are those groups of plastics that can be melted and thus reused. They are either amorphous or semi-crystalline in nature. Amorphous resins are structurally disordered compared to semi-crystalline resins which have relatively ordered structures. Examples of amorphous resins include polycarbonate (PC), polyvinylchloride (PVC), and polystyrene (PS). Semi-crystalline resins are polypropylene (PP) and polyamide (PA). Whereas, thermosets and elastomers cannot be melted and thus are non-recyclable [10]. The former has hard elasticity, while later has soft elasticity. Detailed classification of plastics with examples are shown in Fig. 2 [9].

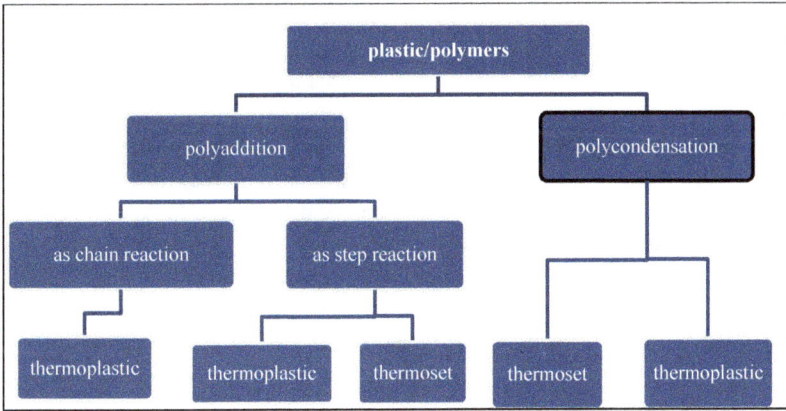

Fig. 1 Process for plastic generation [9]

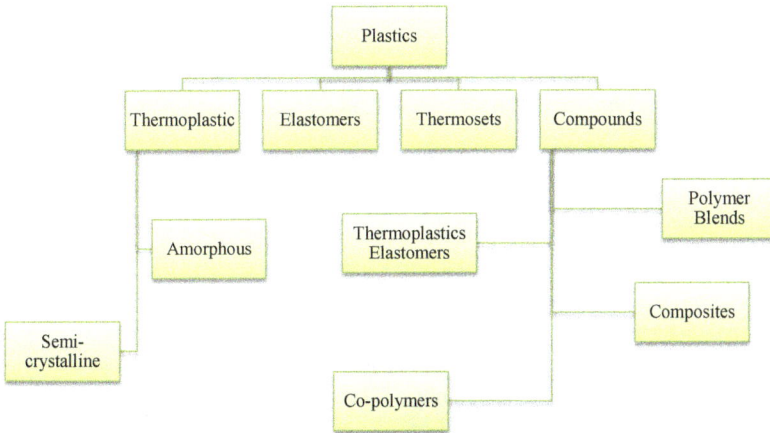

Fig. 2 Classification of plastics [9]

4. Problems associated with plastics

As outlined earlier, there has been a significant increase in plastic production over the years due to the obvious advantages. Global production of these polymers which includes thermoplastics, thermoset plastics and adhesives, and coatings has been estimated to be

260 million metric tonnes annually. This data indicates a growth rate of 9% per annum [11]. The plastic industry has been an ever changing sector due to constant changes in demand. Presently, polyethylene is the polymer that has the highest share of production followed by polyethylene terephthalate (PET) (20%), polypropylene (PP) (18%), polyvinyl chloride (PVC) (15.3%), and polystyrene/expanded polystyrene (PS/EPS) (7.8%). Plastic has been used in various sectors over the years, wherein 72% of plastic demand has been in packaging, construction, and electrical and electronic equipment. Plastic has also been employed in household furniture, agriculture, and also medical applications. Approximately 50% of the plastics generated are single-use disposable plastic which thereby increases the amount of solid waste in the environment. Plastics have been reported to be the second largest component in waste from electric and electronic gadgets [12].

Hence, waste generated through plastic utilization is a global problem, however, there exists regional variation. Currently, a major proportion of plastics utilized are non-biodegradable. Small fractions employ bio-based plastics, also these plastics even though being bio-based vary in the extent of degradation. Extend of biodegradation primarily depends on various geographical conditions such as, ambient temperature, UV exposure, and also oxygen availability, furthermore it also depends on the presence of potential microorganisms for bioremediation [11]. Therefore, the rates of degradation vary significantly amongst the terrestrial and marine environments and also landfills. Thus, it can be stated that plastic remains in the environment for a longer duration without undergoing significant change/ degradation. One major source of air pollution is the burning of plastic waste in the open environment and warming up of the surrounding air. Plastic burning results in the emission of toxic substances such as furans, dioxins and polychlorinated biphenyls. This also holds for plastic waste in the marine environment leading to water pollution. Presently, landfills are overloaded with waste dumps and wastes being burnt along with plastic bags are known to cause health hazards. Plastic wastes, such as, microplastics has the ability to attract contaminants, such as, persistent organic pollutants (POPs). Most of these contaminants being hydrophobic in nature, plastics can be a potential sink for these contaminants. Biomass accumulation on the plastic or biofouling is likely to increase the density of plastic. This observation was used to identify whether the plastics are from the sea or from a landfill [13]. Various factors and mechanisms are involved in the degradation of polymers. Major degradation methods for plastic degradation briefly are:

I. Mechanical degradation: Mechanical degradation basically reduces the molecular weight of the polymer by subjecting the polymer to powerful stress. Extrusion, agitation, and grinding are major causes for the mechanical degradation of polymers.

II. Thermal degradation: Degradation of polymers at elevated temperature or heat. The degradation can occur in the presence of air or oxygen. This process results in the generation of toxic volatile gases [14].

III. Hydrolytic degradation: The polymers are degraded through hydrolysis wherein the water diffuses into the amorphous areas of the polymer structure. It was observed that polymers, such as, polyethylene terephthalate (PET) were found to degrade through hydrolysis when the copolymer has higher nitrate units. The rate was found to be higher than those structures which were either semi-crystalline or amorphous [8].

IV. Ultrasonic degradation: Ultrasound is employed for the degradation of polymers, wherein the most susceptible bond is targeted. This method is known to affect the mechanical, mechano-chemical, and morphological properties of the polymer [15].

V. Biological degradation: The polymers are degraded by microorganisms that specifically target certain functional groups and thereby, break the polymer structure into simpler molecules. These molecules are further mineralized completely with the aid of enzymes. However, a very small proportion of synthetic plastics can be biodegraded due to their recalcitrant nature [12].

VI. Chemical degradation: Degradation of polymers assisted by chemical substances such as esters of H-phosphonic and phosphoric acids etc. to respective monomers or oligomers [16].

VII. Photochemical degradation: Polymer chains are broken down into monomers or oligomers in the presence of UV rays combined with different chemicals. Heterogeneous photocatalysis is one such type of degradation.

5.   Heterogeneousphotocatalysis

Heterogeneous photocatalysis is a promising, economical, and environmentally friendly technology in which a semiconductor catalyst is used for the mineralization of a wide range of organic pollutants (pesticides, pharmaceutical compounds, and persistent organic pollutants). Upon irradiation of the catalyst with an energy higher than its bandgap energy, the electrons from the valence band move to the conduction band of the catalyst creating positive holes. The positive holes are strong oxidants that directly oxidize the compounds or reacts with electron donors to form hydroxyl radicals. The photogenerated electrons and holes at their excited states are unstable and can recombine resulting in inhibiting the photodegradation [17]. The most commonly used semiconductor photocatalyst are TiO_2, ZnO, CdS, Fe_2O_2, etc. Among them, TiO_2 is the most widely used photocatalyst for photocatalytic degradation due to its easy availability, high photocatalytic activity, lower cost, and less toxic nature [15,16]. Several factors such as catalyst loading, light sources, pH of the solution affect the photocatalytic degradation efficiencies [18]. In order to improve the light absorption efficiency of conventional photocatalyst, several techniques such as metal or non-metal doping, dye sensitization, coupling semiconductors are used. These surface modifications help them in avoiding immediate charge recombination, decrease the bandgap energy and increase the overall quantum efficiency. An optimum amount of catalyst particle in the reaction medium increases the available sites for photon absorption, beyond which the scattering and screening phenomena commences [19]. The change in pH of the solution can influence the electron distribution in the target contaminant molecule, the absorbance of light, and hence the degradation potential [20]. The photocatalytic efficiency primarily depends upon the ability of the catalyst to enhance the quantum yield of the electron transfer processes through improvement in charge separation in the semiconductor. In broad terms, the degradation efficiency of the pollutants during photocatalysis is dependent on different operational parameters.

Plastics have become an inevitable daily life commodity all over the world due to its excellent physical properties and low-cost. They became one of the most ubiquitous materials which are used widely all over the world, are cheaper, the imperviousness of water and chemicals, have resistance to temperature and light [21]. However, a major fraction is constantly discarded on the field or burnt uncontrollably causing serious problems. A large number of plastic commodities cause serious environmental problems, called "white pollution' [22]. The higher stability and non-biodegradability of the plastic results in its resistance to microbial or enzymatic degradation [6]. Partial decomposition or incineration of plastics releasing strong carcinogenic dioxins or toxic byproducts into the atmosphere [23,24]. Degradation studies that emphasize preventing the generation of toxic emissions and complete mineralization are required for the removal of plastics in the environment. Heterogeneous photocatalytic oxidation can be an attractive and efficient decomposition technique for plastics polymers. Photons with higher energy than the bandgap of photocatalyst which forms electron-hole pairs will react with surface adsorbed molecules leading to the degradation of organic molecules [22]. The photocatalytic degradation of plastics can occur at moderate conditions, some of the plastics require additional sensitizers due to the non-polar nature [23]. When the catalysts are uniformly distributed with a large interface area with the polymer, the photocatalytic degradation efficiency will also be higher. The photocatalytic degradation of polymers in liquid-phase reactions mainly occurs by the generation of electron-hole pairs under the illumination of light. TiO_2 is commonly used as the catalyst for photocatalytic degradation of plastic The degradation mainly results in the reduction of molecular weight, introducing new functional groups in the polymer chain, etc. Enhanced degradation of polyethene oxide and polyacrylamide was observed by Vijayalakshmi and Madras (2006) [26] for combustion synthesized nano-sized TiO_2 catalyst due to its size, high surface area, and the presence of hydroxyl groups.

5.1 Studies on photocatalytic degradation of plastics

The degradation of polymers under irradiation occurs by the direct absorption of photons by the polymers to create excitation states and further undergoing oxidation reaction and cross-linking [6]. The diffusion of reactive species in the polymer matrix extent the degradation processes, successive reactions lead to chain cleavage. Li et al., 2010 [25] studied the photocatalytic degradation of polyethylene using polypyrrole (PPy) /TiO_2 photocatalyst. For the study, the PPy/TiO_2 photocatalyst was spread on both sides of the rectangular plastic piece (6×8 cm) and the sample was exposed to sunlight for a total duration of 240 h. A maximum weight reduction of 35.4% was observed for PPy/TiO_2 photocatalyst whereas ~11.7% and 3.2% weight reduction was reported with only PPy and TiO_2, respectively. The photoresponse properties of TiO_2 particles in solid-phase photodegradation can be modified by dye sensitization [26]. In a study conducted by Ali et al., 2016 [27], photocatalytic efficiency of titanium nanoparticles (TNPs), titanium nanotubes (TNTs), dye sensitize TNPs and dye-sensitized TNTs were compared for the photocatalytic degradation of LDPE. Under UV light, the weight loss in LDPE film was found to increase with photocatalyst dose. The maximum weight loss of ~78% and 67% in the LDPE film (10% photocatalyst) was observed with TNPs and TNTs, respectively. However, under visible light, ~50% weight reduction in LDPE film was reported with dye-sensitized TNTs. In a photocatalytic study Tofa and co-workers [28] performed with ZnO nanorod on LDPE of 1×1 cm film using 50 W halogen lamp at a fixed distance of 10 cm. An increased value of storage modulus (E_s) and hence increased stiffness in the plastic was observed. A significantly higher value of E_s was observed with an increase in catalyst dose (ZnO) from 3 to 10 mM. With further increase in catalyst dose to 20 mM resulted in the rapture of LDPE films. In a follow-up study Tofa and co-workers [29], researchers further modified the ZnO catalyst by depositing platinum nanoparticles on ZnO nanorods. The carbonyl index (CI) and vinyl index (VI) values for LDPE degradation were increased by 13% and 15% for modified catalyst i.e., ZnO-Pt compared to ZnO.

A well dispersed and uniformly mingled polymer-catalyst composite with a larger interface area between the polymer matrix and catalyst has been proven to be a suitable method to degrade polymers. In a photocatalyst embedded polymer composite, the plastic is dissolved in an organic solvent, and to the solution photocatalyst is added. The obtained plastic composite is treated under different light sources. In this solid-phase photocatalytic degradation process, the hydroxyl radicals and reactive oxidation species formed during the photocatalytic process initiate the degradation by attacking the polymer chains. The degradation process extends to the neighbouring polymer matrix through the diffusion of the reactive radical species [4]. Zhao et al. [6] (2007), evaluated TiO_2 efficacy for the photocatalytic degradation of PE. The TiO_2 content in PE was varied at three concentrations of 0.02, 0.1, and 1 wt %. The PE-TiO_2 films were irradiated under solar or UV light for 300 h duration. Weight loss of PE samples was found to increase with the increase in higher TiO_2 content. Maximum weight loss of ~85% was reported with Pe-TiO_2 (1 wt%) with UV light however, the weight loss was 42% with solar irradiation. The key challenge faced during the solid phase photocatalytic degradation of plastics is the absorption spectrum range of nanomaterials and the type of chromophores present within polymer [30]. The degradation rates of well-dispersed microstructures of composites are much faster than the pure polymers due to the higher absorption of light [31]. Cho and Choi [32] (2001) studied the photocatalytic degradation of PVC-TiO_2 polymer composite by varying the TiO_2 concentration in the composite from 0.5 to 2% weight basis. It was reported 27% of weight loss was observed for PVC-TiO_2 (1.5 wt. %) in the air compared to ~3% weight loss in N_2 atmosphere after 300 h of irradiation. The study confirmed that in the absence of oxygen environment, the generation of radical species is suppressed resulting in lower degradation rate. Chakrabarti et al. [33] (2008) used ZnO for the photocatalytic degradation of PVC. The catalyst dose in the PVC film was varied from 0-30% (wt. basis). With an increase in catalyst up to 20% dose, the weight loss of PVC-ZnO film was increased. The maximum weight loss of ~20% was observed within 2 h of UV irradiation. However, a further increase in catalyst dose to 30% resulted in lower weight loss, which could be due to the blanketing effect or agglomeration of ZnO particles.

Li et al. [25] (2010) carried out photocatalytic degradation of polyethylene plastic directly under sunlight with polypyrrole/TiO_2 nanocomposite synthesized by sol-gel and emulsion polymerization method. It was reported that the nanocomposite enhanced visible light capturing ability, resulted in the formation of cavities on polyethylene surface and molecular weight reduction by 54%. Improved photocatalytic degradation of polyethylene under solar radiation was observed by Thomas and Sandhyarani [34] (2013) using a polyethene-TiO_2 nanocomposite prepared by a solution casting technique. With 0.1 wt% of TiO_2, a remarkable improvement of 68% degradation of polyethene with the formation of pores at the interface of the composites were observed after 200 h of solar irradiation. In another study, an enhanced photodegradation of polypropylene film was observed by Verma et al. [30] (2017) using TiO_2-reduced graphene oxide (rGO) based nanomaterials under solar irradiation for 130 hours. The TiO_2-rGO composites extended the absorption spectra to the visible region and the decreased recombination rate of the composite due to the 2D π- conjugation structure of graphene helped in fragmenting the polymer.

Shang and co-workers [35] (2003) explored the potential of TiO_2 and TiO_2/CuPc composite for the degradation of polystyrene under fluorescent lamps. The ratio of catalyst (i.e., TiO_2 or TiO_2/CuPc) to polystyrene was kept constant to 2 wt%. After 250 h of irradiation, ~7 and ~4 % weight reduction was reported with TiO_2/CuPc and TiO_2 composite, respectively. In a similar kind of study, Thomas and co-workers [36] (2013) studied the effect of TiO_2 particle size on photocatalytic degradation of polyethylene. Two different sizes of TiO_2 nanoparticle i.e. 50 (TiO_2-50) and 200 nm (TiO_2-200) was synthesized using the sol-gel method. For the photocatalytic study, the concentration of catalyst was kept 0.1wt% to PE. It was reported that TiO_2-50 resulted in ~18% weight loss compared to 7.5% with TiO_2 under UV light for 300 h duration. Further SEM images revealed the formation of cavities in PE film.

The modification of photocatalyst shows higher catalytic activity and enhances the degradation of the solid-state polymer. Zan et al. [37] (2004) modified the TiO_2 by grafting polystyrene on nanoparticle (PS-G-TiO_2) in order to reduce their agglomeration during the synthesis of polystyrene (PS) –TiO_2 composite film. On comparing the weight loss of composite film after 300 h of UV illumination, around 30 % weight reduction was reported for PS-G-TiO_2 (1 wt%), whereas 13% weight reduction was reported for PS-TiO_2 (1 wt%). In a study conducted by Kamrannejad et al. [38] (2014), carbon-coated TiO_2 and polypropylene were prepared by melt blending, and the degradation of the composite was studied under UV light. It was observed that by adjusting the carbon content in the composite, Young's modulus and elongation at break of the composites were increased through chain scission reactions and crosslinking. Solid-phase photocatalytic degradation of PE using multi-walled carbon nanotubes (MWCNTs) and TiO_2 nanocomposites photocatalyst synthesized by the sol-gel solvothermal method under UV irradiation was studied by An et al. [39] (2014). They have observed that the higher surface area and hindered electron-hole recombination in the composites enhanced the degradation of PE resulting in 35% weight loss in 180 min irradiation. Photocatalytic oxidation studies conducted on polyethylene-boron-goethite composite film under ultraviolet and visible light was investigated by Liu et al. [40] (2010) and observed that with 0.4wt% of boron, 12.6% weight loss of polymer was observed under UV irradiation of 300h. The narrowing of bandgap due to boron doping extended the light absorption to longer wavelengths resulting in formation of reactive oxygen species and enhanced the photocatalytic degradation

Sensitizing photocatalyst with dyes lowers the bandgap energy of the catalyst and generates singlet electrons upon excitation and facilitates the electron-hole pair separation.

Photocatalytic degradation of polyvinyl chloride on a dye-sensitized PVC- ZnO composite was studied by Chakrabarti et al. [33] (2008). Eosin Y dye used as the sensitizer upon irradiation excites electron and enhanced degradation of PVC to a large extent. In another study conducted by Sil and Chakrabarti [41] (2010), solid-phase photocatalytic degradation of PVC-ZnO composite in air and the aqueous medium were studied under UV and solar radiation. They have observed 6.2 and 14.2% weight loss under UV and solar radiation respectively in 90 minutes which was a significant amount of degradation of the polymer. In another study, a novel photodegradable polyvinyl chloride-vitamin C(VC)–TiO_2 nanocomposite synthesized by Yang et al. [42] (2010) resulted in 2 times higher degradation when compared to PVC-TiO_2 composite films under UV irradiation. The enhanced degradation was due to the formation of a five-member chelate ring structure of Ti^{vi}- VC complex promoting solid-phase photocatalytic degradation of PVC.

Asghar et al. [43] (2011) prepared four different types of polythene films: TiO_2 –PE, Fe doped TiO_2-PE, Ag doped TiO_2-PE and Fe/Ag doped TiO_2-PE. Doped TiO_2 nanoparticles were prepared by liquid impregnation method such as 1% (molar) doping agent to TiO_2. Finally, four polythene films were prepared with 1% photocatalytic weight with respect to PE. The prepared films were kept under UV and artificial light for 300 h duration to compare the efficiency prepared catalyst. Under UV light maximum weight loss of ~14% was observed with Fe/Ag doped TiO_2-PE whereas under artificial light Ag doped TiO_2-PE showed similar reduction.

Conclusion

Plastic is a wonder material due to its various inherent chemical properties such as inert nature, stability, and wide range of application. These properties resulted in the production of huge amounts of plastics however, poor management and non-biodegradable nature of plastic have led to environmental problems. Plastic products use different kinds of materials hence using a single treatment method is not feasible. Recycling plastics is also difficult as segregation of plastic in different categories is required. Due to the dependence of various industries on plastics, it is not easy to put a complete ban on its production. Amongst various disposal methods for plastic waste, photocatalytic degradation can be a viable option. A combination of suitable photocatalyst and plastic can help in a significant reduction in plastic waste. The majority of research on photocatalytic degradation is focused on composite film however, the application of such process is limited to labs. More detailed studies are required on heterogeneous photocatalytic degradation of plastic because of its wide scale applicability. Furthermore, the reusability of the residual catalyst after plastic degradation needs to be studied in detail.

References

[1] V. Koushal, R. Sharma, M. Sharma, R. Sharma, V. Sharma, Plastics: issues challenges and remediation, Int. J. Waste Resour. 04 (2014) 1–6. https://doi.org/10.4172/2252-5211.1000134

[2] J.G.B. Derraik, The pollution of the marine environment by plastic debris: A review, Mar. Pollut. Bull. 44 (2002) 842–852. https://doi.org/10.1016/S0025-326X(02)00220-5

[3] M.C. Krueger, H. Harms, D. Schlosser, Prospects for microbiological solutions to environmental pollution with plastics, Appl. Microbiol. Biotechnol. 99 (2015) 8857–8874. https://doi.org/10.1007/s00253-015-6879-4

[4] J. Shang, M. Chai, Y. Zhu, Solid-phase photocatalytic degradation of polystyrene plastic with TiO_2 as photocatalyst, J. Solid State Chem. 174 (2003) 104–110. https://doi.org/10.1016/S0022-4596(03)00183-X

[5] W. Fa, L. Zan, C. Gong, J. Zhong, K. Deng, Solid-phase photocatalytic degradation of polystyrene with TiO_2 modified by iron (II) phthalocyanine, Appl. Catal. B Environ. 79 (2008) 216–223. https://doi.org/10.1016/j.apcatb.2007.10.018

[6] X. u. Zhao, Z. Li, Y. Chen, L. Shi, Y. Zhu, Solid-phase photocatalytic degradation of polyethylene plastic under UV and solar light irradiation, J. Mol. Catal. A Chem. 268 (2007) 101–106. https://doi.org/10.1016/j.molcata.2006.12.012

[7] A.L. Andrady, Microplastics in the marine environment, Mar. Pollut. Bull. 62 (2011) 1596–1605. https://doi.org/10.1016/j.marpolbul.2011.05.030

[8] K. Fotopoulou, H. Karapanagioti, Degradation of various plastics in environment, Hazard. Chem. Assoc. with Plast. Mar. Environ., Springer, Cham, 2017: pp. 71–92. https://doi.org/10.1007/698

[9] R.K. Laser, Laser welding of plastics:Materials, processes and industrial applications. John Wiley & Sons, 2012

[10] Hisham A. Maddah, Polypropylene as a Promising Plastic: A Review, Am. J. Polym. Sci. 6 (2016) 1–11. https://doi.org/10.5923/j.ajps.20160601.01

[11] J. Hopewell, R. Dvorak, E. Kosior, Plastics recycling: Challenges and opportunities, Philos. Trans. R. Soc. B Biol. Sci. 364 (2009) 2115–2126. https://doi.org/10.1098/rstb.2008.0311

[12] O. Nkwachukwu, C. Chima, A. Ikenna, L. Albert, Focus on potential environmental issues on plastic world towards a sustainable plastic recycling in developing countries, Int. J. Ind. Chem. 4 (2013) 34. https://doi.org/10.1186/2228-5547-4-34

[13]   R. Verma, K.S. Vinoda, M. Papireddy, A.N.S. Gowda, Toxic Pollutants from Plastic Waste- A Review, Procedia Environ. Sci. 35 (2016) 701–708. https://doi.org/10.1016/j.proenv.2016.07.069

[14]   Y. Wang, J.M. Holden, X. xin Bi, P.C. Eklund, Thermal decomposition of polymeric C60, Chem. Phys. Lett. 217 (1994) 413–417. https://doi.org/10.1016/0009-2614(93)E1409-A

[15]   A. V Mohod, P.R. Gogate, Ultrasonic degradation of polymers : Effect of operating parameters and intensification using additives for carboxymethyl cellulose (CMC) and polyvinyl alcohol (PVA), Ultrason. - Sonochemistry. 18 (2011) 727–734. https://doi.org/10.1016/j.ultsonch.2010.11.002

[16]   V. Mitova, G. Grancharov, C. Molero, A.M. Borreguero, K. Troev, J.F. Rodriguez, Chemical degradation of polymers (polyurethanes, polycarbonate and polyamide) by esters of H-phosphonic and phosphoric acids, J. Macromol. Sci. Part A Pure Appl. Chem. 50 (2013) 774–795. https://doi.org/10.1080/10601325.2013.792667

[17]   M. Ahmad, E. Ahmed, Z.L. Hong, W. Ahmed, A. Elhissi, N.R. Khalid, Photocatalytic, sonocatalytic and sonophotocatalytic degradation of Rhodamine B using ZnO/CNTs composites photocatalysts, Ultrason. Sonochem. 21 (2014) 761–773. https://doi.org/10.1016/j.ultsonch.2013.08.014

[18]   C.H. Chiou, C.Y. Wu, R.S. Juang, Influence of operating parameters on photocatalytic degradation of phenol in UV/TiO$_2$ process, Chem. Eng. J. 139 (2008) 322–329. https://doi.org/10.1016/j.cej.2007.08.002

[19]   T. Velegraki, E. Hapeshi, D. Fatta-Kassinos, I. Poulios, Solar-induced heterogeneous photocatalytic degradation of methyl-paraben, Appl. Catal. B Environ. 178 (2015) 2–11. https://doi.org/10.1016/j.apcatb.2014.11.022

[20]   D. Avisar, Y. Lester, H. Mamane, pH induced polychromatic UV treatment for the removal of a mixture of SMX, OTC and CIP from water, J. Hazard. Mater. 175 (2010) 1068–1074. https://doi.org/10.1016/j.jhazmat.2009.10.122

[21]   J.C. Prata, J.P. da Costa, A.C. Duarte, T. Rocha-Santos, Methods for sampling and detection of microplastics in water and sediment: A critical review, TrAC - Trends Anal. Chem. 110 (2019) 150–159. https://doi.org/10.1016/j.trac.2018.10.029

[22]   L. Zan, S. Wang, W. Fa, Y. Hu, L. Tian, K. Deng, Solid-phase photocatalytic degradation of polystyrene with modified nano-TiO$_2$ catalyst, Polymer (Guildf). 47 (2006) 8155–8162. https://doi.org/10.1016/j.polymer.2006.09.023

[23]   A. Bandyopadhyay, G.C. Basak, Studies on photocatalytic degradation of polystyrene, Mater. Sci. Technol. 23 (2007) 307–314. https://doi.org/10.1179/174328407X158640

[24] S.P. Vijayalakshmi, G. Madras, Photocatalytic degradation of poly(ethylene oxide) and polyacrylamide, J. Appl. Polym. Sci. 100 (2006) 3997–4003. https://doi.org/10.1002/app.23190

[25] S. Li, S. Xu, L. He, F. Xu, Y. Wang, L. Zhang, Photocatalytic degradation of polyethylene plastic with polypyrrole/TiO_2 nanocomposite as photocatalyst, Polym. - Plast. Technol. Eng. 49 (2010) 400–406. https://doi.org/10.1080/03602550903532166

[26] X. Zhao, Z. Li, Y. Chen, L. Shi, Y. Zhu, Enhancement of photocatalytic degradation of polyethylene plastic with CuPc modified TiO_2 photocatalyst under solar light irradiation, Appl. Surf. Sci. 254 (2008) 1825–1829. https://doi.org/10.1016/j.apsusc.2007.07.154

[27] S.S. Ali, I.A. Qazi, M. Arshad, Z. Khan, T.C. Voice, C.T. Mehmood, Photocatalytic degradation of low density polyethylene (LDPE) films using titania nanotubes, Environ. Nanotechnology, Monit. Manag. 5 (2016) 44–53. https://doi.org/10.1016/j.enmm.2016.01.001

[28] T.S. Tofa, K.L. Kunjali, S. Paul, J. Dutta, Visible light photocatalytic degradation of microplastic residues with zinc oxide nanorods, Environ. Chem. Lett. 17 (2019) 1341–1346. https://doi.org/10.1007/s10311-019-00859-z

[29] T.S. Tofa, F. Ye, K.L. Kunjali, J. Dutta, Enhanced visible light photodegradation of microplastic fragments with plasmonic platinum/zinc oxide nanorod photocatalysts, Catalysts. 9 (2019). https://doi.org/10.3390/catal9100819

[30] R. Verma, S. Singh, M.K. Dalai, M. Saravanan, V. V. Agrawal, A.K. Srivastava, Photocatalytic degradation of polypropylene film using TiO_2-based nanomaterials under solar irradiation, Mater. Des. 133 (2017) 10–18. https://doi.org/10.1016/j.matdes.2017.07.042

[31] L. Zan, L. Tian, Z. Liu, Z. Peng, A new polystyrene-TiO_2 nanocomposite film and its photocatalytic degradation, Appl. Catal. A Gen. 264 (2004) 237–242. https://doi.org/10.1016/j.apcata.2003.12.046

[32] S. Cho, W. Choi, Solid-phase photocatalytic degradation of PVC-TiO_2 polymer composites, J. Photochem. Photobiol. A Chem. 143 (2001) 221–228. https://doi.org/10.1016/S1010-6030(01)00499-3

[33] S. Chakrabarti, B. Chaudhuri, S. Bhattacharjee, P. Das, B.K. Dutta, Degradation mechanism and kinetic model for photocatalytic oxidation of PVC-ZnO composite film in presence of a sensitizing dye and UV radiation, J. Hazard. Mater. 154 (2008) 230–236. https://doi.org/10.1016/j.jhazmat.2007.10.015

[34]   R.T. Thomas, N. Sandhyarani, Enhancement in the photocatalytic degradation of low density polyethylene-TiO_2 nanocomposite films under solar irradiation, RSC Adv. 3 (2013) 14080–14087. https://doi.org/10.1039/c3ra42226g

[35]   J. Shang, M. Chai, Y. Zhu, Photocatalytic degradation of polystyrene plastic under fluorescent light, Environ. Sci. Technol. 37 (2003) 4494–4499. https://doi.org/10.1021/es0209464

[36]   R.T. Thomas, V. Nair, N. Sandhyarani, TiO_2 nanoparticle assisted solid phase photocatalytic degradation of polythene film: A mechanistic investigation, Colloids Surfaces A Physicochem. Eng. Asp. 422 (2013) 1–9. https://doi.org/10.1016/j.colsurfa.2013.01.017

[37]   L. Zan, L. Tian, Z. Liu, Z. Peng, A new polystyrene-TiO_2 nanocomposite film and its photocatalytic degradation, Appl. Catal. A Gen. 264 (2004) 237–242. https://doi.org/10.1016/j.apcata.2003.12.046

[38]   M.M. Kamrannejad, A. Hasanzadeh, N. Nosoudi, L. Mai, A.A. Babaluo, Photocatalytic degradation of polypropylene/TiO2 nano-composites, Mater. Res. 17 (2014) 1039–1046. https://doi.org/10.1590/1516-1439.267214

[39]   Y. An, J. Hou, Z. Liu, B. Peng, Enhanced solid-phase photocatalytic degradation of polyethylene by TiO_2-MWCNTs nanocomposites, Mater. Chem. Phys. 148 (2014) 387–394. https://doi.org/10.1016/j.matchemphys.2014.08.001

[40]   G. Liu, D. Zhu, W. Zhou, S. Liao, J. Cui, K. Wu, D. Hamilton, Solid-phase photocatalytic degradation of polystyrene plastic with goethite modified by boron under UV-vis light irradiation, Appl. Surf. Sci. 256 (2010) 2546–2551. https://doi.org/10.1016/j.apsusc.2009.10.102

[41]   D. Sil, S. Chakrabarti, Photocatalytic degradation of PVC-ZnO composite film under tropical sunlight and artificial UV radiation: A comparative study, Sol. Energy. 84 (2010) 476–485. https://doi.org/10.1016/j.solener.2009.09.012

[42]   C. Yang, C. Gong, T. Peng, K. Deng, L. Zan, High photocatalytic degradation activity of the polyvinyl chloride (PVC)-vitamin C (VC)-TiO_2 nano-composite film, J. Hazard. Mater. 178 (2010) 152–156. https://doi.org/10.1016/j.jhazmat.2010.01.056

[43]   W. Asghar, I.A. Qazi, H. Ilyas, A.A. Khan, M.A. Awan, M. Rizwan Aslam, Comparative solid phase photocatalytic degradation of polythene films with doped and undoped TiO_2 nanoparticles, J. Nanomater. 2011 (2011). https://doi.org/10.1155/2011/461930

Degradation of Plastics
Materials Research Foundations **99** (2021) 179-192

Materials Research Forum LLC
https://doi.org/10.21741/9781644901335-8

Chapter 8

Overview of the Degradable Plastic Market

Nadia Akram[1*], Khalid Mahmood Zia[1], Muhammad Saeed[1], Waheed Gul Khan[2]

[1] Department of Chemistry, Government College University Faisalabad, Faisalabad-38000, Pakistan

Department of Chemistry, Quaid-i-Azam University, Islamabad-45320, Pakistan

* nadiaakram@gcuf.edu.pk

Abstract

Degradable plastic manufacturing has emerged as an eminent industry due to multiple range of products it offers to the consumers. The diversity induced by the usage of biomaterials lures the customer, making it even more popular for consumption. Degradable plastic industry is a market of multifarious products. The backlash on the massive plastics consumption is expected to be eradicated in the wake of the large business market of degradable plastics in the coming years. The supremacy of the degradable plastic market is not easy to evade by any means. The degradable plastic market is not a solitary market; instead, it encompasses the production, consumption and recycling industry as well. In order to triumph the status of a flourished market a joint venture by the leading companies need to be in harmony. The circular economy of the world is indispensable without a degradable plastic market in future.

Keywords

Bioplastics, Feed Stock, Circular Economy, Global Plastic Market, Degradable

Contents

1.   Introduction

The necessity and utility of the plastic goods are always undeniable. However, the global concern towards environment induced a huge change in the infrastructure of the plastic market. When the plastics were introduced to the world in "twentieth century", the popularity of plastic was due to easy and economical access to everyone. It is logical, this easy availability is due to convenient production, cost reduction and availability of manpower. These plastics provide multiple competitive features; these are resilient, lightweight, transparent, gauzy, and flamboyant. These materials exclusively display the design representation, performance and mass production capacity, in contrast to other alternative materials such as steel, aluminum or glass etc. The availability of plastic goods is beyond our envision. We just contemplate about a product, and the plastic industry makes it a reality. It is surly the diligence and commitment of the plastic manufacturers thriving this industry to the place where it is today.

Its vibrant features make them superlative for all sort of packaging, construction, horticulture, textile sanitation, agriculture, transportation, furniture, electrical products, consumer product, electrical products, cosmetics, etc. Despite the excessive availability of plastic products and its easy availability, consumers are now more conversant and alarmed about the environmental problems instigated by plastics. The prime anxiety is developed due to health and environmental concern in connection with proper plastic litter disposal. This awareness has revolutionized this industry into a degradable plastic market. The consumer's choices to use this new "degradable" or "bioplastic" are in fact booming the industry without affecting its production as shown in Fig. 1. It would be bigoted not to give credit to the plastic manufacturer to covert conventional plastics products into degradable plastic according to the customer and environmental demand. There are numerous examples of non-optimized, low-cost and short-life cycle plastic products that are typically lucrative for the manufacturer but eventually end up as undisposed waste on the credit of consumer. However, the strategic planning of the plastic manufacturer led the foundation of a new plastic industry in the form of degradable plastics. Now this innovative plastic industry elevates a big question of how these degradable plastics are produced apart from conventional plastics. Undoubtedly, if not all, yet a few ingredients must diverge in these degradable plastic products. When the customer thinks about the degradable plastic, it doesn't mean anexorbitant disposal method instead it is looking for an eco-friendly method, or even with the possibility of recycling the products without negotiating the price of the product. Contemporary environment of competition in the global economy helps to ensure the quality of product. The smart manufacturer has actually created a smart market of degradable plastics for consumers [1].

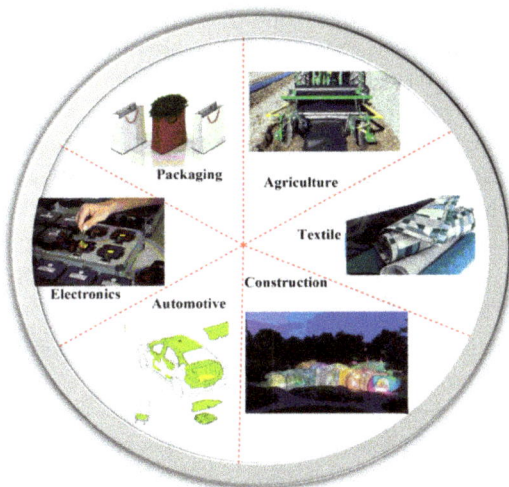

Fig. 1 Emergence of degradable plastic industry in all practical fields.

2.   Conversion of nondegradable to degradable plastic market

Plastics are one of the most ubiquitous products of our routine life. Their extensive applicability has widened their market inducing a number of advantages in various sectors, engendering numerous benefits for society and ecosystem. The voluminous plastics market is thrashing the other markets of paper, wood, aluminum. Their extraordinary properties have introduced them as an important part of various other sectors as well, this is one of those industry which is bifurcated into multiple industries. The plastics are serving two most important sectors of food and automotive industry [2]. As a matter of fact, the whole world is using the plastics however a few countries are leading in the market. China is one of the major plastic producers in the world, its plastic industry started to develop in early 1950s, it was the era of the discovery of crude oil and its significant industrial consumption for synthetic polymers. Within forty years China had more than 200 huge plastics manufacturing amenities. It utilized linear low-density polyethylene, polystyrene, acrylonitrile-butadiene-styrene as major plastic products. It introduced the plastic products in agriculture, packaging, construction and automotive. The China's plastic industry is mature now and can afford to turn towards a degradable plastics market. Hence, China which is usually considered one of the largest polluters globally; actually introduced a ban on non-degradable plastics in 2008. Fig. 2 shows a systematic evolution of biodegradable plastics from nondegradable plastics [3]. Surprisingly, the hazards related to plastics were not taken seriously by the developed countries rather every country played a vital role. Bangladesh was the first country to ban certain kinds of plastic bags in 2002. California was the first US state to endorse a plastic bag prohibition in 2014, followed by Idaho, Arizona, and Missouri between 2015 and 2017. In 2017, the European Union decided to make all plastic packaging recyclable or reusable by 2030; lastly Italian law has banned plastic bags for fruit, vegetables and baked goods. These initiatives were the prime reason to develop and promote the bioplastic industry [4].

The bio-based degradable plastics have experienced revitalization in the recent years. Primarily there are three groups of biobased polymers which are used as bioplastics as well, it includes: **"biochemosynthetic polymers"**, such as polylactic acid (PLA), polybutylene succinate (PBS), polyvinyl alcohol, and polyglycolic acid; biosynthetic polymers, which are also known as bioplastics; **"naturally occurring polymers"**, such as polyhydroxyalkanoate (PHA), and **"modified natural polymers"** such as starch polymers and cellulose derivatives [5]. Starch and polylactic acid (PLA) emerged as renewable feedstocks. This plastic is an excellent option for switching from conventional plastic to bioplastic due to nominal CO_2 emission contributing to global warming. The spectacular transition of non-degradable plastics to degradable plastics has changed the degradable market scenario [6]. The second most abundantly used degradable plastics are soybean based bioplastics. The US is major soybean producer in the world with 82.05 million metric tons. In the European plastic market, large supermarkets such as Sainsbury and Tesco took the initiatives to use biodegradable shopping bags [5]. Numerous renowned consumer and multinational companies have started using bio based materials as part of their sustainability programs. Some foremost examples in this connection are:

- Johnson & Johnson, started using biobased polyethylene packaging for sunscreen products in Brazil in 2009.

- Procter & Gamble, unveiled its long- term vision on 2010 by the use of 100% renewable materials or recycled for all products while eliminating the use of all petroleum-based virgin plastic and PVC-based plastic. P&G perceived the challenges of bioplastics due to lower availability of resources and difference in prices. In the market it is actually not easy to promote bioplastics as mainstream consumers will not pay more for bioplastics.

- Coca Cola also used biobased materials [5]. The world's consumption of bioplastics has increased from 15,000 tons (in 1996) to 225,000 tons (in 2008). With increasing demand for the world's plastic consumption, it was expected that the demand for biodegradable plastics will grow by 30% each year. These growing demands can only be met by exploring new bioplastic materials [7].

Fig. 2 *Transition of global non biodegradable plastic to biodegradable plastic industry ;(A) Representing the discovery of petroleum based non degradable plastics, (B) representing the issues created by the non degradable plastics (C) Struggle to find alternate resources (D) Successful development of degradable bioplastics.*

3. Associated markets for degradable plastics

The bioplastic market is not a single unit rather it comprises two other industries as well.

- Production industries of degradable plastics
- Consumption industries of degradable plastics

3.1 Production industries of degradable plastics

Developing a new market for bioplastics was very challenging because their competitors were well developed petrochemicals, which had a trusted fame in the entire polymer supply chain. It became more challenging as the Middle East was the hub of the petrochemicals. Despite these factors, major investments were made for new bio-based plastics which revealed a significant potential of bioplastics transforming the production of nondegradable plastics to a new field. The convenience of conversion of nondegradable plastics to the degradable plastics was in the hope of availability and utilization of biomass. Hence several biomaterials have been used for the production of bioplastics.

There method can be adopted to produce bio-based plastics, i.e.

i) Modification of the natural biomaterial for the conversion to bioplastics, e.g. starch plastics

ii) Conversion of monomers to bioplastics by the process of fermentation followed by polymerization such as: polylactic acid

iii) Production of bioplastics from monomers by micro-organisms such as; genetically modified crops.

When the production of any product achieves the highest position in the market, then it is not the product market only, actually there are multiple associated markets which develop with the main product as well. The most important plastic raw materials are: "**Starch plastics**". These starch plastics are leading the degradable plastics for two decades of evolution of bio-based plastics. These plastics are enjoying the experience of massive growth in its production. This is one of the widely used commercialized, degradable plastic in the market, used with the intention to replace non-biodegradable plastics. This inexpensive raw material can be blended with other copolymers to achieve the desired characteristics. The European countries optimized the starch production to approximately 130,000 metric tons from 2003 to 2007 representing an annual growth rate of nearly 50%. Five main types of starch can be obtained from the processing step, including: *Moderately fermented starch*; used primarily for those applications where the mechanical properties can be compromised. 2. *thermoplastic starch;* used for blending with other polymers to produce starch blends such as foamed-trays and boxes especially in food packaging, water soluble products and ingestible products. 3. *chemically modified starch*; expensive and not in excessive use, 4. *starch blends*; high water vapour permeability of starch plastics as fog-free packaging of warm foodstuffs. These are also used in catering service such as cups, food trays, knives and forks etc. 5 *Starch composites.* A native starch based composite blended with one or more constituents to enhance the properties of the main starch stream, coupled with the spark in the properties of the product [3]. The "Cellulose Plastics" industry is in a mature state with moderate growth because of China's remarkable role in flourishing its market. The main types of cellulosic plastics are *Organic cellulose esters* and *Regenerated cellulose*. The organic cellulose esters were developed to supplement cellulose nitrate to avoid its flammability resulting in its market today. Approximately 20% chemical grade pulp produces organic cellulose ester. Which is used in packaging films, cigarette filters, textile fibers, surface coatings, pharmaceutical (sustained release) and many other specialty industrial applications? The *Regenerated cellulose* is a widely used product. Almost 60% of the worldwide use of chemical grade pulp is consumed to produce regenerated cellulose fibers and films, suggesting fibers economically more favorable [8]. *Polylactic acid (PLA) is* an aliphatic polyester production was started to be used in 2002 by NatureWorks' (formally Cargill Dow) production plant and became the third type of bio-based polymer that was commercialized and is now produced on a mega scale. Dupont, Coors Brewing (Chronopol) and Cargill all these companies have significantly contributed in large research and development programs to investigate the possible technical applications for lactic acid, lactide and PLA. The important applications of this plastic are; extruded sheet, biaxially oriented film, blow moulded bottles, injection moulded products, packaging products including (cups, bottles, films, trays), textiles (shirts, furniture), nonwovens (diapers), electronics etc. **Poly(trimethylene terephthalate) (PTT);** This material was developed with the aid of processability of polybutylene terephthalate (PBT) in order to overcome the rigidity of polyethylene terephthalate (PET). The resiliency, wearability, dyeability, static and resistance made it a favorite choice for carpets and industrial textiles. In 2001, PTT Poly Canada, a joint venture between Shell Chemical and SGF Chimie JV, was formed to produce petrochemical PTT on a scale of 95 kt.p.a. in Montreal, Canada; the plant was operational in 2004. **Bio-based polyamides (nylon),** This is a product of castor oil, bio based adipic acid, *based azelaic acid, bio-based caprolactam dealing in automotive industry* **Arkema** (formally AtoFina) is the only producer of only producer of PA from castor oil. The production plant is located at Marseille Saint Menet (France). The manufacturing capacity is 22,000 tons/year in Europe. Arkema's PA11 has the tradename Rilsan. It is produced in France (Serquigny), in China (Changshu) and in the USA (Birdsboro, PA). **Soy-based**

Polyhydroxyalkanoates (PHA) Soybean oil is an attractive carbon source for the production of PHA as it supports the yield of PHA from soybean oil ranged from 0.72 to 0.76 g/g and the PHA productivity was around 1 g/L/h [6].

3.2   Consumption industries of degradable plastics

There are various consumption sectors of the biobased materials and many companies took the initiatives and played leading role in order to promote the degradable plastics. Procter & Gamble pledged to reduce petroleum- derived materials by 25%. Eventually, it intended to develop 100% packaging and products with renewable or recycled materials. Similarly, the world famous company "Coca-Cola" company established a global office of sustainability. The appropriate packaging of biodegradable material has a significant advantage over conventional packaging. In the event of the Olympics 2012 organized in London large amounts of starch and cellulose- based packaging were utilized [5]. Various renowned industrial sectors factors have started consuming degradable plastics such as; the food industry. Packaging industry utilizes 41% for plastic packaging for different items among which 20% is earmarked for food. Bioplastics are committed for sustainability and biodegradability to protect the environment and the consumer. A few distinguished companies which are developing bioplastics include: Dow chemicals (EcoPLA), DuPont (Sorona and Hytrel), BASF (Ecoflex and Ecovio) etc. Particularly BASF a famous face of polymers and chemicals has established numerous biobased and biodegradable plastics utilizing starch, PLA, PBAT etc. Ecoflex and Ecovio (a blend of Ecoflex and PLA) are both made of biodegradable plastics are extensively used materials in packaging specifically in shopping bags. Additionally, many other companies such as; Nature Works, Environmental Polymers, Novamont, Mirel, Tianan, Innovia etc. are developing bioplastics from renewable sources. Despite of exceptional features of bioplastics it still is a big task to replace petroleum based non degradable plastics entirely from the market [8]. "**Engineering Applications of Bioplastics**", bioplastics also deliver an ecological substitute for building and construction market. These materials are converted into biocomposites to get the maximum benefits. "**Biomedical Applications of Bioplastics**", the revolutionary use of bioplastics in medical industry is bringing more advantage than we have ever imagined. New materials are being reconnoitered for usage in the broad spectrum of biomedical applications for instance; implants, tissue engineering, drug delivery etc. [7-9].

4.   Challenges of degradable plastic market

Whenever there is emergence of a new market, it is not devoid of hurdles, struggles and challenges. It is not a one day practice to think and to create especially when it is a global market issue. It is a real blessing for mankind that the proprietors of the leading companies took the special initiative to convert their non degradable products into degradable products. All the countries specified the legislations to control the hazards associated with the non degradable issue. Currently, these industries are flourishing rapidly, however as a matter of fact it is not easy to convert the nondegradable plastic market to degradable plastic market rapidly. The most important challenges faced by this industry is given below:

4.1   Consumer acceptance

The market relies on two important figures; in the market there is a strong relationship between the consumer and the product. The bioplastic industry is yet not as flourished as plastic industry. In the beginning of the plastic industry, its popularity was associated with its low cost and long shelf life. Both are conversed features of bioplastics, where there is greater investment and the ultimate degradation. Hence, it is actually difficult to convince people to buy the products wrapped in bioplastic and to accept not only its higher price but to make them aware with the responsibility to dispose of the wrapper according to prescribed protocol as well. This also involves using only right packaging otherwise, the customer will be reluctant to buy products due to limited shelf life and probability of quick degradation of packaging. It is actually the maturity and preference of the customer to buy the product with degradable packaging. Corporate's social responsibility also urges them to promote various activities in all market areas with the specific concerns to aware the customer about environmental protection and to switch towards their new product. Awards are given to companies who have worked remarkable in the field of environmental awareness. The UK provides special training regarding the control of climate change. By the interest of many countries and companies, the environmental policies can be adopted easily. The product price plays a significant role to its acceptability to the consumer, this difference is viable in plastic goods, both natural and synthetic plastics follow approximately the same prices even then it's a serious challenge for the bioplastic market to convince customer to ignore the minor difference in the price to create a big difference in the society. The difficulty is the poverty in the world, in the developing and underdeveloped countries it is difficult for the customer to even pay some extra coins to prefer degradable plastic over non degradable plastic. It is a very critical situation where the bioplastic industry has to measures the business opportunity as well. To fix a market price is essential to prevent the market decline. Though, the companies are playing a key role in making the bioplastic products popular among consumer, still they need strategic planning. To compete with petro plastics the bioplastics have to be easily available in the market and it should be really ecofriendly as well [10-13]

Degradation of Plastics   Materials Research Forum LLC
Materials Research Foundations **99** (2021) 179-192   https://doi.org/10.21741/9781644901335-8

4.2   Inadequate availability of biomass feedstock

The bioplastic industry is dependent on the availability of biomaterials. The variable cost of grains impacts the price of bioplastic, which is influenced by grain's stocks and estimation of grain's annual crop price. Inflation in grain's prices due to diminutive production, stock or supply can reduce the production of bioplastic. Hence, it is a real challenge to keep an eye on feedstock all the time and to plan production accordingly. Hence it is significantly different to make a sole synthetic product with the use of biomaterial. Obviously, in this situation the petroleum based non degradable plastic will face less completion, promoting a more intense and challenging situation for bioplastics. The low wastage of biomaterials and the promotion to harvesting is only the right solution for this problem which can help to control the rates of bioplastics. This challenge can also be coped with recycling which however a tedious process. Apart from this, the real challenge is to fill the gap of demand and supply. Once the consumer is convinced to use the bioplastic product there should be no gap in the demand and supply chain [14-16].

4.3   New market of degradable plastics for circular economy

The global economy is oscillating due to drastic changes in international market in recent years, including escalation in demand and prices of crude oil. The linear 'take-make-dispose' economic model is fading, and initiatives to develop alternative economic models are evolving. Circular economy is in limelight. **Fig. 3** represents a relationship between linear and circular economy of degradable plastics. The price change in the stock market is always an issue especially for manufacturing companies. This change is usually associated with the oil price change which could be best situation where bio plastic can replace crude oil products. Similarly, the increase in the price of sugar for the production of bioplastics is also important which can be replaced with more economical corn production. Bioplastic is becoming a vital component in circular economy which can help in transformation of static economy into circular economy. It keeps the resources to circulate continuously. It is degenerated, recycled and reformed again in circular economy. It enhances system effectiveness by enhancing and optimizing the products. [10,17] The global resource ingesting, political, industrial and civil society organizations are progressively solving the puzzle to the resource 'limits to growth. In the European Union, the 'circular economy' has been presented as a high-level approach to move the societies beyond these limits. European policy makers consider the European business community very important to launch such processes. It is assumed that by applying new business models based on circular economy principles, the society will adopt circular economy. However, this assumption has yet to be scrutinized empirically. We know very little about how the business community takes up the circular economy as a potential strategy and whether and how debates and initiatives towards a circular economy contribute to business innovation. The bio-based sector, as one of the largest producers and consumers of natural resources in Europe, has become a vocal sector in the context of the circular economy particularly in Northern and Western Europe. Despite its relative importance to the European economy, evidence on the impact of the bio-based sector on the circular economy is still residual [11,18].

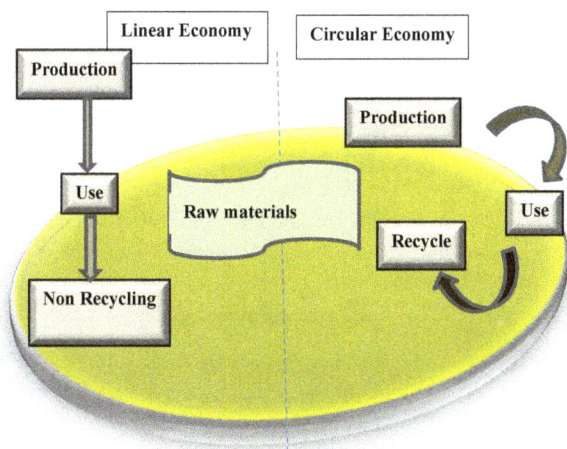

Fig. 3 Economy cycle of degradable bioplastics.

Conclusion and future prospects

In order to cope with the competitive environment in the petroleum based plastic industry, the launched product should be optimized enough to attract and satisfy the customer. The appropriate resource and raw materials consumption are the key to ensure profit. Degradable plastics are in the struggle to occupy the maximum share of the plastic industry due to its salient features. The degradable plastic and especially the bioplastics have covered a long passage to attain its present status, though it was very challenging to compete the petrochemical plastic industry but it is the global effort with the help of renowned companies to introduce it to the plastic consumer. After the successful production of bioplastic products, the next challenge is to spread the awareness about bioplastic. It is a million dollars industry but the consumer is still reluctant to embrace it fully due to its high cost and low shelf life over conventional plastics. The promising future of bioplastic will try to eradicate petroleum plastics. Bioplastic industry is committed to diminish the issues of marine litter and fossil fuel waste spreading by using green credentials. However, there is dire need of behavioral changes on bioplastics. Bioplastics have the potential to alter the shape of the economy from linear to circular but it should be in the limit of ethical boundaries. The climate change information suggests that the global carbon and greenhouse gases emission will be worst in condition until 2050. In order to maximize the potential of bio plastics a systematic approach is required through the controlled life cycle of bio plastics to promote it by then. A long path is ahead to realistically achieve this target. This dream can be true by the mass production of feed stock, its preservation, public awareness and only by the ethical constraints worldwide.

References

[1]   V.J. Romero, A. Sanchez, Methodology for multi-criteria design optimization of plastic products with a focus on highly competitive markets, Procedia. Manuf. 41 (2019) 1087–1094. 8th Manufacturing engineering society international conference, (http://creativecommons.org/ licenses/by-nc-nd/4.0/)

[2]   A.A. Rodriguez, M. Bosnjak, M.J. Sirgy, Moderators of the self-congruity effect on consumer decision-making: A meta-analysis, J. Bus. Res. 65 (2012) 1179–1188. https://doi.org/10.1016/j.jbusres.2011.07.031.

[3]   Z. Babutsidze, A. Chai, Look at me saving the planet! The imitation of visible green behavior and its impact on the climate value-action gap, Ecol. Econ. 146 (2018) 290–303. https://doi.org/10.1016/j.ecolecon.2017.10.017.

[4]   I. Confentea, D. Scarpib, I. Russoa, Marketing a new generation of bio-plastics products for a circular economy: The role of green self-identity, self-congruity, and perceived value, J. Bus. Res.112 (2020) 431–439. https://doi.org/10.1016/j.jbusres.2019.10.030

[5]   T.A. Cooper. Developments in bioplastic materials for packaging food, beverages and other fast- moving consumer goods, in: N. Farmer (Ed.), Trends in packaging of food, beverages and other fast- moving consumer goods (FMCG) , Woodhead Publishing, 1518 Walnut Street, Suite 1100, Philadelphia, 2013, pp. 108-238.

[6]   L. Shen, J. Haufe, M.K. Patel, Product overview and market projection of emerging bio-based plastics PRO-BIP 2009 Final report Group Science, Technology and Society (STS) Copernicus institute for sustainable development and innovation Utrecht University, 2009, 1-243. www.chem.uu.nl/nws www.copernicus.uu.nl

[7]   S. Pilla, Engineering applications of bioplastics and biocomposites - An overview, in: S. Pilla, (Ed), Handbook of bioplastics and biocomposites engineering applications. Scrivener ublishing LLC. Co-published by John Wiley & Sons, Inc. Hoboken, New Jersey, and Scrivener Publishing LLC, Salem, Massachusetts, 2011, pp. 1-14.

[8]   C. Barbarossa, P. de Pelsmacker, Positive and negative antecedents of purchasing eco-friendly products: A comparison between green and non-green consumers, J. Bus. Ethics. 134(2016) 229–247. https://doi.org/10.1007/s10551-014-2425-z.

[9]   H. Chen, W. Jiang, Y. Yang, Y. Yang, X. Man, State of the art on food waste research: A bibliometrics study from 1997 to 2014, J. Clean. Prod. 140 (2016) 840–846. https://doi.org/10.1016/j.jclepro.2015.11.085.

[10]   M.S. Jorgensena, A. Remmenb, A methodological approach to development of circular economy options in businesses, Procedia CIRP 69 (2018) 816 – 821, 25th CIRP Life Cycle Engineering (LCE) Conference, 30 April – 2 May 2018, Copenhagen, Denmark (http://creativecommons.org/licenses/by-nc-nd/4.0/).

[11]   L.Sina, A.P. Boix, The circular economy and the bio-based sector - Perspectives of European and German stakeholders, J. Clean. Prod. 201 (2018) 1125-1137 ttps://doi.org/10.1016/j.jclepro.2018.08.019

[12]   Y.N. Cho, E. Baskin, It's a match when green meets healthy in sustainability Labeling, J. Bus. Res. 86(2018) 119–129. https://doi.org/10.1016/j.jbusres.2018.01.050.

[13]   S.Y. Ong, K. Sudesh, Soy-based and plant oil-based polyhydroxyalkanoates, in: V. K. Thakur, M. K.Thakur, M.R. Kessler (Eds.),Soy-based bioplastics. Woodhead Publishing Limited, 80 High Street, Sawston, Cambridge CB22 3HJ, UK, 2013, pp,167-191

[14]   J. Ying, Z. Li-jun, Study on green supply chain management based on circular Economy, Physics Procedia. 25(2012) 1682-1688. https://doi.org/10.1016/j.phpro.2012. 03.295.

[15]   D. Andrews, The circular economy, design thinking and education for sustainability, Local Econ. 30 (2015.) 305-315. https://doi.org/10.1177/0269094215578226.

[16]   N.M.P. Bocken, S.W. Short, P. Rana, S. Evans, A literature and practice review to develop sustainable business model archetypes, J. Clean. Prod. 65(2014) 42-56. https://doi.org/10.1016/j.jclepro.2013.11.039.

Degradation of Plastics   Materials Research Forum LLC
Materials Research Foundations **99** (2021) 179-192   https://doi.org/10.21741/9781644901335-8

[17] S. Seuring, M. Müller, From a literature review to a conceptual framework for sustainable supply chain management, J. Clean. Prod. 16 (2008) 1699-1710. https://doi.org/10.1016/j.jclepro.2008.04.020.

[18] G. Roos, Business model innovation to create and capture resource value in future circular material chains, RES. 3 (2014) 248-274. https://doi.org/10.3390/resources3010248.

Degradation of Plastics
Materials Research Foundations **99** (2021) 193-237

Materials Research Forum LLC
https://doi.org/10.21741/9781644901335-9

Chapter 9

Plastics Versus Bioplastics

Faizan Muneer[1], Sabir Hussain[2], Sidra-tul-Muntaha[1], Muhammad Riaz[3], Habibullah Nadeem*[1]

[1]Department of Bioinformatics and Biotechnology, Government College University Faisalabad, Pakistan

[2]Department of Environmental Sciences and Engineering, Government College University Faisalabad, Pakistan

[3]Department of Food Sciences, University College of Agriculture, Bahauddin Zakariya University, Multan, Pakistan

*habibullah@gcuf.edu.pk

Abstract

Plastics are polymers of long chain hydrocarbons based on petrochemicals. Due to their physiochemical properties these are almost non-degradable and their complete recycling is impossible. High production rate and less disposal capacities have made plastic environmental pollutant resulting in severe impacts on the health of organisms and destruction of habitats thus effecting the biosphere in different ways. Biodegradation, thermal and catalytic degradation of plastics is widely studied to ensure a sustainable disposal of plastic waste with limited results until the present however, a new field where ecofriendly polymers obtained from natural biomass are used to make materials is flourishing. Bioplastics are polymers derived from biomass such as cellulose, starch, chitin and microbial polyhydroxyalkanoates that have the ability to produce products of daily use that can replace their counter parts made from the synthetic plastics. Bioplastics degrade easily in natural environment and replace the petrochemical based plastic polymers, thus saving the natural environment from plastic pollution and ensuring a sustainable environment.

Keywords

Polyhydroxyalkanoates, Biopolymers, Biodegradation, Polyethylene, Plastic Pollution

Contents

1.   Introduction to plastics

Plastics are large variety of materials or products that are synthetic or semisynthetic organic compounds and are basically the petrochemical derivatives [1]. The term plastic came from the Greek word "plastikos", means 'ability to change into different shapes [2]. Plastics consist of various elements such as carbon, oxygen, nitrogen, hydrogen, chlorine, and sulfur. In 1907, Leo Baekeland invented the world's first fully synthetic plastic-bakelite $(C_6H_6O.CH_2OH)_n$ in New York [3]. Since the discovery of bakelite, plastic industry got a sharp jump in its production and different types of plastics were discovered that are available today in different forms. Some most common examples include polyvinyl chloride (PVC), polyethylene (PE), polyethylene terephthalate (PET), polyurethane (PU) and a lot more. The invention of the first synthetic plastic at the very beginning of 20th century was seen as a great development, which opened the road for a variety of new plastic developments and established a completely new industry. Unfortunately, plastic in 21st century is the biggest environmental concern which has taken up the planet earth as an invading pollutant [4]. The use of plastic in almost all of the consumer products has produced piles of plastic waste which stands high above the ground like a mountain peak and floats in oceans and rivers like living creatures [5].

The use of plastic in the US is approximately 109 Kg/person/year, Europeans use 65 Kg of plastic per person/year, and every Indian uses at least 9.7 kg of plastic annually [6]. A long term study in the North Atlantic has shown that some of the water sample collected had 5,80,000 pieces of plastic per square kilometer of the ocean [7]. This is a huge environmental concern not only for the marine life and eco system but also for the human beings who hold the highest rank in the food chain of biosphere. It is found that more than 5 million tons of plastic either microplastic or macroplastic, is lost to the environment every year [8]. The main reason behind plastics turning from a marvelous product (which is used in almost all of the aspects of modern life) to an alarming concern for the environment is its slow degradation. The synthetic plastic may take more than 1000 years to degrade, not completely but partially and this again depends upon the type, environmental conditions and molecular mass of the polymer used for its synthesis [9]. The global production of plastic is currently ~320 billion tons per annum [10]. The current global production of plastics is estimated to triple by 2050 [11]. The global plastic industry was worth 1,722 billion Euros in 2015 showing its mightiness as a giant global industry [8]. The synthetic plastics like PVC, PE and PET cannot be recycled at the rate at which these are being produced. There is no method that can effectively remove these plastic wastes from the environment without negative effects on the environment and disturbing the human health. Burning plastic waste will lead to the production of toxic gases which will directly result in a number of health problems and will deplete ozone layer causing more ultra violet (UV) radiations to reach the ground which will increase the global warming and other health related issues (skin cancer) [11]. If production of synthetic plastic will rise, plastic pollution will follow suit. There is a dire need to concentrate and focus research on some natural and sustainable ways to produce plastics from renewable organic sources like biomass from plants, agricultural waste residues and microorganisms that produce natural polymers (biopolymers) that can be used for the production of plastic, i.e. bioplastic.

1.1   Types of plastics

Plastic to be straightforward is the most abundantly used material used in our daily life commodities, goods and other articles. Depending upon their physiochemical behaviour and recyclability plastics can be of seven different types as discussed below.

1.1.1 Polyethylene terephthalate (PETE or PET)

PET was introduced in 1940 by J. Rex Whinefield and James T. Dickson. It is the most widely used plastic on the planet, however, it took almost three decades to produce crystal-clear beverage bottles from it. In United States up to 96% of all plastic bottles are PET based and only 25% of them are recycled [12].

1.1.2 High-density polyethylene (HDPE)

High-density polyethylene was first introduced by Karl Ziegler and Erhard Holzkamp in 1953. They used a number of catalysts and low pressure to create high-density polyethylene. Primarily it was used for pipes in drains and culverts. In modern world high-density polyethylene is used for the production of a wide variety of products [13]. HDPE is recyclable, Environmental protection agency (EPA) suggests that almost 12% of all high-density products are recycled each year however, it still a very small dent in the world's carbon footprint [14].

1.1.3 Polyvinyl chloride (PVC)

If we call polyvinyl chloride an accidental discovery it will be true. In 1838 French physicist Henri Victor and in 1872 German Chemist Eugen Baumann discovered Polyvinyl chloride inside the vinyl chloride flasks that were somehow exposed to sunlight. The major use of the polyvinyl chloride is in the production of pipes and related goods [15]. It is the least recyclable plastic type. Less than 1% of PVC plastic is recycled per year. Due to its toxic nature and harmful impacts on health and the environment, it is known as "Poison Plastic".

1.1.4 Low-density polyethylene (LDPE)

Low-density polyethylene has less molecular mass as compared to high-density polyethylene, hence, these are considered as separate recycling type of plastics. LDPE was the first polyethylene to be ever produced [14]. Almost 56% of the plastic waste on planet earth is due to the packaging material produced from low-density polyethylene.

1.1.5 Polypropylene (PP)

Polypropylene was discovered in 1951 at Phillips Petroleum Company by J. Paul Hogan and Robert L. Banks when they were trying to convert propylene into gasoline. A very small fraction of this plastic is recycled [16]. In United States only about 3% of the products made from polypropylene are recycled each year.

1.1.6 Polystyrene or Styrofoam (PS)

Polystyrene too like polyvinyl chloride was discovered accidently. A German scientist Eduard Simon while working on a medication in 1839 accidently came across polystyrene. Another German Chemist Hermann Staudinger studied this polymer and expanded its uses. Due to its lightweight it is used as a multipurpose plastic however, as it breaks apart easily it contributes very badly polluting our water resources and endangering marine life. In United states almost 35% of plastic materials in landfills are made up of styrofoam [17].

1.1.7 Miscellaneous plastics

All other plastic materials that do not fit in the above mentioned categories due to their diverse properties and varying conditions of recycling and degradability in the environment such as polycarbonate, polylactide, acrylic, styrene and nylon are placed in this group.

2.   Recycling codes for plastics

In order to have more information about the plastics and make the consumer aware of the type and its possible impacts on the health and environment recycling codes for plastics are used [18]. Resin identification numbers that are more commonly known as SPI (society of the plastic industry) codes are used to sort plastics more effectively for the purpose of recycling.

2.1   SPI codes

In order to properly manage, organize, recycle and dispose of plastics they must be labeled with information pertaining to their use and properties. The Society of the Plastic Industry (SPI) established a classification system in 1988, which we now call as SPI code. Codes are given to different classes of plastics which explains the recyclability of the plastic material and its specific type [19]. Manufacturers around the world follow this coding system and label their specific plastic products with it so that the consumer or the recycler can have the basic knowledge of the type and properties of the plastic material they are dealing with. Different plastic types with SPI code, their symbols, and uses are summarized in Table 1 [6-14].

Table 1 *Classification of different plastics based on the SPI codes [6-14]*

SPI Code	Symbolic Representation	Polymeric Name (Abbreviation)	Melting temperature range (°C)	Major Use(s)	Recycled to make
1.	PETE	Polyethylene terephthalate (PET or PETE)	163-200	Beverage bottles, Medicine jars, Rope, Fiberfill in winter clothing	Storage containers, Carpets, Auto parts, Life jackets
2.	HDPE	High density polyethylene (HDPE)	65-140	Grocery bags Detergent containers, Toys	Lumber, fencing
3.	PVC	Polyvinyl chloride (PVC)	215	Pipes for plumbing purposes, Ducts, Window frames	Flooring
4.	LDPE	Low density polyethylene (LDPE)	65-135	Squeezable bottles, Frozen food bags,	Lumbers and Garbage canes
5.	PP	Polypropylene (PP)	65-170	Disposable utensils like cups and plates, Bottle caps, Kitchenware	Battery cables
6.	PS	Polystyrene or Styrofoam (PS)	65-240	Plastic cutlery, Packing foam	Insulations
7.	OTHER	Miscellaneous Plastics (polycarbonate, styrene, nylon etc.)	85-225	Multipurpose uses depending upon type	Plastic lumber often used in outdoor decks and park benches

Materials Research Forum LLC
https://doi.org/10.21741/9781644901335-9

3.   Global industry and production rates of synthetic plastics

With the population surge, in recent years the demand for plastics is increasing on daily basis. All sectors use plastic based products directly or indirectly to manufacture different products. Approximately 395 million metric tons of plastics is being manufactured each year with an approximate net worth of 1900 billion Euros [10]. Polyethylene (PE), polyvinyl chloride (PVC), styrofoam, polyethylene terephthalate (PET) constitutes the major plastic polymers used [20]. Data used from different sources and international forums such as United Nations Environmental Programme, World Economic Forum 2019 suggest that the seven major types of plastics are being produced on different scales according to their demands and use in the industry. Fig. 1 [8-11] shows a graph exhibiting the average production rates of different types of plastic for past five years from 2014-2019.

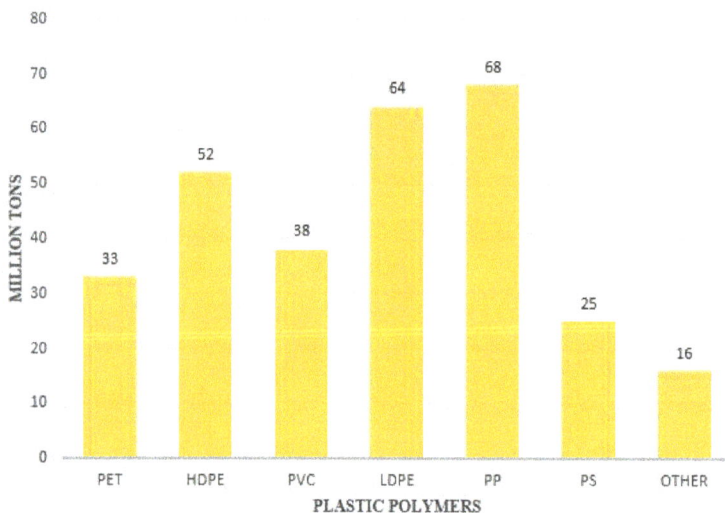

Fig. 1 *Production rates for different types of plastics [8-11].*

The plastics find different applications in our daily life; it follows us like a shadow, from a motorbike to a jumbo jet all have plastic components of some types employed [21]. It will be very difficult to discuss the percentage of use of different plastic materials, however; we can easily summarize the facts in a graph, which will give us a sharp estimation of the facts that which particular sector is using more plastic.

Depending upon the physiochemical properties and suitability for manufacturing and use of plastic consider, Fig. 2 [12-18] which shows estimated use of plastics in different sectors [5]. It can be seen from the graph that the largest use of plastic materials is packaging. Plastic packaging constitutes 36% of the total use of plastics. Polyethylene either high-density polyethylene or low-density polypropylene is used most commonly for this purpose. Thus the demand for polyethylene and its consumption rates are higher than the other types of plastic materials [22].

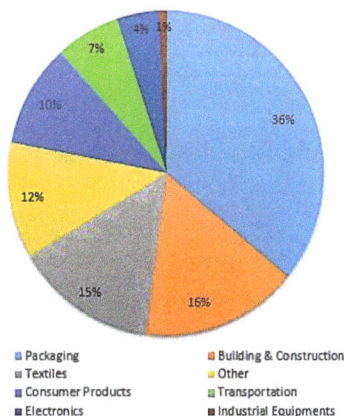

Fig. 2 *Chart showing the percentage use of plastics in different sectors of daily life [12-18].*

4.   Plastics: From a marvelous synthetic material to a nightmare

With the industrial revolution, the plastic polymers have got prominence in our daily life. It is almost impossible to create a new product without incorporation of some sort of plastic material in it. The world stands indebted to the greatness of plastics and its support in creating a developed and modern world hence no one can underestimate its importance in anyway [23]. Although they have countless uses in our modern world but due to some of their characteristics they are now turning invaders of the biosphere and are close to destroying the normal functioning of the food chain. Plastic pollution has got more prominence in recent years due to the drastic effects we are experiencing now. Plastics are turning into a nightmare as they are destroying our habitat and natural environment [24]. The non-degradability, recycling limitations, more production and less disposal of plastics have turned them environmental pollutants. High temperature requirements for molding or

Degradation of Plastics
Materials Research Foundations **99** (2021) 193-237

Materials Research Forum LLC
https://doi.org/10.21741/9781644901335-9

melting them is another environmental problem in addition to energy. The temperatures required for molding and melting different types of plastics are shown in Fig. 3 [25-29]. Plastic debris in oceans and on land is visible without any minute speculation and its impacts are harsh both for the environment and life. Some of the major environmental problems caused by plastics are discussed to get an idea about its destructive effects on environment.

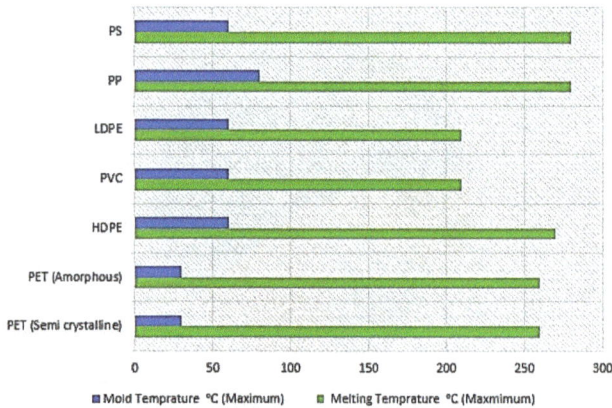

Fig. 3 *Maximum temperature for molding and melting different plastics [25-29]*

4.1 Destruction of water reservoirs due to plastics

Aquatic reservoirs such as ponds, lakes, and oceans are worst affected by plastic pollution as more and more land based plastic enters them. Marine organisms such as different species of fish, and turtles are facing alarming situations and awful scenarios due to the water pollution caused by plastics. Plastic products that enter these water bodies changes to smaller and smaller fragments of plastic upon mechanical disruption of water currents and other factors with the passage of time [24]. These smaller plastic fragments depending upon their varying size are nano, micro and macroplastics. Water polluted with these plastic particles generates different health issues and are a direct threat to the destruction of natural habitat to marine and fresh water flora and fauna. Plastics that end up in ocean and fresh water bodies usually flows from the land based sites. Marine environment is currently acting as a plastic sink. A number of different studies suggest that approximately over 250,000 tons afloat at sea [25]. It has been well estimated that 2/3rd of the surface of planet earth is covered by water while the remaining is land. Oceans make up 98% of the total water of the globe which cannot be used for drinking purposes without treatments such as filtration at water filtration facilities which no doubt cost a lot to fulfil the growing demand of population. Today, as much as 40% of our seas are intensely affected by human activities, fresh water constitutes about 2% of the total water available on earth. This water however, is still limited to only 0.4% in freely usable form as the rest 1.6% is locked as glaciers and polar ice caps, which are beyond our access to use which indicates us for a dark future if we don't save these water resources. Today, human activities such as plastic pollution has intensely affected as much as 40% of our seas and oceans [26].

4.2 Effects of plastics on marine life and human health

United Nations department for the environment estimated that approximately 800 species are affected by ocean debris globally and nearly 75-80% of which is plastic or products derived from plastics. It is measured and evaluated that almost 67% of marine plastic comes from land-based sources while the remaining come from ocean-based sources [23]. It is calculated that nearly 13 million metric tons of plastics enter the ocean every year. Seabirds, fishes, sea turtles and mammals can ingest plastic debris, which ultimately results in suffocation, starvation, and drowning of these marine creatures. Scientists have estimated that nearly 60% of all the seabird species have ingested fragments of plastics at some point in their lives: the point of concern here is a study that foresees this figure to increase to 99% by 2050 [27]. A number of long term scientific studies in oceanic environment has shown that the plastic particles can get ingested by a number of different marine organisms including fish and tortoises which can pass from the intestines reaching into the animal's circulatory system eventually generating an immune response [18,19]. Bioaccumulation as a result of ingesting these nano size particles is possible. As a result of bioaccumulation of plastic particles in different organisms inhabiting a wide range of water bodies the food web can get disturbed. Bioaccumulation usually starts from simpler microorganisms and reaches higher animals in the food web for example algae to zooplankton and then to fish [28]. It has been studied that the nano size plastic particles that are accumulated inside the bodies of different organisms can get access to the brain and incite behavioral disorders [29-31]. In nature, however, many animals throughout their lifetime are rarely exposed to nanoparticles. Most commonly used plastic types are enlisted in Table 1 [6-14]. Egg hatching and algal feeding was shown to diminish in copepods due to polystyrene ingestion with negative consequences for its survival. Polyvinyl chloride microplastics have effected soil-dwelling worms by reducing their energy reserves and normal physiology and functions [20, 21]. It has been evaluated through studies that 8.7 % of the turtles that died due to plastic pollution had a plastic or polyethylene bag blocking the esophagus and preventing the food passage to stomach which eventually resulted in the death [32]. Almost 19% of the global population depends on sea food in order to meet their animal protein intake [21]. Human beings hold the highest level in food web and trophic levels hence, they consume plastics indirectly for example by consuming seafood more commonly shellfish and small fish which are eaten as a whole including their gut on the other hand larger fish has more flesh hence human consume only their meat and are therefore, less effected by them. It has been estimated and evaluated through experiments that nano plastics have more mobility rates in the tissues of living organisms as compared to other types hence their larger surface to volume ratio increases their concentration in the tissues causing mass concerns of health and reproduction. If such sea food is taken it may have unknown and uncharacterized effects on humans [33].

4.3   Economic losses due to plastic pollution

Pollution creates challenges on different frontiers including health, environment and common day life. A bad environmental condition can result in the shutdown of business and health problems, need more medical treatments and accurate tests to diagnose the disease which means that more money from the budget and savings will be spent on such issues which otherwise could have been used in a more promising way such as education or personal development. Plastic pollution in sea and oceans can render the quality as well as quantity of seafood low effecting the fish market and seafood market. Plastic results in a loss of 13 billion dollars to marine ecosystem each year [34] suggesting huge economic losses due to plastic pollution alone.

5.   Possible solutions for plastic pollution

Researchers, innovators, environmentalists, and other stakeholders have put forward different solutions to counter plastic pollution. These solutions are important to be met with immediate effect as the plastics are continuously damaging the planet. Some of the major proposed solutions are discussed here.

5.1   Recycling and conversion to other useful and efficient products

Waste plastic or plastic debris can be recycled depending upon its type and conditions that are needed for its recycling. However, apart from recycling difficulties and expenses, only 12-20% of the plastic is recycled every year [35]. Plastic can be converted into fuels by catalytic and thermal cracking. Plastic is degraded into fractions like gas, crude oil, and solid residue upon thermal degradation. Parameters such as the catalyst, design of the reactor, and pyrolysis temperature plays an important role in the generation of gasoline and diesel grade fuel. Despite all these results this is not usually cost effective apart from reasons that the process generates toxic byproducts in gaseous and solid residual forms [13].

5.2   Ban on the single use of plastics

It has been proposed on different forums that a ban on single use of plastics such as polyethylene shopping bags that are used once and thrown away results in plastic pollution on a faster rate hence ban on single use can effectively control plastic pollution. If a plastic is used for multi times it will not become debris until thrown away as a used one.

5.3   Biodegradation of synthetic plastics

Palm et al. [36] 2019 has recently shown that polyethylene terephthalate debris that has been known as an environmental pollutant because of its extreme durability can now be degraded using a microbial strain *Ideonella sakaiensis* 201-F6 [37,38]. The polyethylene terephthalate degradation is a two-step simple process; during the first step, the bacterial enzyme PETase converts polyethylene terephthalate to mono-(2-hydroxyethyl) terephthalate. In the second step another enzyme known as MHETase hydrolyzes mono-(2-hydroxyethyl) terephthalate to the ethylene glycol and terephthalic acid both of which are nontoxic and can be used for other useful purposes.

The conversion of polyethylene terephthalate to its simple residual form which are non-toxic to the environment using *Ideonella sakaiensis* is of great importance in plastic waste management [36]. From this simple example of plastic degradation by the microorganisms, we can now say that there are a lot of opportunities out there in environment to degrade plastics using different microbes or enzymes isolated from them for this purpose. Some important bacterial strains that are shown to degrade plastics (without concerning the fact that how efficient these strains are in the process of degradation). The microbial strains mentioned in Table 2 [39-45] are a few examples where microbes are used to degrade plastics. One important thing to bear in mind is that these microbial strains that have been studied until now are not available on commercial scale to degrade large garbage dump sites in land, or to degrade plastic liter in oceans. These studies are just a hope that in near future we might discover some strains or modify the existing strains using biotechnology, recombinant DNA technology or our knowledge of protein and enzyme engineering. Scientists are working to enhance the speed of catalytic activity of enzymes that degrade plastics so that these modified enzymes can break the strong bonds between the synthetic polymer chains of the plastics and convert them into simpler residues that might be less toxic or can be used in making a diverse range of other useful products [46]. Hence, there is a hope that in near future it is possible to come up with new versions of plastic degrading enzyme that will help to counter plastic pollution not only in land but also in ocean and other water bodies like rivers, lakes and ponds [47].

Table 2 *Microbial degradation of synthetic plastics [39-45].*

Microbial strain	Type of plastic degraded	References
Pseudomonas sp.	Polyethylene, Polypropylene, Polyvinyl chloride.	[39]
Aspergillus flavus, Mucor circinilloides,	Low density polythene, Polyvinyl chloride films	[40]
Alcaligenes faecalis Clostridium botulinum	Polycaprolactone	[41]
Micrococcus sp.	Polyvinyl chloride	[42]
Ideonella sakaiensis Flavobacterium sp.	Polyethylene terephthalate, Nylon	[43]
Thermophilic Bacillus	Low-density polyethylene, High density polyethylene	[44]
Bacillus magaterium strain B1	Polyvinyl chloride	[45]

6.   The future of plastics

Due to revolution in information technology and recent developments of scientific research, people are becoming more and more aware of the negative impacts of the plastics. We have a society that is more conscious of its activities on the globe than ever before. Researchers, innovators and international forums are in continuous struggle to find solutions and better alternatives to plastics. Scientists at Berkeley Lab have found a brand new plastic material that may become efficient without becoming a pollutant for the environment as it can be made over and over again without losing its characteristic properties. They have called this plastic as poly diketoenamine (PDK) plastic.

6.1   What is poly diketoenamine or PDK?

Plastics create problems due to non-degradable behaviour and harsh requirements to partially degrade and even then, releasing harmful and toxic gases and byproducts. Plastic is also known for being tricky to recycle. PDK or poly(diketoenamine) is an up-and-coming plastic that can be recycled again and again without creating any waste ever having to go into landfills. As of 2019, research is still under progress with the goal to produce new

Materials Research Forum LLC
https://doi.org/10.21741/9781644901335-9

polymeric material, which can be used effectively for the production of like textile materials, foams, and 3D printing.

7.   Bioplastics

Bioplastics are ecofriendly, naturally derived polymers or monomers that can be used to synthesis materials that have properties and uses almost similar to synthetic plastics but differ in the sense that these are the derivatives of biomass and are renewable while plastics are the derivatives of petroleum compounds [38]. Bioplastic is not a single material or substance rather it is a family of materials that are either bio-based, biodegradable or both [39]. We can divide bioplastic family in to three main groups namely:

a) Bio-based or partly bio-based non-biodegradable bioplastics

b) Bio-based and biodegradable bioplastics

c) Fossil based biodegradable bioplastics [40]

Bioplastics, depending upon their direct or indirect biomass source are divided as first-generation bioplastics and second generation bioplastics. The first generation bioplastics are manufactured directly from natural polymers such as starch or cellulose while those which are obtained as monomers like lactic acid or extracted as polyesters from renewable sources such as polyhydroxyalkanoates (PHAs) from microbes are second generation bioplastics [41]. Each of the above group of bioplastic family is discussed here comprehensively.

7.1   Bio-based or partly bio-based non-biodegradable bioplastics

Some most commonly used plastics such as polyethylene (PE), polyvinyl chloride (PVC) and polypropylene (PP) are known as commodity plastics which are usually made from petrochemical derivatives but these can be made from renewable resources too, for example from bioethanol. Approximately 200000 tons per annum (p.a) of Bio-PE (Bio-based polyethylene) is being produced by some industries like Braskem, Brazil. Similarly, Bio-PP (Bio-based polypropylene) and Bio-PVC (Bio-based polyvinyl chloride) are in the pipeline for production [42]. The world renowned beverage industry icon Coca-Cola uses partially bio-based polyesters polyethylene terephthalate (Bio-PET) for beverage bottles. These bio-based or partially bio-based variants of the synthetic commodity plastics are bioplastics by definition but these retain most of the properties of their parent types i.e. these are non-biodegradable like their fossil based parent types but are called bioplastics due to their bio-material composition [43].

7.2   Bio-based biodegradable plastics

Bioplastics in this group includes biopolymers like starch or thermos-plastically modified starch polymers as well as polyesters like polyhydroxyalkonates (PHAs) or polylactic acid (PLA). These are not only derived from biomass (e.g. starch from plants, vegetables and PHAs like polyhydroxybutyrate from microorganisms) but are also biodegradable [44]. Bio-based biodegradable plastics are in real sense the ecofriendly bioplastics that are desired in real sense however, they have been available on commercial scale only for a few years [45]. Until now their use have been limited to short-lived packaging but still this dynamic and flourishing and even the future of plastic industry continues to grow smoothly due to the recent introductions of some new bio-based monomers like fatty acid derivatives, propane diol, succinic acid etc. Their renewable basis for production, biodegradability to reduce the concern of becoming a pollutant and environmental safety has made them a sustainable group of bioplastic family. Hence, these are at the prime focus of research to create ecofriendly plastics for daily use.

7.3   Biodegradable, fossil-based plastics

These constitute a comparatively small bioplastic group because these are used in combinations with other biomass polymers such as starch. These improve and enhance the performance of the plastic in respect to its biodegradability and mechanical properties. Most of the biodegradable fossil based plastics are usually produced in petrochemical production processes [46]. Partially biobased versions of theses plastics will be available for commercial use in the near future.

7.4   Bioplastic: Plastic from biomass and its effectiveness to counter plastic problems

Bioplastics or biopolymers can be derived from sources like plants and microorganisms i.e. bacteria and algae [47]. Some macroalgae like seaweeds has also been studied for the production of bioplastics [48]. There are monomers like carbon dioxide (CO_2), terpenes, vegetable oils and carbohydrates, which can be effectively used and employed as feedstock for the manufacture of a variety of sustainable materials and products, including elastomers and plastics. The global bioplastic industry is growing at more than 20 to 25% each year. In 2007 the bioplastic industry worth 7 billion US\$ which is estimated to reach more than 10 billion US\$ by 2020 [7]. The early developing history of some biomaterials that lead to the modern research and industry of bioplastic is given in Table 3 [48-51].

Table 3 *A brief history of biobased and biodegradable plastic materials [48-51].*

Year	Biomaterial	Characteristics	References
1862	Parkesine: Derived from cellulose.	Biobased	[48]
1897	Galalith: derived from casein (milk)	Biobased and biodegradable	[49]
1912	Cellophane: derived from cellulose	Biodegradable	[50]
1926	Polyhydroxybutyrate (PHB): polyester derived from bacteria	Biobased and biodegradable	
1947	Rilsan or Nylon 11: Produced by the polymerization of 11-aminoundecanoic acid from castor beans.	Biobased and biodegradable	[51]

8. Sources of bioplastic production

Bioplastics, as discussed earlier are either bio-based, biodegradable or both hence any monomer, polymer or polyester that is either obtained from biomass (organic source) like plant and microbes or which might be degraded by them will be considered as a bioplastics [49]. Plants such as switch grass (*Panicum virgatum* L.) and *Arabidopsis thaliana* [50] and microorganisms like *Bacillus megaterium* NCIM 2475, *Bacillus cereus* SPV, *Bacillus flexus, Sinorhizobium meliloti* [9,51] are widely studied for this purpose. Table 4 [52-61] shows some important microbial strains that have been studied and used for the purpose of production of different biopolymers.

Table 4 *Microbial sources of bioplastic polymers [52-61]*

Microbial source(s)	Bioplastic Polymer(s)	References
Alcaligenes latus	PHB	[52]
Bacillus sp.	PHB	[53]
Bacillus megaterium	P3HB	[54]
Pseudomonas oleovorans	P3HB, P3HV	[55]
Burkholderia sacchari	P3HB	[56]
Cupriavidus necator	PHB, PHA	[57,58]
Hydrogenophilus thermoluteolus TH-1	PHB	[59]
Hydrogenophaga sp.	P3HB	[60]
Pseudomonas putida KT2440	mcl-PHA	[61]

It is evident from the Table 4 [52-61] that the most commonly produced biopolymer from the microbial strains is polyhydroxyalkanoates, especially the short chain length polymer known as polyhydroxybutyrate or (PHB). Plants on the other hand are a good source of lignocellulosic biomass which includes starch, cellulose and lignin which can be used to produce biopolymers for the production of bioplastics. Some important plant sources that have been previously studied for this purpose are enlisted in Table 5 [62-69].

There is an enormous amount of biomass available in different forms and from different sources, which can be somehow used directly or indirectly to produce bioplastics, which can help us to replace the synthetic plastics with these naturally produced and environmental friendly bioplastics. In order to have a quick look at the different polymers, proteins and polyesters that have been used and are still being extensively studied for the production of bioplastics consider Fig. 4 [23-32] and Table 6 [69-81].

Table 5 *Important plant sources that have been studied for the production of bioplastic materials [62-69].*

Plant sources	References
Switch grass	[62]
Wheat	[63]
Maize and Potato	[64]
Kenaf (*Hibiscus cannabinus L.*)	[65]
Flax (*Linum usitatissimum*)	[66]
Water hyacinth(*Eichhornia crassipes*)	[67]
Banana	[68]
Rice straw	[69]

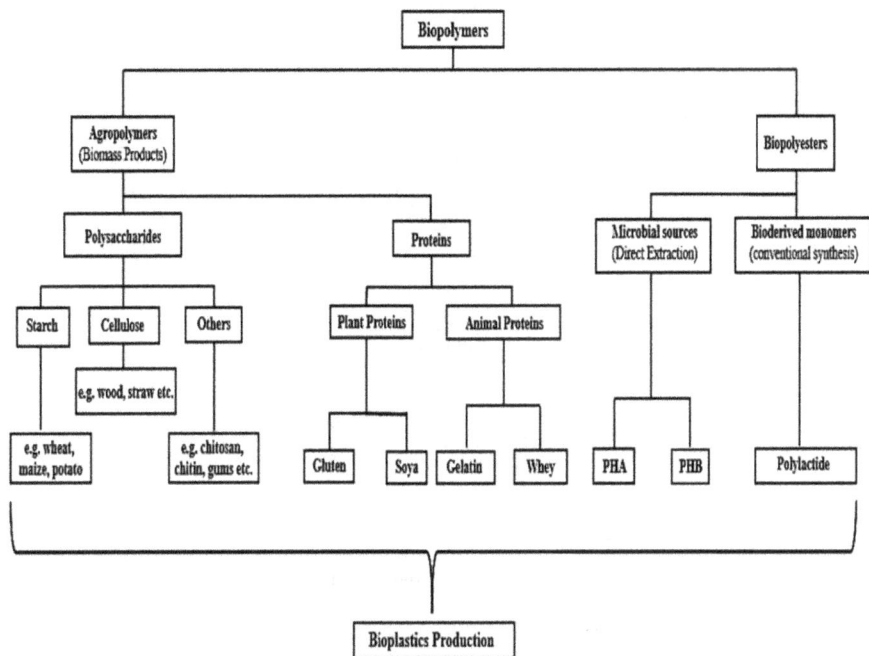

Fig. 4 *A wide variety of sources of biopolymers that can be used to produce bioplastics [23-32].*

Materials Research Forum LLC

https://doi.org/10.21741/9781644901335-9

Table 6 *Natural polymers and proteins from different sources like plants, animals and microbes that can be used for the production of bioplastic products [69-81].*

Natural polymers & proteins	Structure	Summary of the properties	Uses	References
Polysaccharides				
Starch		Used for the production of biocompatible, rigid & elastic plastic. It yields a thermoplastic and can act as a gas barrier. Degradable in soil and water.	Packaging & trash bags	[69]
Cellulose		Cellulose triacetate & cellulose acetate are two biopolymers derived from cellulose. These molecules yield thermostable and rigid bioplastic which is degradable in soil and water.	Air filters, Wound dressing and textiles	
Chitin		It is the second most abundant polysaccharide in nature, which is used to form ultrathin & labile nanostructured films.	Construction of bioplastic for bone repair.	[70]
Chitosan		It is derived from chitin and is structurally similar to it.	It is used as bioplastic film for enhancing structural characteristic of other polymers.	[71,72]

Plant proteins				
Gluten		It is plasticized with glycerol yielding bioplastic film. This wheat gluten derived bioplastic is elastic and resistant to oxygen and water but is completely biodegradable in nature.	Biopolymers formed from wheat gluten are used as nanofibers in many medical applications such as drug delivery.	[73]
Animal Proteins				
Whey		Whey protein is used with the polyethylene glycol or PEG, glycerol and water to form bioplastic films that have good mechanical properties	Food packaging and development of other biomaterials	[74]
Keratin		Keratin protein is present in abundant quantities in hair and nails. It is non-burning and hydrophilic in nature.	Used for the development of bioplastic films with a wide range of properties	[75]
Gelatin		It is derived from collagen from different parts of an animal. It is odourless and translucent.	Used as a biomaterial to produce edible biocomposite films.	[76]

Lactic acid and Lactides				
PLA		Polylactic acid is an aliphatic polyester that can be derived from a wide range of petrochemicals but the most common and obvious sources of its production are starch or sugar rich plants and crops. High transparency and surface gloss that make it a replacement for synthetic plastics like PET and PVC.	Used in food packaging.	[77]
PHAs (Polyhydroxyalkanoates) or Polyesters				
P(3HB)		Poly 3 hydroxybutyrate is a polyester produced in microbial strains under limited nutrient conditions as granules in the cytoplasm.	Is widely used in biopolymer studies for the production of bioplastics.	[78]
P(4HB)		Poly-4-hydroxybutyrate is a short chain length PHA (Polyhydroxyalkanoates) that is produced by different microbes as intracellular granules.	Approved by the FDA for use in medical devices.	[79]
PHV		Polyhydroxyvalerate is a biopolyester that has been produced by bacteria. It has special characteristics that enable it to be used in bioplastic production.	PHV is used as blends and copolymers along with other PHAs	[80]
PHBV		Poly(hydroxybutyrate-co-hydroxyvalerate) is a copolymer of PHB and PHV. This copolymer is formed to get a better version of polyester material with dual characteristics of both the individual molecules.	It enhances the mechanical performance of the bioplastic products.	[81]

9.   Polysaccharides for the production of bioplastic materials

9.1   Starch

Starch is widely associated with plant based biomass and is most common among polysaccharides found in nature [82]. The monomer that forms this polysaccharide is a simple monosaccharide- glucose. The glucose monomers are joined by 1,4-linkage. There are two forms of starch available in nature, a linear polymer known as amylose and a branched polymer called as amylopectin [83]. Starch has been used for centuries for different purposes with paper manufacturing being the most common. Large quantities of starch are also used in textile industry where it is used to impart strength to the thread during weaving process. Starch has been investigated for the development of bioplastics as a natural derivative of plant based biomass. Maulida et al. [84] (2016) investigated the potential of starch based bioplastic production using cassava peel [85]. They used sorbitol as plasticizer during their experiments and found that the starch from cassava peel reinforced with microcrystalline cellulose and sorbitol as plasticizer increased the tensile strength of the bioplastic that was formed. Their study confirms that starch, as a biopolymer is a potential candidate for the synthesis of bioplastics. Starch based bioplastic films are not only degradable in water and soil but are also used to form plastic films that are biocompatible, rigid and elastic. These bioplastic films can act as gas barriers and can be used in food packaging with some slight modifications in their tensile strength properties by preparing their composites with some other biobased materials [86].

9.2   Cellulose

Cellulose is the most common polysaccharide associated with the cell wall of plant cells [55]. Being a polymer made from simple repeating monomer units of glucose it is widely used in paper production and as cotton in manufacturing clothing. Cellulose triacetate and cellulose acetate are two biopolymers derived from cellulose. These molecules yield thermostable and rigid bioplastic, which is degradable in soil and water. Lignocellulosic biomass sources like wood and rice straw can be effectively used as non-food mass for the sustainable development of bioplastics however, the conversion of this biomass to simple residues to be used sustainably in biopolymeric research needs cost effective routes for developing bioplastics [56]. Bioplastic production from cellulose of oil palm empty fruit bunch has been investigated by Isroi et al. [57] (2017), nearly 40% cellulose is present in the empty fruit bunch of the oil palm, although cellulose has no direct properties to become a potential bioplastic however, its composites with plasticizers is a key success in the development of bioplastics [87]. Bioplastic composite of cellulose-starch was prepared by solution casting and evaporation process using cassava starch as the polymers matrix and glycerol as plasticizer. Cellulose side chain was oxidized to reduce hydroxyl group and increase the carboxyl group. Their research successfully produced bioplastic sheet. It is believed that further research can make bioplastic from EFB cellulose which can be used to develop plastic bags and can be used in food packaging [57].

9.3 Pectin and chitin

Pectin is a hetero polysaccharide that is found primarily in the cell walls of the terrestrial plants. On chimerical scale it is produced as a white to light brown powder most commonly from citrus fruits [58]. Pectin is the main agent behind the thickening properties of jams and jellies. The pectin bioplastics film can be produced by adding 5-gram pectin obtained from banana peel extract in distilled water and heating the mixture at 60 °C. The film obtained will have a thickness of 0.00387 cm, water resistance of 63.63%, the tensile strength value of 10.562 MPa, and the elongation value of 58.33% [59,86]. Chitin is the second most abundant polysaccharide in nature. It is the main structural component of the exoskeleton of arthropods like insects and crustaceans. Crab, shrimps and fish scale are the ideal biomass for the production of chitin. Pandharipande and Bhagat [60] (2016) used chitin extracted from the crab shall to develop bioplastic film with properties that help us to conclude that chitin obtained from natural sources can be used sustainably to develop biomaterials like bioplastics [60,87]. It is also used to form ultrathin and labile nanostructured films [60]. He et al. [61] (2017) constructed a biocompatible bioplastic from chitin that was used to aid bone repair. It is completely evident that chitin derived biodegradable and biocompatible products can be used for the sustainable development of different eco-friendly biobased products [61].

9.4 Animal and plant derived proteins as bioplastic sources

With more understanding of the sustainable use of biological materials scientists have now developed new methods to use proteins as an excellent source for the manufacturing of bioplastic materials and products as a replacement to conventional plastics [62]. Proteins are complex heteropolymers, which offer a number of different functional side groups capable of forming strong intermolecular bonds. The most important aspects of protein processing include denaturing, cross-linking and plasticization. All these properties give proteins a unique characteristic to be used as a biomass resource in the development of sustainable products such as bioplastics [62]. Proteins such as casein, whey, gelatin, zein, soya and gluten can be used as precursors for the development of biomaterials such as bioplastics [63-66]. Chalermthai et al. [67] (2019) has shown the synthesis of protein-based thermoset elastomers using whey protein isolates through free radical polymerization. The whey protein based copolymers produced in their study have shown the potential to bear a high tensile strength, which make a future possibility to use them as packaging material and thus diminishing the role of conventional plastics. Whey protein is also used with the polyethylene glycol or PEG, glycerol and water to form bioplastic films that have good mechanical properties[67]. Bioplastic containers for planting have been produced by Helgeson et al. [68] 2009 using zein protein extracted from corn as a bio-renewable component [68]. Gluten is a protein obtained from cereals. Wheat gluten derived bioplastic is elastic and is resistant to oxygen and water but is completely biodegradable in nature [69]. Aziz et al. [69] (2019) has shown that wheat gluten can be used efficiently for the manufacturing of electrospun nanofibers based on wheat gluten containing azathioprine for biomedical applications [88]. In nutshell there are a lots of possibilities available or can be derived further to employ plant and animal based protein biomass for the sustainable and ecofriendly development of biobased materials. A large part of which can be used as bioplastics that will not only counter the pollution caused by the synthetic or conventional plastics that are non-degradable but will also ensure the sustainable development of the ecosystem and the global environment effected very badly by the plastic wastes.

10.   Polyhydroxyalkanoates (PHAs): Biopolyesters from microbial sources used as bioplastics

Prokaryotes such as bacteria produce polyesters from renewable resources like lipids, carbohydrates, organic acids and alcohols under insufficient or limited supply of nutrients, one such type of polyester is polyhydroxyalkanoates (PHA) [70]. PHA is actually a diverse class of polymers which have at least 100 different types of monomers which are either obtained from plant sources such as wheat, corn, sugar beet, potatoes and vegetable oils or from different microbial sources [89]. More than 300 species of microorganism have been reported to produce PHA [72]. The diversity and versatility in PHA is due to the difference in molecular structure and chain lengths of its members.

10.1   Structural classification of PHAs

Polyhydroxyalkanoates are classified into three different types depending upon the carbon chain length which include short chain length PHAs (scl-PHAs), medium chain length PHAs (mcl-PHAs) and long chain length PHAs (lcl-PHAs) [61]. The general structure of PHAs is shown in Fig. 5 [38].

Scl-PHAs have 3-4 carbon atoms such as poly(3-hydroxybutyrate) or P(3HB) and poly(4-hydroxybutyrate) or P(4HB). Scl-PHAs can be used in food packaging and disposable products [90-92]. Polyhydroxyalkanoates with 6-14 carbon atoms are termed as mcl-PHAs these can either be homopolymers like poly(3-hydoxyhexanoate), simplified as P(3HHx) and poly(3-hydroxyoctanoate) or P(3HO). The scl-PHAs homopolymers or mcl-PHAs homopolymers have quite different properties [74]. Fig. 6 [33,36, 45-49] provides a comparison between the different properties of the two main types of PHAs i.e. scl-PHAs and mcl-PHAs.

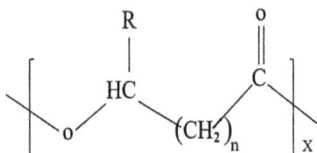

Fig. 5 *General structure of polyhydroxyalkanoates [38].*

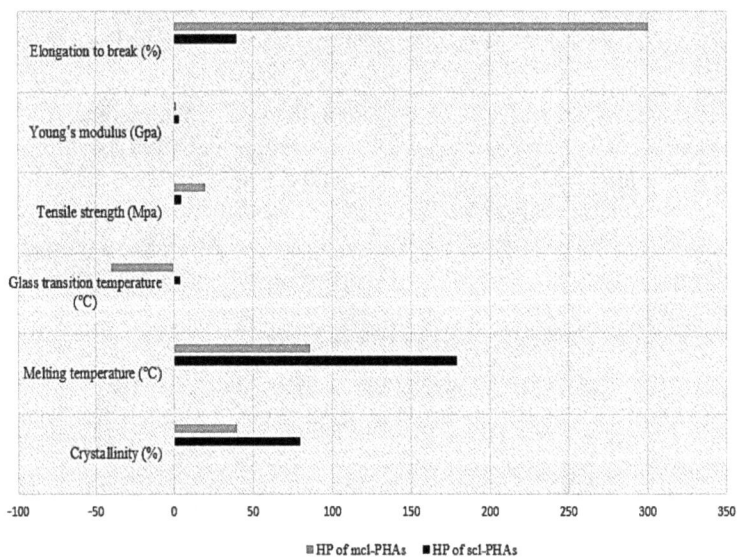

HP= homopolymers

Fig. 6 *Comparison between the properties of two main types of PHAs i.e. scl-PHAs and mcl-PHAs [33,36, 45-49]*

The heteropolymers include P(3HHx-co-3HO). These can be used to manufacture high value added products such as drug delivery systems, medical implants and stitching [93-95]. Lcl-PHAs are less frequent and are therefore, rarely studied [61].

PHAs, depending upon these differences can be polyhydroxybutyrate (PHB), polyhydroxyvalerate (PHV) or the copolymer of the PHB and PHV i.e. PHBV [76]. PHAs have been studied for their application in the development of bioplastics, medicines, fine chemicals and implantable biomaterials [96]. Different microbial strains accumulate PHAs in the presence of abundant carbon sources. The simple hydroxyalkanoates (HAs) monomers are polymerized in to long chain polyesters with the help of PHA synthase – a key enzyme in the biochemical conversion of hydroxyalkanoates into PHAs. The PHA accumulation in the cytoplasm is due to its insolubility in water. Hence, several bacteria can therefore, store 80% of PHAs of their cell dry mass [97]. PHA has been in the focus of research for bioplastic development due to its near identical properties to that of conventional plastics that are nonrenewable and non-biodegradable and possess serious environmental threats [77]. PHAs are not only environmental friendly but can be applied in different fields where its conventional counterparts can't be applied [78]. The commercial production of bioplastics from PHAs has grown in recent years but their market share is still marginal due to higher manufacturing costs, which make them an economic challenge. Despite using all expertise in creating good microbial strains using biotechnology, high production facilities, and better fermentation technologies, only reduction of the cost of PHAs production up to US$ 5 per Kg has been possible in 2009, which is still three times the rate of its closest synthetic version of plastic-polypropylene (PP) [98].

11. Polylactic acid (PLA): A bioderived monomer used for bioplastic production

Polylactide most commonly termed as polylactic acid is one of the most promising polymer that has the ability to become a focused biomolecule for the bioplastic production [80]. Polylactic acid is an aliphatic polyester that can be derived from a wide range of petrochemicals but the most common and obvious sources of its production are starch or sugar rich plants and crops [81]. PLA has characteristics like high transparency and surface gloss that make it a replacement for synthetic plastics like PET and PVC [82].

Polylactic acid can be produced from lactide or lactic acid through the reaction as shown in Fig. 7 [34-36]. Dehydrative condensation of the lactic acid will yield polylactic acid while polylactic acid and lactide are interconvertible through a reverse reaction i.e. the ring opening polymerization of lactide will yield polylactic acid whereas the depolymerisation of the PLA will yield lactide molecule [40]. PLA has more recently got more attention due

Materials Research Forum LLC
https://doi.org/10.21741/9781644901335-9

to its ability to be used in food packaging where it is blended with other biopolymers to enhance its required characteristics [83].

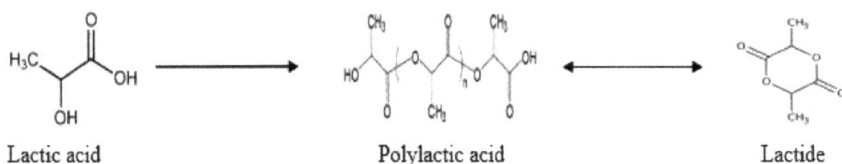

| Lactic acid | Polylactic acid | Lactide |

***Fig. 7** Conversion of lactic acid to polylactic acid through Dehydrative condensation and interconversion of PLA and Lactide through ring opening polymerization [34-36].*

12. Applications of bioplastics

Bioplastics are actually meant to replace the synthetic versions of conventional plastics that are accumulated in environment causing pollution and effecting thousands of species living in aquatic and terrestrial environments [84]. Synthetic plastics such as polyethylene (PE), polypropylene (PP), Polyvinyl chloride (PVC), polyethylene tetraphalate (PET) are used in daily life. Their multipurpose uses have increased their consumer demand, hence industry is producing millions of metrics of tones of plastic each year which ultimately results to environmental pollution and is also associated with health effects [85]. Bioplastics are primarily meant to replace conventional plastics therefore, they are desired to replace the applications as are provided by the petrochemically derived plastics. However due to their ecofriendly behaviour, biocompatibility bioplastics have found a wide range of application ranging from home appliance to food packaging material and from industrial products to pharmaceuticals see Table 7 [99-108] and Fig.8 [46,48,55-64].

Apart from the applications mentioned in the Table 7, protein based bioplastics which are derived from casein, gelatin and keratin impart special characteristics to the bioplastic product which include flexibility, strength and toughness [109]. Protein based biopolymers that can be used as bioplastic materials, have the potential applications in formation of food packaging, and as biomaterial for the reconstructive surgery and tissue engineering [110].

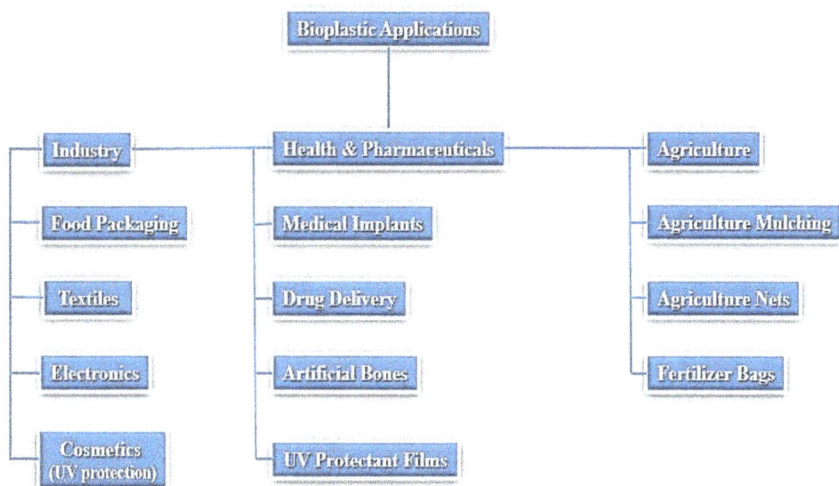

Fig. 8 *Three main application sectors of bioplastic i.e. Health, Industry and Agriculture [46,48,55-64].*

Table 7 *Some important applications of bioplastics derived from different derivatives [99-108].*

Bioplastic Derivative	Applications	References
Starch	Agriculture foils, Textiles, Food Packaging	[99-101]
Cellulose	Packaging, Medical devices, Electronic devices, Bio Films	[102-104]
Polyhydroxyalkanoates (PHAs)	Coating, Food packaging, Medical implant, Agricultural grow bags and fertilizer bags, Agriculture nets	[105-107]
Polylactic acid (PLA)	Food packaging	[108]

12.1 Antimicrobial films for food wrapping (fruits & vegetables)

One of the most important role of the non-toxicity which enables the development of bioplastic films for packaging food material is the use of antimicrobial films that can save the food such as fresh fruits and vegetables from being spoiled by bacterial or microbial attacks [111]. Chitosan based nano silver and Ag-ion incorporated bioplastic films have antimicrobial properties that enhance post processing food quality and safety [112].

Padil et al. [89] (2019) has analyzed the sustainable use of gum arabic as a potential bioplastic material by using it as a food wrapping biomaterial [89]. Gum arabic (GA)-based electrospunbioplastic fibers were fabricated using GA blended with PVA (2:1 wt. % of GA/PVA) via electrospinning process the potential of the electrospun fiber were further evaluated as a food wrapper [89].

12.2 Agricultural mulching

PHA made biofilms or plastic material can be used for agricultural mulching. Mulching is used to not only protect the crops but also increase its yield. It prevents soil structure and retain its moisture [113]. Natural and synthetic mulching are two types that are available but the expensiveness and inadequateness of natural material for the synthesis of natural mulch has increased and favored the use of synthetic mulching which are made up of high density polyethylene (HDPE) [91,114]. The increased rate of synthetic mulch has increased the rate of landfill and causes environmental pollution when buried or burnt. Bioplastics have addressed the very issue as PHA based mulch films are being developed on research level and are expected to help in reduction of use of synthetic mulch films [115].

12.3 Removal of heavy metals

Hungund et al. [93] (2018) developed and used polyhydroxyalkanoates (PHAs) composites such as PHB-Starch and PHB-PEG composites for the removal of heavy metals from the aqueous solution. These composites have very positively showed successful adsorption of Pb^{2+} ion.

13. Comparison of plastics & bioplastics

Although plastic and bioplastics are derived from completely distinct sources they can replace each other due to some of their physiochemical properties [94,116]. Replacing synthetic plastics with bioplastics in everyday life can save the planet earth from a number of environmental problems, including the continuous destruction of our water reservoirs and harm to sea life. Bioplastics are still under investigation on different levels of research and there have been studies around the world where bioplastic products have replaced the synthetic plastic materials however, this is still on very small scale [117-119]. Experts are of the view that decades of research is needed to enable bioplastics to replace the synthetic plastics used today and even then plastics will remain due to their large scale applications and their properties that cannot be completely replaced with bioplastics [120-122]. However, the adverse effects that are imposed by the plastics on the environment by the petrochemical-based plastics must be addressed in order to save the world from growing environmental issues such as pollution. In the present scenario, bioplastics can effectively help us to reduce plastic related pollution and its impacts on the ecosystem due to their invading and non-degradable behaviour. Bioplastic films for packaging can replace their synthetic counterparts [123]. Some medical devices, auto parts, and a large number of electronic equipment and parts can be replaced with bioplastics to an extent which will lower the use of synthetic plastics for the manufacturing of such products [124]. Therefore, the petrochemically derived, non-degradable and toxic plastic materials can be replaced by ecofriendly materials derived from biomass. A brief and comprehensive comparison between the properties of plastics and bioplastics is given in Table 8 [125-134].

Summary and conclusion

Plastics being the synthetic derivatives of the petrochemicals have evolved too much in the modern world where their production rates are far greater than their disposal rates. The non-degradability of the synthetic plastics is due to their physiochemical properties that enable them to become environmental pollutants. The plastics are divided into seven large categories depending upon their recyclability and some other properties. Researchers around the world have developed and tried different methods to degrade, recycle or dispose plastics in a sustainable way but the results have not been so positive. Biodegradation, thermal and catalytic degradation of plastic materials to convert them into some useful products is the hot topic nowadays but results need to be improved. Bioplastics which are derived from a large variety of biomass such as cellulose, starch, chitin, keratin protein and microbial polyhydroxyalkanoates are a new horizon to tackle the over production of plastic materials. Plastic materials can be effectively replaced with bio-derived plastic polymers that are not just environment friendly but will help to reduce the plastic pollution as these polymers actively degrade in natural environment without leaving toxic byproducts. It is believed that eventually in the future bioplastic research may help to produce environmental friendly and sustainable products.

Table 8 *Comparative properties of plastics and bioplastics [125-134]*

Property	Plastic	Bioplastic
Main constituents	Hydrocarbons from Petrochemical sources	A wide range of Biomass e.g. Cellulose, Starch, Chitin, Polyhydroxyalkanoates from bacteria etc.
Major categories	7	3
Degradability in natural environment	Non-degradable	Mostly Degradable
Melting temperatures	Mostly High	Moderate
Recycled for	Fuel production (By catalytic and thermal decomposition). Other products of daily use.	Different monomers that may be used again for developing the same plastic. Can be used as fertilizers.
Toxicity	Yes	Usually non-toxic
Applications for consumer product development	Very High	Low and under developed
Life Cycle Analysis (LCA) or Cradle to grave analysis		
Energy requirements MJ/kg	77-120 (For different plastics)	25-57 (For different biobased plastics)
Global warming Kg CO_2eq/Kg	2.70- 7.64 (For different plastics)	1.14-3.84

References

[1] S. Thakur, A. Verma, B. Sharma, J. Chaudhary, S. Tamulevicius, V.K. Thakur, Recent developments in recycling of polystyrene based plastics, Curr. Opin. Green Sustain. Chem. 13 (2018) 32–38. https://doi.org/10.1016/j.cogsc.2018.03.011

[2] R. Wei, W. Zimmermann, Microbial enzymes for the recycling of recalcitrant petroleum-based plastics: how far are we?, Microb. Biotechnol. 10 (2017) 1308–1322. https://doi.org/10.1111/1751-7915.12710

[3] R. Porta, The plastics sunset and the bio-plastics sunrise, Coatings. 9 (2019) 526. https://doi.org/10.3390/coatings9080526

[4] A. Malizia, A.C. Monmany-Garzia, Terrestrial ecologists should stop ignoring plastic pollution in the anthropocene time, Sci. Total Environ. 668 (2019) 1025–1029. https://doi.org/10.1016/j.scitotenv.2019.03.044

[5] A.M. Brandon, C.S. Criddle, Can biotechnology turn the tide on plastics?, Curr. Opin. Biotechnol. 57 (2019) 160–166. https://doi.org/10.1016/j.copbio.2019.03.020

[6]   S. Kumar, K. Thakur, Bioplastics - classification, production and their potential food applications, J. Hill Agric. 8 (2017) 118. https://doi.org/10.5958/2230-7338.2017.00024.6

[7]   E.B. Arikan, H.D. Ozsoy, A Review: Investigation of bioplastics, J. Civ. Eng. Archit. 9 (2015) 188–192. https://doi.org/10.17265/1934-7359/2015.02.007

[8]   M.W. Ryberg, A. Laurent, M. Hauschild, Mapping of global plastics value chain and plastics losses to the environment, United Nations Environ. Program. (2018) 5–99

[9]   F. Alshehrei, Biodegradation of synthetic and natural plastic by microorganisms, J. Appl. Environ. Microbiol. Vol. 5, 2017, Pages 8-19. 5 (2017) 8–19. https://doi.org/10.12691/JAEM-5-1-2

[10]   (WEF) World Economic Forum, Top 10 Emerging Technologies 2019, World Econ. Forum Annu. Meet. 2019. (2019) 4–15. www.weforum.org

[11]   R. Devi, V. Kannan, K. Natarajan, D. Nivas, K. Kannan, S. Chandru, A. Antony, The role of microbes in plastic degradation, 2015. https://doi.org/10.1201/b19243-13

[12]   X. Jia, C. Qin, T. Friedberger, Z. Guan, Z. Huang, Efficient and selective degradation of polyethylenes into liquid fuels and waxes under mild conditions, Sci. Adv. 2 (2016) 1–8. https://doi.org/10.1126/sciadv.1501591

[13]   B. Kunwar, H.N. Cheng, S.R. Chandrashekaran, B.K. Sharma, Plastics to fuel: a review, Renew. Sustain. Energy Rev. 54 (2016) 421–428. https://doi.org/10.1016/j.rser.2015.10.015

[14]   S. Skariyachan, A.A. Patil, A. Shankar, M. Manjunath, N. Bachappanavar, S. Kiran, Enhanced polymer degradation of polyethylene and polypropylene by novel thermophilic consortia of *Brevibacillus sps.* and *Aneurinibacillus sp.* screened from waste management landfills and sewage treatment plants, Polym. Degrad. Stab. 149 (2018) 52–68. https://doi.org/10.1016/j.polymdegradstab.2018.01.018

[15]   L. Giacomucci, N. Raddadi, M. Soccio, N. Lotti, F. Fava, Polyvinyl chloride biodegradation by *Pseudomonas citronellolis* and *Bacillus flexus*, N. Biotechnol. 52 (2019) 35–41. https://doi.org/10.1016/j.nbt.2019.04.005

[16]   C. Andreeßen, A. Steinbüchel, Recent developments in non-biodegradable biopolymers: Precursors, production processes, and future perspectives, Appl. Microbiol. Biotechnol. 103 (2019) 143–157. https://doi.org/10.1007/s00253-018-9483-6

[17]   M.C.M. Blettler, M.A. Ulla, A.P. Rabuffetti, N. Garello, Plastic pollution in freshwater ecosystems: macro-, meso-, and microplastic debris in a floodplain lake, Environ. Monit. Assess. 189 (2017). https://doi.org/10.1007/s10661-017-6305-8

[18]   C. Wabnitz, W.J. Nichols, Editorial: Plastic pollution: An ocean emergency, Mar. Turt. Newsl. (2010) 1–4. http://search.proquest.com/docview/924334169?accountid=27795

[19]   M. Vert, Y. Doi, K.-H. Hellwich, M. Hess, P. Hodge, P. Kubisa, M. Rinaudo, F. Schué, Terminology for biorelated polymers and applications (IUPAC Recommendations 2012), Pure Appl. Chem. 84 (2012) 377–410. https://doi.org/10.1351/pac-rec-10-12-04

[20]   L. Godfrey, Waste plastic, the challenge facing developing countries—Ban it, change it, collect it?, Recycling. 4 (2019) 3. https://doi.org/10.3390/recycling4010003

[21]   N.J. Beaumont, M. Aanesen, M.C. Austen, T. Börger, J.R. Clark, M. Cole, T. Hooper, P.K. Lindeque, C. Pascoe, K.J. Wyles, Global ecological, social and economic impacts of marine plastic, Mar. Pollut. Bull. 142 (2019) 189–195. https://doi.org/10.1016/j.marpolbul.2019.03.022

[22]   R. Scalenghe, Resource or waste? A perspective of plastics degradation in soil with a focus on end-of-life options, Heliyon. 4 (2018) e00941. https://doi.org/10.1016/j.heliyon.2018.e00941

[23]   S. Karbalaei, P. Hanachi, T.R. Walker, M. Cole, Occurrence, sources, human health impacts and mitigation of microplastic pollution, Environ. Sci. Pollut. Res. 25 (2018) 36046–36063. https://doi.org/10.1007/s11356-018-3508-7

[24]   J.R. Jambeck, R. Geyer, C. Wilcox, T.R. Siegler, M. Perryman, A. Andrady, R. Narayan, K.L. Law, Plastic waste inputs from land into the ocean, Science 80;347 (2015) 768–771. https://doi.org/10.1126/science.1260352

[25]   O.S. Ogunola, O.A. Onada, A.E. Falaye, Mitigation measures to avert the impacts of plastics and microplastics in the marine environment (a review), Environ. Sci. Pollut. Res. 25 (2018) 9293–9310. https://doi.org/10.1007/s11356-018-1499-z

[26]   M. Valentukevičienė, E. Brannvall, Marine pollution: an overviewJūros Geologija. 50 (2008) 17–23. https://doi.org/10.2478/v10056-008-0002-9

[27]   S. Sharma, S. Chatterjee, Microplastic pollution, a threat to marine ecosystem and human health: a short review, Environ. Sci. Pollut. Res. 24 (2017) 21530–21547. https://doi.org/10.1007/s11356-017-9910-8

[28]   L. Ivanova, K. Sokolov, G. Kharitonova, Plastic pollution tendencies of the Barents sea and adjacent waters under the climate change, Arct. North. 32 (2018) 121–145. https://doi.org/10.17238/issn2221-2698.2018.32.121

[29]   L.G.A. Barboza, A. Dick Vethaak, B.R.B.O. Lavorante, A.K. Lundebye, L. Guilhermino, Marine microplastic debris: An emerging issue for food security, food safety and human health, Mar. Pollut. Bull. 133 (2018) 336–348. https://doi.org/10.1016/j.marpolbul.2018.05.047

[30]   Y.H. Yang, C.J. Brigham, C.F. Budde, P. Boccazzi, L.B. Willis, M.A. Hassan, Z.A.M. Yusof, C. Rha, A.J. Sinskey, Optimization of growth media components for polyhydroxyalkanoate (PHA) production from organic acids by *Ralstonia eutropha*, Appl. Microbiol. Biotechnol. 87 (2010) 2037–2045. https://doi.org/10.1007/s00253-010-2699-8

[31]   Y. Yu, D. Zhou, Z. Li, C. Zhu, Advancement and challenges of microplastic pollution in the aquatic environment: A review, Water. air. soil Pollut. 229 (2018) 2–18. https://doi.org/10.1007/s11270-018-3788-z

[32]   A.C. Vegter, M. Barletta, C. Beck, J. Borrero, H. Burton, M.L. Campbell, M.F. Costa, M. Eriksen, C. Eriksson, A. Estrades, K.V.K. Gilardi, B.D. Hardesty, J.A. Ivar do Sul, J.L. Lavers, B. Lazar, L. Lebreton, W.J. Nichols, C.A. Ribic, P.G. Ryan, Q.A. Schuyler, S.D.A. Smith, H. Takada, K.A. Townsend, C.C.C. Wabnitz, C. Wilcox, L.C. Young, M. Hamann, Global research priorities to mitigate plastic pollution impacts on marine wildlife, Endanger. Species Res. 25 (2014) 225–247. https://doi.org/10.3354/esr00623

[33]   A.L. Andrady, Microplastics in the marine environment, Mar. Pollut. Bull. 62 (2011) 1596–1605. https://doi.org/10.1016/j.marpolbul.2011.05.030

[34]   S.G. Tetu, I. Sarker, V. Schrameyer, R. Pickford, L.D.H. Elbourne, L.R. Moore, I.T. Paulsen, Plastic leachates impair growth and oxygen production in *Prochlorococcus*, the ocean's most abundant photosynthetic bacteria, Commun. Biol. 2 (2019) 1–9. https://doi.org/10.1038/s42003-019-0410-x

[35]   D. Danso, J. Chow, W.R. Streita, Plastics: Environmental and biotechnological perspectives on microbial degradation, Appl. Environ. Microbiol. 85 (2019) 1–14. https://doi.org/10.1128/AEM.01095-19

[36]   G.J. Palm, L. Reisky, D. Böttcher, H. Müller, E.A.P. Michels, M.C. Walczak, L. Berndt, M.S. Weiss, U.T. Bornscheuer, G. Weber, Structure of the plastic-degrading *Ideonella sakaiensis* MHETase bound to a substrate, Nat. Commun. 10 (2019) 1–10. https://doi.org/10.1038/s41467-019-09326-3

[37]   M. Furukawa, N. Kawakami, A. Tomizawa, K. Miyamoto, Efficient degradation of Poly(ethylene terephthalate) with *Thermobifida fusca*cutinase exhibiting improved catalytic activity generated using mutagenesis and additive-based approaches, Sci. Rep. 9 (2019) 1–9. https://doi.org/10.1038/s41598-019-52379-z

[38]   T. Volova, E. Kiselev, N. Zhila, E. Shishatskaya, Synthesis of polyhydroxyalkanoates by hydrogen-oxidizing bacteria in apilot production process, Biomacromolecules. (2019). https://doi.org/10.1021/acs.biomac.9b00295

[39]   A. Iles, A.N. Martin, Expanding bioplastics production: Sustainable business innovation in the chemical industry, J. Clean. Prod. 45 (2013) 38–49. https://doi.org/10.1016/j.jclepro.2012.05.008

[40]   H. Karan, C. Funk, M. Grabert, M. Oey, B. Hankamer, Green bioplastics as part of a circular bioeconomy, Trends Plant Sci. 24 (2019) 237–249. https://doi.org/10.1016/j.tplants.2018.11.010

[41]   S. Beucker, F. Marscheider-Weidemann, Potentials and challenges of bioplastics – Insights from a German survey on " GreenFuture Markets", Borderstep Inst. Innov. Sustain. (2007) 1–7. http://www.borderstep.de/wpcontent/uploads/2014/12/beucker_marscheider_potentials _and_challenges_of_bioplastics_2007.pdf

[42]   M.V. Reddy, Y. Mawatari, R. Onodera, Y. Nakamura, Y. Yajima, Y.-C. Chang, Bacterial conversion of waste into polyhydroxybutyrate (PHB): A new approach of bio-circular economy for treating waste and energy generation, Bioresour. Technol. Reports. 7 (2019) 100246. https://doi.org/10.1016/j.biteb.2019.100246

[43]   M.. Paridah, A. Moradbak, A. Mohamed, F. abdulwahab taiwo Owolabi, M. Asniza, S.H.. Abdul Khalid, Prospective biodegradable plastics from biomass conversion processes. Intech. 13 (2016)https://doi.org/http://dx.doi.org/10.5772/57353

[44]   R.J.K. Helmes, A.M. López-Contreras, M. Benoit, H. Abreu, J. Maguire, F. Moejes, S.W.K. van den Burg, Environmental impacts of experimental production of lactic acid for bioplastics from *Ulva spp*, Sustain. 10 (2018) 1–15. https://doi.org/10.3390/su10072462

[45]   S. Spierling, C. Röttger, V. Venkatachalam, M. Mudersbach, C. Herrmann, H.J. Endres, Bio-based Plastics - A building block for the circular economy?, Procedia CIRP. 69 (2018) 573–578. https://doi.org/10.1016/j.procir.2017.11.017

[46]   M. He, X. Wang, Z. Wang, L. Chen, Y. Lu, X. Zhang, M. Li, Z. Liu, Y. Zhang, H. Xia, L. Zhang, Biocompatible and biodegradable bioplastics constructed from chitin

via a "green" pathway for bone repair, ACS Sustain. Chem. Eng. 5 (2017) 9126–9135. https://doi.org/10.1021/acssuschemeng.7b02051

[47] P. Trivedi, A. Hasan, S. Akhtar, M. Haris Siddiqui, U. Sayeed, M. Kalim, A. Khan, Role of microbes in degradation of synthetic plastics and manufacture of bioplastics, J. Chem. Pharm. Res. 8 (2016) 211–216.

[48] C. Rajendran, N. , Sharanya Puppala, Sneha Raj M., Ruth Angeeleena B., and Rajam, Seaweeds can be a new source for bioplastics, J. Pharm. Res. 5 (2012) 1476–1479

[49] O. Valerio, J.M. Pin, M. Misra, A.K. Mohanty, Synthesis of glycerol-based biopolyesters as toughness enhancers for polylactic acid bioplastic through reactive extrusion, ACS Omega. 1 (2016) 1284–1295. https://doi.org/10.1021/acsomega.6b00325

[50] M.N. Somleva, K.D. Snell, J.J. Beaulieu, O.P. Peoples, B.R. Garrison, N.A. Patterson, Production of polyhydroxybutyrate in switchgrass, a value-added co-product in an important lignocellulosic biomass crop, Plant Biotechnol. J. 6 (2008) 663–678. https://doi.org/10.1111/j.1467-7652.2008.00350.x

[51] B.T. Ho, T.K. Roberts, S. Lucas, An overview on biodegradation of polystyrene and modified polystyrene: the microbial approach, Crit. Rev. Biotechnol. 38 (2018) 308–320. https://doi.org/10.1080/07388551.2017.1355293

[52] Maulida, M. Siagian, P. Tarigan, Production of starch based bioplastic from cassava peel reinforced with microcrystalline cellulose avicel PH101 using sorbitol as plasticizer, J. Phys. Conf. Ser. 710 (2016). https://doi.org/10.1088/1742-6596/710/1/012012

[53] S. Mohapatra, S. Maity, H.R. Dash, S. Das, S. Pattnaik, C.C. Rath, D. Samantaray, Bacillus and biopolymer: Prospects and challenges, Biochem. Biophys. Reports. 12 (2017) 206–213. https://doi.org/10.1016/j.bbrep.2017.10.001

[54] J. Gonzalez-Gutierrez, P. Partal, M. Garcia-Morales, C. Gallegos, Development of highly-transparent protein/starch-based bioplastics, Bioresour. Technol. 101 (2010) 2007–2013. https://doi.org/10.1016/j.biortech.2009.10.025

[55] I. Reiniati, A.N. Hrymak, A. Margaritis, Recent developments in the production and applications of bacterial cellulose fibers and nanocrystals, Crit. Rev. Biotechnol. 37 (2017) 510–524. https://doi.org/10.1080/07388551.2016.1189871

[56] M. Brodin, M. Vallejos, M.T. Opedal, M.C. Area, G. Chinga-Carrasco, Lignocellulosics as sustainable resources for production of bioplastics – A review, J. Clean. Prod. 162 (2017) 646–664. https://doi.org/10.1016/j.jclepro.2017.05.209

[57] Isroi, A. Cifriadi, T. Panji, N.A. Wibowo, K. Syamsu, Bioplastic production from cellulose of oil palm empty fruit bunch, IOP Conf. Ser. Earth Environ. Sci. 65 (2017). https://doi.org/10.1088/1755-1315/65/1/012011

[58] R. Villa-Rojas, A. Valdez-Fragoso, H. Mújica-Paz, Manufacturing Methods and Engineering properties of pectin-based nanobiocomposite films, Food Eng. Rev. 10 (2018) 46–56. https://doi.org/10.1007/s12393-017-9163-9

[59] S. Chodijah, A. Husaini, M. Zaman, Hilwatulisan, Extraction of pectin from banana peels (*Musa Paradiasica Fomatypica*) for biodegradable plastic films, J. Phys. Conf. Ser. 1167 (2019). https://doi.org/10.1088/1742-6596/1167/1/012061

[60] S.L. Pandharipande, P.H. Bhagat, Synthesis of chitin from crab shells and its utilization in preparation of nanostructured film, Int. J. Sci. Eng. Technol. Res. 5 (2016) 2278–7798

[61] Z.A. Raza, S. Abid, I.M. Banat, Polyhydroxyalkanoates: Characteristics, production, recent developments and applications, Int. Biodeterior. Biodegrad. 126 (2018) 45–56. https://doi.org/10.1016/j.ibiod.2017.10.001

[62] C.J.R. Verbeek, L.E. van den Berg, Recent developments in thermo-mechanicalprocessing of proteinous bioplastics, recent patents Mater. Sci. 2 (2010) 171–189. https://doi.org/10.2174/1874465610902030171

[63] M.P. Ryan, G. Walsh, The biotechnological potential of whey, Rev. Environ. Sci. Biotechnol. 15 (2016) 479–498. https://doi.org/10.1007/s11157-016-9402-1

[64] S. Sharma, I. Luzinov, Whey based binary bioplastics, J. Food Eng. 119 (2013) 404–410. https://doi.org/10.1016/j.jfoodeng.2013.06.007

[65] A. Jerez, P. Partal, I. Martínez, C. Gallegos, A. Guerrero, Protein-based bioplastics: Effect of thermo-mechanical processing, Rheol. Acta. 46 (2007) 711–720. https://doi.org/10.1007/s00397-007-0165-z

[66] B. Chalermthai, W.Y. Chan, J.R. Bastidas-Oyanedel, H. Taher, B.D. Olsen, J.E. Schmidt, Preparation and characterization of whey protein-based polymers produced from residual dairy streams, Polymers (Basel). 11 (2019). https://doi.org/10.3390/polym11040722

[67] M. Javanmard, Biodegradable whey protein edible films as a new biomaterials for food and drug packaging, Iran. J. Pharm. Sci. 5 (2009) 129–134.

[68]   M. Helgeson, W. Graves, D. Grewell, G. Srinivasan, Degradation and ntrogen release of zein-based bioplastic containers, J. Environ. Hortic. 27 (2009) 123–127. http://www.hriresearch.org/docs/publications/JEH/JEH_2009/JEH_2009_27_2/JEH 27-2-123-127.pdf

[69]   S. Aziz, L. Hosseinzadeh, E. Arkan, A.H. Azandaryani, Preparation of electrospun nanofibers based on wheat gluten containing azathioprine for biomedical application, Int. J. Polym. Mater. Polym. Biomater. 68 (2019) 639–646. https://doi.org/10.1080/00914037.2018.1482464

[70]   M. Koller, A. Salerno, M. Dias, A. Reiterer, G. Braunegg, Modern biotechnological polymer synthesis: A review, Food Technol. Biotechnol. 48 (2010) 255–269

[71]   R. Jain, S. Kosta, A. Tiwari, Polyhydroxyalkanoates: A way to sustainable development of bioplastics, Chronicles Young Sci. 1 (2010) 10–15. https://doi.org/10.4103/4444-4443.76448

[72]   M.E. Grigore, R.M. Grigorescu, L. Iancu, R.M. Ion, C. Zaharia, E.R. Andrei, Methods of synthesis, properties and biomedical applications of polyhydroxyalkanoates: a review, J. Biomater. Sci. Polym. Ed. 30 (2019) 695–712. https://doi.org/10.1080/09205063.2019.1605866

[73]   B. Johnston, I. Radecka, D. Hill, E. Chiellini, V.I. Ilieva, W. Sikorska, M. Musioł, M. Zięba, A.A. Marek, D. Keddie, B. Mendrek, S. Darbar, G. Adamus, M. Kowalczuk, The microbial production of Polyhydroxyalkanoates from waste polystyrene fragments attained using oxidative degradation, Polymers (Basel). 10 (2018). https://doi.org/10.3390/polym10090957

[74]   J. Mozejko-Ciesielska, K. Szacherska, P. Marciniak, Pseudomonas species as producers of eco-friendly polyhydroxyalkanoates, J. Polym. Environ. 27 (2019) 1151–1166. https://doi.org/10.1007/s10924-019-01422-1

[75]   T. Keshavarz, I. Roy, Polyhydroxyalkanoates: bioplastics with a green agenda, Curr. Opin. Microbiol. 13 (2010) 321–326. https://doi.org/10.1016/j.mib.2010.02.006

[76]   M. Kootstra, H. Elissen, S. Huurman, PHA's (Polyhydroxyalkanoates): General information on structure and raw materials for their production, 2017. https://doi.org/http://library.wur.nl/WebQuery/wurpubs/fulltext/414011

[77]   T.V. Ojumu, J. Yu, B.O. Solomon, Production of polyhydroxyalkanoates , a bacterial biodegradable polymer, African J. Biotechnol. 3 (2004) 18–24

[78]   L. Favaro, M. Basaglia, J.E.G. Rodriguez, A. Morelli, O. Ibraheem, V. Pizzocchero, S. Casella, Bacterial production of PHAs from lipid-rich by-products, Appl. Food Biotechnol. 6 (2019) 45–52. https://doi.org/10.22037/AFB.V6I1.22246

[79]   K. Dietrich, M.J. Dumont, L.F. Del Rio, V. Orsat, Producing PHAs in the bioeconomy — towards a sustainable bioplastic, Sustain. Prod. Consum. 9 (2017) 58–70. https://doi.org/10.1016/j.spc.2016.09.001

[80]   K.M. Nampoothiri, N.R. Nair, R.P. John, An overview of the recent developments in polylactide (PLA) research, Bioresour. Technol. 101 (2010) 8493–8501. https://doi.org/10.1016/j.biortech.2010.05.092

[81]   R. Coles, M. Kay, J. Song, Bioplastics, Adv. Biochem. Eng. Biotechnol. 166 (2010) 427–468. https://doi.org/10.1007/10_2016_75

[82]   S.M. Emadian, T.T. Onay, B. Demirel, Biodegradation of bioplastics in natural environments, Waste Manag. 59 (2017) 526–536. https://doi.org/10.1016/j.wasman.2016.10.006

[83]   P. Scarfato, L. Di Maio, L. Incarnato, Recent advances and migration issues in biodegradable polymers from renewable sources for food packaging, J. Appl. Polym. Sci. 132 (2015). https://doi.org/10.1002/app.42889

[84]   N. Benn, D. Zitomer, Pretreatment and Anaerobic Co-digestion of Selected PHB and PLA Bioplastics, Front. Environ. Sci. 5 (2018) 1–9. https://doi.org/10.3389/fenvs.2017.00093

[85]   R. Geyer, J.R. Jambeck, K.L. Law, Production, use, and fate of all plastics ever made., Sci. Adv. 3 (2017) e1700782. https://doi.org/10.1126/sciadv.1700782

[86]   S. Mangaraj, A. Yadav, L.M. Bal, S.K. Dash, N.K. Mahanti, Application of Biodegradable polymers in food packaging industry: A comprehensive review, J. Packag. Technol. Res. 3 (2019) 77–96. https://doi.org/10.1007/s41783-018-0049-y

[87]   L.A.M. Van Den Broek, R.J.I. Knoop, F.H.J. Kappen, C.G. Boeriu, Chitosan films and blends for packaging material, Carbohydr. Polym. 116 (2015) 237–242. https://doi.org/10.1016/j.carbpol.2014.07.039

[88]   J.W. Rhim, H.M. Park, C.S. Ha, Bio-nanocomposites for food packaging applications, Prog. Polym. Sci. 38 (2013) 1629–1652. https://doi.org/10.1016/j.progpolymsci.2013.05.008

[89]   V.V.T. Padil, C. Senan, S. Waclawek, M. Černík, S. Agarwal, R.S. Varma, Bioplastic fibers from gum arabic for greener food wrapping applications, ACS

Sustain. Chem. Eng. 7 (2019) 5900–5911.
https://doi.org/10.1021/acssuschemeng.8b05896

[90]   J. Rydz, W. Sikorska, M. Kyulavska, D. Christova, Polyester-based (bio)degradable polymers as environmentally friendly materials for sustainable development, Int. J. Mol. Sci. 16 (2015) 564–596. https://doi.org/10.3390/ijms16010564

[91]   H.Y. Sintim, M. Flury, Is biodegradable plastic mulch the solution to agriculture's plastic problem?, Environ. Sci. Technol. 51 (2017) 1068–1069. https://doi.org/10.1021/acs.est.6b06042

[92]   T.S.M. Amelia, A.M. Sharumathiy Govindasamy, Tamothran, and K.B. Sevakumaran Vigneswari, Biotechnological applications of polyhydroxyalkanoates. Applications of PHA in Agriculture, (2019) 347–361. https://doi.org/10.1007/978-981-13-3759-8

[93]   B.S. Hungund, S.G. Umloti, K.P. Upadhyaya, J. Manjanna, S. Yallappa, N.H. Ayachit, Development and characterization of polyhydroxybutyrate biocomposites and their application in the removal of heavy metals, Mater. Today Proc. 5 (2018) 21023–21029. https://doi.org/10.1016/j.matpr.2018.6.495

[94]   F. Alshehrei, Biodegradation of synthetic and natural plastic by microorganisms, J. Appl. Environ. Microbiol. Vol. 5, 2017, Pages 8-19. 5 (2017) 8–19. https://doi.org/10.12691/JAEM-5-1-2

[95]   N.A. Mostafa, A.A. Farag, H.M. Abo-dief, A.M. Tayeb, Production of biodegradable plastic from agricultural wastes, Arab. J. Chem. 11 (2018) 546–553. https://doi.org/10.1016/j.arabjc.2015.04.008

[96]   T. Ahmed, M. Shahid, F. Azeem, I. Rasul, A.A. Shah, M. Noman, A. Hameed, N. Manzoor, M. Muhammad, Irfan Manzoor Sher, Biodegradation of plastics: current scenario and future prospects for environmental safety, Environ. Sci. Pollut. Res. 25 (2018) 7287–7298. https://doi.org/10.1007/s11356-018-1234-9

[97]   R.A. Wilkes, L. Aristilde, Degradation and metabolism of synthetic plastics and associated products by *Pseudomonas* sp.: capabilities and challenges, J. Appl. Microbiol. 123 (2017) 582–593. https://doi.org/10.1111/jam.13472

[98]   A.K. Urbanek, W. Rymowicz, A.M. Mirończuk, Degradation of plastics and plastic-degrading bacteria in cold marine habitats, Appl. Microbiol. Biotechnol. 102 (2018) 7669–7678. https://doi.org/10.1007/s00253-018-9195-y

[99] G. Caruso, Plastic degrading microorganisms as a tool for bioremediation of plastic contamination in aquatic environments, J. Pollut. Eff. Control. 03 (2015). https://doi.org/10.4172/2375-4397.1000e112

[100] R. Patil, U.S. Bagde, Isolation of polyvinyl chloride degrading bacterial strains from environmental samples using enrichment culture technique, African J. Biotechnol. 11 (2012) 7947–7956. https://doi.org/10.5897/ajb11.3630

[101] T.C.H. Dang, D.T. Nguyen, H. Thai, T.C. Nguyen, T.T.H. Tran, V.H. Le, V.H. Nguyen, X.B. Tran, . T. P. T. Pham, T. G Nguyen, Plastic degradation by thermophilic Bacillus sp. BCBT21 isolated from composting agricultural residual in Vietnam, Adv. Nat. Sci. Nanosci. Nanotechnol. 9 (2018) 015014. https://doi.org/10.1088/2043-6254/aaabaf

[102] P. Sriyapai, K. Chansiri, T. Sriyapai, Isolation and characterization of polyester-based plastics-degrading bacteria from compost soils, Microbio. 87 (2018) 290–300. https://doi.org/10.1134/s0026261718020157

[103] S. Pathak, C.L.Rp. Sneha, B.B. Mathew, Bioplastics : Its timeline based scenario &challengesJ. Polym. Biopolym. Chem. 2 (2014) 84–90. https://doi.org/10.12691/jpbpc-2-4-5

[104] R. Mülhaupt, Green polymer chemistry and bio-based plastics: Dreams and reality, Macromol. Chem. Phys. 214 (2013) 159–174. https://doi.org/10.1002/macp.201200439

[105] N. Jabeen, I. Majid, G.A. Nayik, Bioplastics and food packaging: A review, Cogent Food Agric. 1 (2015) 1–6. https://doi.org/10.1080/23311932.2015.1117749

[106] F.I. Khan, L. Aktar, T. Islam, M.L. Saha, Isolation and Identification of Indigenouspoly-β-Hydroxybutyrate (PHB) producing bacteria from different waste materials, Plant Tissue Cult. Biotechnol. 29 (2019) 15–24. https://doi.org/10.3329/ptcb.v29i1.41975

[107] S. Rohner, J. Humphry, C.M. Chaléat, L.J. Vandi, D.J. Martin, N. Amiralian, M.T. Heitzmann, Mechanical properties of polyamide 11 reinforced with cellulose nanofibres from Triodia pungens, Cellulose. 25 (2018) 2367–2380. https://doi.org/10.1007/s10570-018-1702-x

[108] T.M.M.M. Amaro, D. Rosa, G. Comi, L. Iacumin, Prospects for the use of whey for polyhydroxyalkanoate (PHA) production, Front. Microbiol. 10 (2019) 1–12. https://doi.org/10.3389/fmicb.2019.00992

[109] P. Carlozzi, A. Giovannelli, M.L. Traversi, E. Touloupakis, T. Di Lorenzo, Poly-3-hydroxybutyrate and H_2 production by *Rhodopseudomonas* sp. S16-VOGS3 grown in a new generation photobioreactor under single or combined nutrient deficiency, Int. J. Biol. Macromol. 135 (2019) 821–828. https://doi.org/10.1016/j.ijbiomac.2019.05.220

[110] A.H.M. Fauzi, L.W. Yoon, T. Nittami, H.K. Yeohd, Enrichment of PHA-accumulators for sustainable PHA production from crude glycerol, Process Saf. Environ. Prot. 122 (2019) 200–208. https://doi.org/10.1016/j.psep.2018.12.002

[111] H. Al-Battashi, N. Annamalai, S. Al-Kindi, A.S. Nair, S. Al-Bahry, J.P. Verma, N. Sivakumar, Production of bioplastic (poly-3-hydroxybutyrate) using waste paper as a feedstock: Optimization of enzymatic hydrolysis and fermentation employing Burkholderia sacchari, J. Clean. Prod. 214 (2019) 236–247. https://doi.org/10.1016/j.jclepro.2018.12.239

[112] R.R. Dalsasso, F.A. Pavan, S.E. Bordignon, G.M.F. de Aragão, P. Poletto, Polyhydroxybutyrate (PHB) production by *Cupriavidus necator* from sugarcane vinasse and molasses as mixed substrate, Process Biochem. (2019) 0–1. https://doi.org/10.1016/j.procbio.2019.07.007

[113] A.O. Pérez-Arauz, A.E. Aguilar-Rabiela, A. Vargas-Torres, A.I. Rodríguez-Hernández, N. Chavarría-Hernández, B. Vergara-Porras, M.R. López-Cuellar, Production and characterization of biodegradable films of a novel polyhydroxyalkanoate (PHA) synthesized from peanut oil, Food Packag. Shelf Life. 20 (2019) 100297. https://doi.org/10.1016/j.fpsl.2019.01.001

[114] T.H. Nguyen, F. Ishizuna, Y. Sato, H. Arai, M. Ishii, Physiological characterization of poly-β-hydroxybutyrate accumulation in the moderately thermophilic hydrogen-oxidizing bacterium *Hydrogenophilus thermoluteolus* TH-1, J. Biosci. Bioeng. 127 (2019) 686–689. https://doi.org/10.1016/j.jbiosc.2018.11.011

[115] T. Yamaguchi, J. Narsico, T. Kobayashi, A. Inoue, T. Ojima, Production of poly(3-hydroyxybutylate) by a novel alginolytic bacterium *Hydrogenophaga* sp. strain UMI-18 using alginate as a sole carbon source, J. Biosci. Bioeng. 128 (2019) 203–208. https://doi.org/10.1016/j.jbiosc.2019.02.008

[116] I. Pernicova, Vojtech Enev, Ivana Marova, Stanislav Obruca, Interconnection of waste chicken feather biodegradation and keratinase and mcl-PHA production employing *Pseudomonas putida* KT2440, Appl. Food Biotechnol. 6 (2018) 83–90. https://doi.org/10.22037/afb.v6i1.21429

[117] Z.N. Terzopoulou, G.Z. Papageorgiou, E. Papadopoulou, E. Athanassiadou, E. Alexopoulou, D.N. Bikiaris, Green composites prepared from aliphatic polyesters and

bast fibers, Ind. Crops Prod. 68 (2015) 60–79.
https://doi.org/10.1016/j.indcrop.2014.08.034

[118] K. Preethi, M. Umesh, Water Hyacinth : A potential substrate for bioplastic (PHA) production using *Pseudomonas aeruginosa*, Int. Joural Appl. Res. 1 (2015) 349–354

[119] J. Yaradoddi, V. Patil, S. Ganachari, N. Banapurmath, A. Hunashyal, A. Shettar, Biodegradable plastic production from fruit waste material and its sustainable use for green applications, Int. J. Pharm. Res. Allied Sci. 5 (2016) 55–66

[120] R.G. Saratale, G.D. Saratale, S.K. Cho, D.S. Kim, G.S. Ghodake, A. Kadam, G. Kumar, R.N. Bharagava, R. Banu, H.S. Shin, Pretreatment of kenaf (*Hibiscus cannabinus L.*) biomass feedstock for polyhydroxybutyrate (PHB) production and characterization, Bioresour. Technol. 282 (2019) 75–80. https://doi.org/10.1016/j.biortech.2019.02.083

[121] C. Choi, J.P. Nam, J.W. Nah, Application of chitosan and chitosan derivatives as biomaterials, J. Ind. Eng. Chem. 33 (2016) 1–10. https://doi.org/10.1016/j.jiec.2015.10.028

[122] Y.E. Agustin, K.S. Padmawijaya, Effect of glycerol and zinc oxide addition on antibacterial activity of biodegradable bioplastics from chitosan-kepok banana peel starch, IOP Conf. Ser. Mater. Sci. Eng. 223 (2017). https://doi.org/10.1088/1757-899X/223/1/012046

[123] S. Domenek, P. Feuilloley, J. Gratraud, M.H. Morel, S. Guilbert, Biodegradability of wheat gluten based bioplastics, Chemosphere. 54 (2004) 551–559. https://doi.org/10.1016/S0045-6535(03)00760-4

[124] N. Ramakrishnan, S. Sharma, A. Gupta, B.Y. Alashwal, Keratin based bioplastic film from chicken feathers and its characterization, Int. J. Biol. Macromol. 111 (2018) 352–358. https://doi.org/10.1016/j.ijbiomac.2018.01.037

[125] S. Van Vlierberghe, E. Vanderleyden, V. Boterberg, P. Dubruel, Gelatin functionalization of biomaterial surfaces: Strategies for immobilization and visualization, Polymers (Basel). 3 (2011) 114–130. https://doi.org/10.3390/polym3010114

[126] S. Khalid, L. Yu, L. Meng, H. Liu, A. Ali, L. Chen, Poly(lactic acid)/starch composites: Effect of microstructure and morphology of starch granules on performance, J. Appl. Polym. Sci. 134 (2017) 1–12. https://doi.org/10.1002/app.45504

[127] S.F. Williams, S. Rizk, D.P. Martin, Poly-4-hydroxybutyrate (P4HB): A new generation of resorbable medical devices for tissue repair and regeneration, Biomed. Tech. 58 (2013) 439–452. https://doi.org/10.1515/bmt-2013-0009

[128] I.S. Sidek, S. Fauziah, S. Draman, S. Rozaimah, S. Abdullah, N. Anuar,Current development on bioplastics and its future: An introductory review, I Tech Mag. 1 (2019) 3-8.https://doi.org/http://doi.org/10.26480/itechmag.01.2019.03.08 CURRENT

[129] V. Mehta, M. Darshan, D. Nishith, Can a starch based plastic be an option of environmental friendly plastic?, J. Glob. Biosci. 3 (2014) 681–685.

[130] W.J. Orts, J. Shey, S.H. Imam, G.M. Glenn, M.E. Guttman, J.F. Revol, Application of cellulose microfibrils in polymer nanocomposites, J. Polym. Environ. 13 (2005) 301–306. https://doi.org/10.1007/s10924-005-5514-3

[131] V.K. Modi, Y. Shrives, C. Sharma, P.K. Sen, Review on green polymer nanocomposite and their applications, Int. J. Innov. Res. Sci. Eng. Technol. 2014 (2015) 17651–17656. https://doi.org/10.15680/IJIRSET.2014.0311079

[132] K. Khosravi-Darani, D.Z. Bucci, Application of poly(hydroxyalkanoate) in food packaging: Improvements by nanotechnology, Chem. Biochem. Eng. Q. 29 (2015) 275–285. https://doi.org/10.15255/CABEQ.2014.2260

[133] I.M. Shamsuddin, J.A. Jafar, A.S.A. Shawai, S. Yusuf, M. Lateefah, I. Aminu, Bioplastics as better alternative to petroplastics and their role in national sustainability: A review, Adv. Biosci. Bioeng. 5 (2017) 63.https://doi.org/10.11648/j.abb.20170504.13

[134] F. Muneer, I. Rasul, F. Azeem, M.H. Siddique, M. Zubair, H. Nadeem, Microbial Polyhydroxyalkanoates (PHAs): Efficient Replacement of Synthetic Polymers, J PolymEnviron. (2020)28: 2301–2323. https://doi.org/10.1007/s10924-020-01772-1

Degradation of Plastics
Materials Research Foundations **99** (2021) 238-268

Materials Research Forum LLC
https://doi.org/10.21741/9781644901335-10

Chapter 10

Versatile Applications of Degradable Plastic

Anan Ashrabi Ananno[1], Sami Ahbab Chowdhury[3], Mahadi Hasan Masud[*,2,3], Peter Dabnichki[2]

[1]Department of Management and Engineering, Linköping University, SE-581 83, Linköping, Sweden

[2]School of Engineering, RMIT University, Bundoora Campus, Melbourne, VIC-3083

[3]Department of Mechanical Engineering, Rajshahi University of Engineering and Technology, Rajshahi-6204, Bangladesh

*masud.08ruet@gmail.com

Abstract

In the last 50 years, plastics has become a favorite industry for packaging materials for their ease of manufacture and excellent performance. The advancement of food, electronics, automobile, medical and agricultural industries has increased the demand for packaging and casing materials made of large hydrocarbon polymers. Since plastics show resistance to biodegradation, they pose considerable threats to the environment. Degradable plastics and biopolymers offer promising solutions to this problem. Degradable plastics can be easily absorbed in the environment while exhibiting the properties of conventional plastics. There are three types of biopolymers according to their source: biomass extracted polymers, synthesized from microorganisms and produced from bio-derived monomers. Biodegradable plastics are commonly used in one-off packaging such as crockery, food service containers and cutlery. Although biodegradable plastics can replace conventional plastics in a lot of applications, their performance and cost are sometimes problematic. This chapter analyses the growth of the degradable plastic industry and explores their potential applications.

Keywords

Degradable Plastic, Plastic Waste, Industrial, Agricultural, Consumer, Energy, Waste Management

Contents

1.   Introduction

Public and government awareness of using sustainable products has increased considerably over the last decade. This provided an impetus for researchers to develop sustainable technologies and eco-friendly materials. A combination of climate change and environmental destruction threats pressed governments to apply for strict industry regulations to limit their plastic pollution [1]. This encouraged the research and development of degradable plastics that dissolve in the environment at the end of their life cycle without any negative impact. Understanding the application aspects and sprouting trends of this promising technology will aid investors, policymakers, and researchers in investing further advance the research on degradable plastic.

In 1907, Bakelite - the first synthetic plastic was produced, which paved the way for the plastic industrial revolution in the 1950s. However, within a span of only 65 years, the production of plastic has increased 200 folds and reached an annual production rate of 381 million tons in 2015. Disposing of this enormous amount of plastic waste is near impossible; therefore, it is essentials that the produced plastic can decay naturally or be disposed using inexpensive physio-chemical treatments. The production of plastics by industrial sectors is shown in Fig. 1 [2].

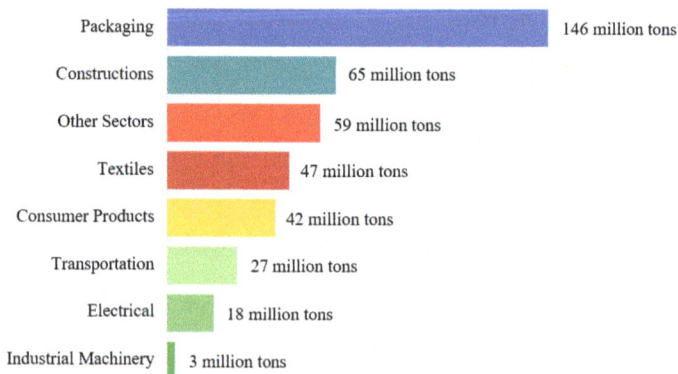

Fig. 1 *Production of plastics by industrial sectors, 2015[2].*

From Fig. 1, it is evident that a considerable amount of plastic is produced for packaging, which is essentially single-use disposable plastic. Research has shown that photodegradable and thermally degradable plastics have the ideal properties for being potential packaging material [3,4]. After their useful life cycle, these plastics degrade naturally in the presence of light and heat.

This study provides a bibliographic analysis of the potential application of degradable plastic in various industrial sectors. Additionally, the chapter highlights existing research landscape and predicts the future developing trends focusing on the contexts of degradable plastics such as polylactic acid (PLA), polycaprolactone (PCL), polyhydroxy alkenoates (PHAs), polybutylene adipate-co-terephthalate (PBAT) and polybutylene succinate (PBS). More specifically, PLA is at the top preference for both industry and academy, and PHAs are in almost all international biodegradable polymer patents. However, in technological terms, PLA is the fastest expanding technology in terms of both technology strength and maturity whereas PBS, PCL and PHAs are uncertain in the matter of technology and PBAT has less developmental potential.

The chapter discusses possible industrial applications based on the physio-chemical properties of each type of degradable plastics. Section 2 highlights the negative impact of plastic waste, providing a detailed view of the plastic waste scenario. Various applications of degradable plastics are discussed in section 3. In section 4, the growth of the degradable plastic industry is studied to assess its potential market share. Section 5 summarizes the findings.

2.   Plastic waste and its adverse effects

2.1   Plastic waste scenario

Different aerospace and automobile parts, health and medical industry, clothing, pipes and fittings in buildings, engineering parts, plastic home appliances, and food packaging, etc. sectors widely use plastic. Due to this immense dependence on plastics, their overall production has reached almost 8 billion tons [5], and the production rate is increasing exponentially. Due to the absence of proper waste management, almost all plastics are accumulated on land or in the ocean. Some of the most common ways of plastic disposal are landfilling, open dumping and burning or littering [6]. These conventional ways of plastic disposal cause severe damage to the environment [5]. The plastic products are also washed into water bodies causing significant harm to marine lives.

The production of plastic is gradually increasing as only a meagre amount isadequately recycled [5]. In developed countries, about 98% of plastic waste is collected, whereas in

Degradation of Plastics   Materials Research Forum LLC
Materials Research Foundations **99** (2021) 238-268   https://doi.org/10.21741/9781644901335-10

developing countries, the share falls to 41%, and the rest is littered or burned. In the USA, the recycling rate for plastics is about 9%, and in Europe, the amount is up to 30%. As packaging is a primary application of plastic, about 24-42% of packaging plastic is recycled in developed countries [5].

2.2 Effects of plastic waste

2.2.1 Landfills

In the whole world, about 22-43% of plastics are disposed in landfills, which is one of the cheapest methods of disposal [7]. Such landfilling practices reduce the capacity of potential land that can be used for other productive purposes. Although landfilling is not a recommended practice but is common among many countries, for example, in the USA most unrecycled plastics (about 90%) are landfilled, and in the EU, this landfill rate is about 31%[8]. In developing countries, open landfills and dumps is a common scenario [9]. This causes severe damage to the environment [10]. Plastic production will increase with urbanization and that will bring upon even more wastes in the future [10, 11].

2.2.2 In ocean

According to Jambeek et al. [12] about 50% of the plastic waste that enters the ocean, comes from five south Asian countries. Plastic waste of at least 100 million tons is predicted to enter the ocean in the period 2010- 2025 [5]. It is predicted that by 2050, the ocean will contain more plastic than fish by weight [13]. Cellulose acetate, polyethylene (PE), polypropylene (PP) are some of the common plastic wastes found in the ocean [14]. The used plastic enters the ocean body through wastewater discharge points, coastal areas and rivers etc. Due to pollution caused in the ocean, the following problems occur:

- The tourism industry is negatively affected
- Marine life is damaged and threatened by extinction
- Coastal species numbers are severely depleted
- Interference with boats cause accidents
- Organic pollution

Fig. 2 shows the impact of several waste materials on marine animals [14].

Plastic waste entanglement and ingestion leads to choking and starvation of marine animals such as sea birds, whales, fish, dolphins etc. The global environmental impact of plastic waste causes an estimated economic loss of about 40 billion USD per annum.

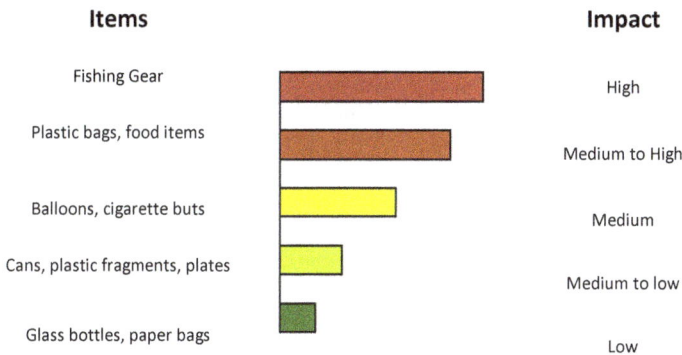

Fig. 2 *Comparative effects of several waste materials on marine animals [14]*

3.   Application of degradable plastic

3.1   Packaging

Plastics are used widely as packaging materials over the last 50 years. With the expansion of the food processing industry, the demand for plastics is also increasing. This caused an alarming ecological issue as conventional plastics resist biodegradation.

Some of the conventional packaging materials, i.e. polyamide (PA), polyvinyl chloride (PVC), polystyrene (PS), polyethylene terephthalate(PET),polypropylene (PP) exhibit specific properties for which they are used including low cost, tensile strength, and O_2 and CO_2 transmission properties [15–17]. These properties increase shelf life and attractiveness of the food to the consumers [18,19]. But these synthetic materials are non-degradable [20,21]. Biopolymers can be divided into three categories [22–26]. The first type is biopolymers derived from biomass, the second type is derived from renewable bio-based monomers and the third type is derived from microorganisms.

The first group of biopolymers extracted from biomass are hydrophilic and crystalline and are unsuitable for moist food products packaging; however, they play significant role in the food packaging industry due to their gas barrier property [23–26]. Conversely, some protein-based biopolymer (e.g., milk, gelatin, whey, egg white, wheat gluten, zein, corn, soy protein) of this group can be used to make edible films to be consumed with the food. Such protein-based films are already used as edible packaging materials [27].

The second group of biopolymers is dominated by polylactic acid (PLA). It is economically viable during processing, and that is why becoming a popular choice for different types of applications [28,29]. PLAs performance was significantly better than synthetic plastics [30]. So, day by day, it is becoming a good alternative as a food packaging material. Resistance to water solubility, large molecular weight, ease of processing through biodegradability and thermoforming established PLA as a food packaging material [22,31,32]. Moreover, PLA exhibits similarities with some conventional plastics like polyethylene, PET, PVC, polystyrene etc. that makes it an ideal alternative of conventional plastics. It can be processed in several ways, which makes the processing pathway more flexible. Additionally, it can be processed into different forms and shapes. It can also be recycled and re-polymerized. Fig. 3 [29] illustrates some of the positive traits and limitations of PLAs.

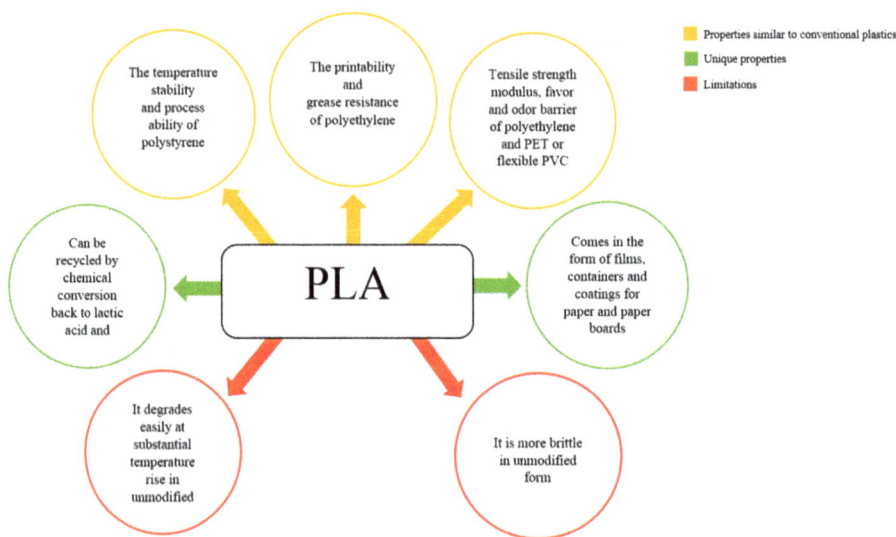

Fig. 3 *Some positive and negative traits of PLA [29]*

Some limitations of PLA include the tendency to degrade rapidly in elevated temperatures and its brittleness can be an issue, but these limitations are likely to be eliminated in the near future.

The third group consists of polyhydroxyalkanoates (PHAs), which is a type of bacterial polyester. The PHA has a small share of the overall biopolymer market as only 453.59 tons of PHAs are produced annually. Fig. 4 [33–40] illustrates the favourable properties of PHAs. Some of these properties are unique while others are similar to conventional plastics.

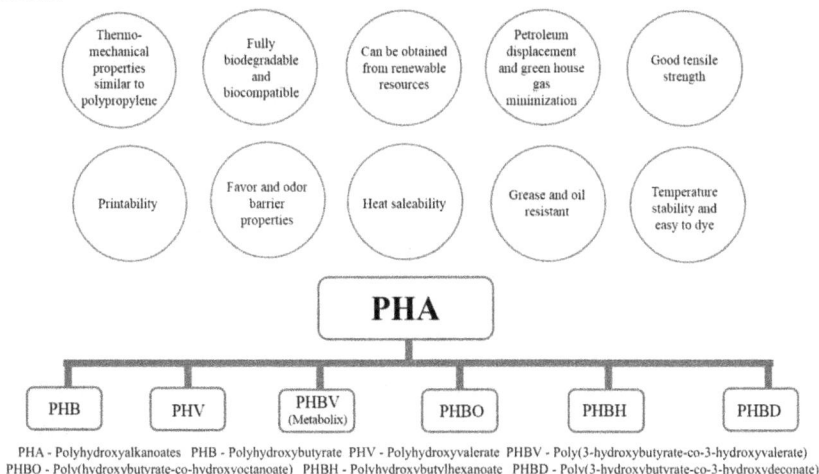

PHA - Polyhydroxyalkanoates PHB - Polyhydroxybutyrate PHV - Polyhydroxyvalerate PHBV - Poly(3-hydroxybutyrate-co-3-hydroxyvalerate)
PHBO - Poly(hydroxybutyrate-co-hydroxyoctanoate) PHBH - Polyhydroxybutylhexanoate PHBD - Poly(3-hydroxybutyrate-co-3-hydroxydeconate)

Fig. 4 *Some favorable properties of PHA and some of its derivatives[33–40].*

PHAs are practical for biodegradable bottles, sheets, containers, coatings, films, fibres, and laminates. Furthermore, "Metabolix PHA" is a blend of poly (3-hydroxyoctanoate) and polyhydroxybutyrate (PHB) which is approved by Food and Drug Administration (FDA) for food packages that match the characteristics and performance of non-degradable plastics.

Several companies from the USA, Japan, Denmark, Canada, Germany, Thailand, France, etc. have been producing biodegradable plastic films. Table 1 [27] shows some of the companies and the type of biodegradable film they manufacture and Fig. 5 [6] shows the different applications of those films.

Materials Research Forum LLC

https://doi.org/10.21741/9781644901335-10

Table 1   *Manufacturers of bio-based films[27]*

Country	Company name, Film name		
	Starch-based films	**PLA based films**	**PLA/Starch-based films**
USA	Plantic technologies, Plantic	Nature works, Ingeo	Nature works, Ecovio
	Biotec, Bioplast		Plantic Co, Plantic
	Novamont, Mater Bi		
Japan	Rodenburg biopolymers, Solanyl	Teijin, BIOFRONT	Japan corn starch, Compole
Germany	Biop, Biopar		BASF, Bioflex
Canada		Revoda, HiSun	
France			Limagrain, Biolice

Biopolymers have a wide variety of commercial use. The major types of biopolymers discussed above can be used as an alternative for PP, PET, PES, and PE. They can be shaped like cups, trays, cutlery, shrink and stretch films, etc. Figure 5 shows a detailed overview of the plastics that biopolymers can alternate [6]. It also shows the form in which they are available for use in the food packaging industry.

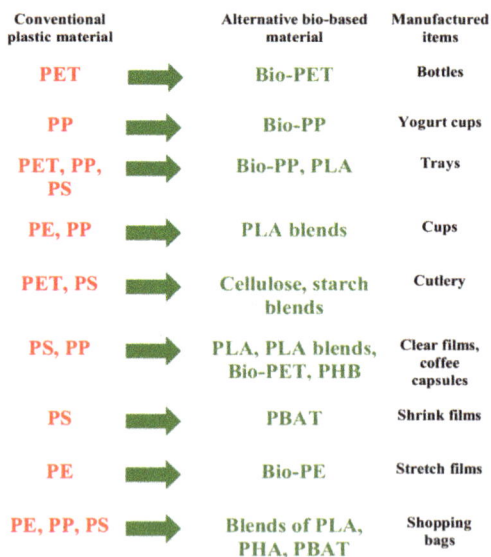

Conventional plastic material		Alternative bio-based material	Manufactured items
PET	→	Bio-PET	Bottles
PP	→	Bio-PP	Yogurt cups
PET, PP, PS	→	Bio-PP, PLA	Trays
PE, PP	→	PLA blends	Cups
PET, PS	→	Cellulose, starch blends	Cutlery
PS, PP	→	PLA, PLA blends, Bio-PET, PHB	Clear films, coffee capsules
PS	→	PBAT	Shrink films
PE	→	Bio-PE	Stretch films
PE, PP, PS	→	Blends of PLA, PHA, PBAT	Shopping bags

Fig. 5 *Alternating potential of biopolymers with conventional plastics [6].*

Degradation of Plastics   Materials Research Forum LLC
Materials Research Foundations **99** (2021) 238-268   https://doi.org/10.21741/9781644901335-10

To increase the shelf life of food materials, a special type of packaging is used that decreases the O_2 level inside the package and increases the CO_2 level. This slows down the overall metabolism rate and hence increases the shelf life. This type of packaging is called modified atmosphere packaging (MAP). Some biodegradable polymers such as chitosan-based films, biofex, PLA, PVC films etc. are used for MAP of some fruits and vegetables, and they are described in Fig. 6 [6].

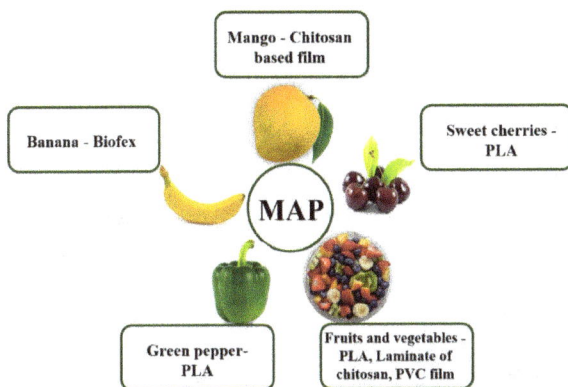

Fig. 6 *Application of MAP in different fruits and vegetables [6].*

Undoubtedly, biopolymers reduce the adverse environmental effect of plastic production and processing, creating a pathway towards a green economy. But surprisingly, only 1% of the total plastics are replaced by the biodegradable polymers. Limitations such as complex processing operation, lack of durability and stability, production cost etc. have prevented biopolymers from offering a competitive alternative to plastics. More research is needed to improve their thermal and physiochemical properties, to reduce cost by simplifying the production process. A comprehensive study about the interaction between biopolymers and the food components should be performed prior to recommending a biopolymer for food packaging. The utilization of agricultural waste, solid waste management, health hazards issues, and greenhouse gas emission can be at least partially addressed by the use of biodegradable polymers contributing to sustainable environment society and economy.

Materials Research Forum LLC
https://doi.org/10.21741/9781644901335-10

3.2   Industrial Application

3.2.1 Paper covers

PHA latex is used to make paper or cardboard coverings. This cover is water-resistant and is much cost-efficient than the cardboard-aluminium mixture that is used at present. It is also used to make diaphragms, foils and films [41,42].

3.2.2 Day to day accessories

A German company "Biomer" makes P(3HB) from the bacteria "Alcaligenes latus". It degrades within two months, and accessories like comb, pen etc. are made with this material [43].

3.2.3 Piezoelectric objects

As PHAs have piezoelectric nature; therefore, they can be used to produce acceleration measuring instruments, stretch sensors, lighters, shockwave detectors, keyboard pressure sensors, etc. [42].

3.2.4 Acoustics

Microphones, sounds pressure sensors, ultrasonic sensors, etc. can also be made by PHAs. This can be done by virtue of their acoustic properties [42].

3.2.5 Oscillators

PHA is used for making headphones, atomizing liquids, ultrasonic therapy, and loudspeakers. Oscillatory properties of PHAs can be useful in this regard [42].

3.2.6 Others

The combined effort of some companies from China, USA, and Korea, allowed P(3HB-3HHx) to be industrially produced from "Aeromonas hydrophila" and to make flexible packaging, synthetic papers, nonwovens, medical devices, thermoformed articles etc. [44].

3.3   Aerospace & automobile

Nearly 27 million tons of plastic are being used in the automotive industry each year. Hence, there is a significant concern for their proper recyclability. To comply with the environmental regulations, various automobile companies are using bio-based degradable plastics such as bio-based polyamides and polyesters to produce the various components. Research has shown that high-performance bio-based polypropylene and polyesters have similar properties to the polybutylene terephthalate. These plastics have high durability, which makes them ideal for designing dashboard components.

Toyota is using degradable plastic polytrimethylene terephthalate (PTT) for manufacturing the interior vent louvre vanes of their latest Toyota Prius vehicle. PTT allows for thinner and stronger vent design. Moreover, Toyota is producing 60% of the interior fabrics of their cars from bio-based polyesters, which makes them a front-runner in adopting degradable plastic in the automobile industry. In their newest hybrid-electric compact vehicle model Lexus CT 200h and Lexus HS 250h, Toyota uses bio-based polypropylene/polylactic acid (PP/PLA) for interior design. Similarly, Röchling automotive are using high heat PLA compounds to design their interior trim parts and air filter box. A selection of applications is in Table 2 below [45].

Table 2 *Use of biopolymers in the automobile industry[45]*

Name	Use
Bio based polyamides	
PA 11	Mono and multi-line fuel lines, flexible tubing, fluid transfer lines, pneumatic brakes, friction parts and quick connectors, oil and fuel filters, steering shafts and coil springs etc
PA 6,10	Air filter, fuel contacting line, special cables etc.
PA 10,10	Turbo air duct, air coolers, oil pans, engine mounts, transmission parts etc.
PA 5,10	Housing of air filter, steer angle sensor cogwheel, cooling fan etc.
Bio based glass reinforced polyamides	
PA 6, 10 30% GF	
PA 10, 10 50% GF	Technical automotive parts
PA 10, 10 65% GF	
PLA based materials	
Standard commercial PLAs	Protective vehicle management wrappings
PLA fabrics and fibers	Roof, floor and carpet mats
Bio-based polyurethane foams	
Raw material: Soybean oil	Seats, door and body panel, head rests, head liners, impact absorbing foams, bumper fascia, etc.
Raw material: Castor oil	Head and arm rests, seats
Raw material: Sunflower oil	Seats

3.4   Medical application

Biodegradable plastics have versatile uses in medical applications being used as biomaterials in tissue engineering, surgical and pharmaceutical practice.

3.4.1 Biomaterials

A number of biomaterials are in essence disposable medical products such as catheters, syringes, and blood bags. Surgery supporting materials (adhesives, sealants, suture), temporary or permanent assisting artificial organs (vascular graft, artificial kidney or heart etc.), tissue replacements materials (dental implants, intraocular lens etc.) are also considered biomaterials. Biodegradable plastics are slowly attracting attention because of their biocompatibility [46], [47].

3.4.2 Surgical use

The most popular use of biodegradable polymers is for suturing. Polyglycolide and glycolide-L-lactide are widely used for this purpose. Hemostasis, adhesion and sealing are the second most important surgical procedures that highly require biodegradable polymers. The processed products are applied in the form of liquid that turns into gel and sticks in the defect area and later on, is absorbed by the body [48].

3.4.3 Pharmaceutical use

Biodegradable polymers are used for controlled release of drugs into the body. For this reason – discs, beads, cylinders, microspheres, nanospheres are made using biopolymers [49,50].

3.4.4 Tissue engineering

Biodegradable polymers are used in scaffolds for differentiation and proliferation of cell that causes tissue construction and regeneration. Also, at the location of regeneration, biodegradable polymers are used for sustained releasing of growth factors. Collagen, copolymers of lactide and glycolide-lactide are commonly used for scaffolds [51].

3.5   Agricultural application

3.5.1 Mulching

Degradable plastics have garnered popularity in agricultural applications as mulches. In its simplest form mulches, degradable plastic films contain plants and assist its growth. Farmers use such plastic films to raise soil temperature, prevent weed growth and maintain adequate moisture. Plastics that undergo photodegradation are highly recommended for this application.

Light sensitive degradable plastic materials such as ethylene-carbon monoxide copolymers, vinyl ketone copolymers, aromatic ketone additives, and Ti/Zr complexes substituted with benzophenones can be used for agricultural mulching [3]. Using a mixture of ferric and nickel dibutyldithio-carbamates, the photodegradation rate can be controlled. This mixture is added in such a ratio that these degradable plastics start decaying once the growing season is over [52].

3.5.2 Planting containers

The use of plastic in various types of containers is a common practice in most agricultural farms and gardening [53]. Polycaprolactone, a type of degradable plastic, can be used for producing small plant containers. One of the fundamental advantages of using polycaprolactone is its longer life cycle. It is found that after six months of application, polycaprolactone begins to undergo considerable biodegradation. Experiments show that 48% of the weight is lost after six months, and 95% of the weight is lost in a year [3]. The most common use of polycaprolactone planting containers is in automated machine seed planting.

3.5.3 Packaging

Packing is a critical section of any industrial production process, including agriculture. Various types of agricultural products are now being transported to long-distance which requires improved packaging. Moreover, adequate packaging improves consumer perception of the end product and enhances shelf life. Photodegradable polymers such as ethylene-carbon monoxide (usually 1% carbon monoxide) copolymers are widely adopted as an alternative to conventional plastic for packaging [41].

Recently, additive approaches for preparing biodegradable packaging plastics have also gained considerable attention. Some of the most commonly used additives are ceric compounds, zirconium/titanium chelates, ferric compounds, and benzophenones that are used in various ratios to control the photodegradation rate. Additionally, biodegradable starch-filled polyethylene is also popular for producing packing plastics and container bags [52].

3.5.4 Seed film cultivation

Seed film cultivation is an innovative approach cultivate crops that are grown in water, such as rice. Experiments have shown that high rice yields can be achieved in aerobic soil biosphere in the absence of water. However, this process is susceptible to aerobic weed growth, which ultimately dominates the field and reduces rice yield. In order to ameliorate this situation, biodegradable films can be used to protect the seed and produce rice in the absence of water. Seed film cultivation (SFC) is the direct sowing of seeds protected using biodegradable film.

Similar to the mulching, various types of photodegradable polymers such as ketone carbonyl copolymers can be used for producing cultivation films. Additionally, biodegradable plastics produced from a mixture of vinyl ketone comonomer with polystyrene (PS) and polyethylene (PE) can also be used to produce the cultivation film for SFC.

3.6 Consumer and institutional products application

Nearly 42 million tons of conventional plastics are produced to manufacture consumer and institutional products. Considering only 9% of these plastic products are recycled - this massive production rate raises grave concerns for environmental sustainability. There are thousands of types of consumer products that use plastics. Since plastic is cheap and easy to mold, finding an alternative material is challenging. However, degradable plastics can be a convenient alternative to reduce plastic pollution significantly. Based on the physio-chemical properties of various types of degradable plastic, certain products can be modelled and assembled. Table 3 [52–55,79–85] discusses the possible application of degradable plastic in consumer and institutional product applications.

3.7 Energy application

Nearly 300 million metric tons of plastics are produced globally [56]. Realizing the potential of plastic to fuel technologies, industries are investing significantly to develop effective methods for producing energy using plastics. Some of the most feasible techniques are described below.

3.7.1 Pyrolysis

Thermally treating and breaking large polymers into smaller hydrocarbons using pyrolysis is a popular method for converting plastic into fuel. In low oxygen and controlled environment with elevated temperatures, plastics can be converted into hydrocarbon-rich compounds that can be used as diesel and gasoline [57,58]. Belgiorno et al. [59] have shown that hydrocarbons with a boiling point between 35 and 185 °C are suitable for motor gasoline application. Similarly, hydrocarbons having a boiling temperature between 350 and 538 °C can be recommended as vacuum gas oil, and between 185 and 290 °C as diesel [59]. Table 4 [4,60–62,86–92] discusses various pyrolysis methods for converting plastic into fuel.

Materials Research Forum LLC
https://doi.org/10.21741/9781644901335-10

Table 3 *Application of degradable plastic for producing consumer and institutional products[52–55, 79–85].*

Type of products	Alternative Degradable Plastic	Method of Treatment	Conventional Plastic	Reference
Food containers, water bottles, soft-drinks bottles and other consumer products	Polyethylene containing pro-oxidant, Hydroxyalkanoates	Partial assimilation by microorganism, pre-aged thermally, activated sludge	PETE-polyethylene and PET	[54]
Laundry and dishwashing detergent, juice and milk bottles, grocery bags	HDPE/ starch with pro-oxidants, Polyhydroxy butirrate	Biodegradable, bioactive soil degradation, artificial weathering using UV and Xenon arc radiation	HDPE-high density polyethylene	[52,79,80]
Electrical insulation, rigid packaging sheets, food trays, irrigation pipes	Isotactic polymers (i-PP), Ethylene-propylene copolymers	Bio-assimilation and microbial attack, UV-irradiation	Polyvinyl Chloride	[81]
Wire insulation, trash bags, grocery bags, storage bags for food, bottles	LDPE comprising pro-oxidants and totally degradable plastic additives (TDPA)	Environmentally degradable, physiochemical treatment, microbial degradation	LDPE-low density polyethylene	[53,82,83]
Drinking straw, packaging of vegetable and fruit, bottle caps	PP/ starch mixed with Mater-Bi	Partially biodegradable	Polypropylene	[84]
Plastic kitchen and tableware, egg cartons, electronics, carryout containers	Polyhydroxyalkanoate (PHA), Polyolefines	Biodegradable, degradation in extremely acidic environment	Polystyrene	[85]
Baby toys, belts, book covers, household items, interior decoration, boots, shoes, slippers	Copolymers styrene-(ethylene-butylene)-styrene, Laprene, Polyolefines	Thermal degradation	Polyvinyl Chloride, polyethylene	[55]

Table 4   *Various pyrolysis methods for converting plastic into fuel [4,60–62,86–92]*

Type of Reactor	Plastic Feed Type	Operating Temperature	Output		Reference
			Gas (wt%)	**Crude oil (wt%)**	
Distillation type reactor	PETE-1	515-795	14.25	12.4	[61]
Batch Reactor with 2 litter capacity	HDPE	440	74	9	[4,86,87]
Natural State Research Inc reactor	Mixed (PE, PP, PS, PET)	370-420	90	5	[88]
Parr mini bench top	PP, PS, PET, PE	500	93	7	[89]
Fixed-bed and Fluidized bed reactor	LEPD and Mixed	330-450	15-87		[90,91]
Continuous reactor with activated carbon	PP, HDPE	520	88-96		[60,92]
Conical shaped spouted bed reactor	HDPE	500	58-70		[62]

In addition to pyrolysis, another thermal treatment method called gasification can also be used to convert plastics into combustible fuel [63]. Using gasification agents or oxidizing agent, large hydrocarbon polymers can be broken into gasoline and diesel limiting the adverse effect on the environment [64]. Moreover, hydrocracking is a potential technology where plastics can be used as feed material for energy. Scott et al. have observed that combustible gas with a hydrocarbon range of C_{5+} can be produced by hydrocracking polyethylene at 600 °C [60].

3.7.2 Cold plasma pyrolysis

Using the advanced cold plasma pyrolysis, biodegradable plastics such as polyolefines, polyesters, polyurethanes, polyvinlyacetate and degradable plastics like HDPE (High-density polyethylene), polypropylene, polystyrene are converted into methane, hydrogen and ethylene. The produced gas is then combusted and used as a clean source of energy [65]. In addition to that, ethylene can be used as the raw material for biodegradable plastic production. Nearly 40% of plastic waste in the United States and 31% in the European Union is used for landfilling instead of being used as feed material for energy

Degradation of Plastics   Materials Research Forum LLC
Materials Research Foundations **99** (2021) 238-268   https://doi.org/10.21741/9781644901335-10

generation [66]. This massive plastic waste dumb causes environmental damage. Clean energy can be generated from plastic waste by utilizing cold plasma pyrolysis.

Although there are numerous advantages of cold plasma pyrolysis, it is an expensive and complicated process. Cold plasma pyrolysis operates at 500 to 600ºC and maintaining the optimal reaction condition requires expert human resources and a sophisticated cooling system [67,68].

3.8 Waste management

A waste hierarchy is devised to outline the most and least preferred methods for plastic waste management (see Fig. 7 [5, 69–72]). Prevention and reuse are at the top of the hierarchy because preventing waste generation and reusing will significantly reduce the use of resources needed to manage waste. This will reduce costs and the requirement of human resources. Also, it enables the system to be a closed cycle and hence lessen further plastic production. On the other hand, landfilling and disposal are at the least as this has many side effects both to mankind and the environment. Fig. 7 [5, 69–72] shows the waste hierarchy of plastic waste management.

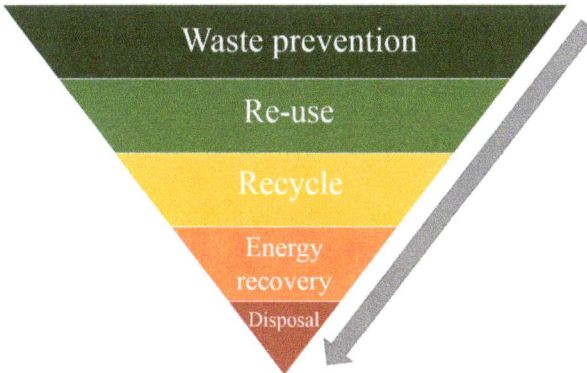

Fig. 7 *Waste hierarchy of plastic waste materials [5, 69–72].*

There are four conventional ways of plastic waste management; mechanical recycling, chemical recycling, incineration, and landfill. Fig. 8 [72–75] illustrates the benefits of biodegradable plastics compared to the conventional plastic in terms of recycling is represented. Some specialized recycling practices for biodegradable plastics are also represented.

Properties	Conventional plastics	Bio based degradable plastic	Comments
Landfill			
Can be landfilled	✔	✔	Production of methane gas is an issue
Degradation	✖	✔	0-85% degradation
Incineration			
Can be incinerated	✔	✔	
Good calorific value	✔	✔	Nearly same energy capacity as conventional plastics
Release of greenhouse gases	✔	✖	
Release of toxic gases	✔	✖	Very low environmental impact
Emission control required	✔	✖	
Mechanical recycling			
Can be mechanically recycled	✔	✔	Need enough material to be recycled or else not viable
Chemical recycling			
Can be chemically recycled	✔	✔	Possible for PLA and PHA but no evidence for starch
Industrial composting			
Can be industrially composed	✖	✔	More than 90% of the material converts into CO_2
Anaerobic digestion			
Can be anaerobically digested	✖	✔	Relatively higher capital cost

Fig. 8 *Comparison of conventional plastics and bio-based polymers[72–75].*

All the benefits of conventional plastics in case of recycling can also be obtained from biopolymers which are indicated by green ticks. But there are some unwanted harmful sides in some of the conventional plastic recycling processes, indicated by red ticks.

These are not present in recycling biopolymers and are indicated by green crosses. Additionally, biopolymers can be recycled by some other methods that are not possible for conventional plastics. These methods are indicated by a red cross for conventional plastics. Anaerobic digestion and industrial composting are special types of recycling techniques only available for biopolymers. It can be seen that biopolymers are more desirable than conventional plastics in terms of flexibility of recycling and environmental issues [5,76].

4.   Growth of plastic degradation industry

Plastic has entered every aspect of human life; nearly every consumable item uses plastic at some point in its supply chain. As a result, its growth has increased exponentially over the past decades and is likely to expand further. In order to evaluate the growth of plastic industries, it is helpful to look at the cumulative growth of plastic production over the years (see Fig. 9 [77, 78]). Back in 1950, the annual production of plastic was only 2 million tons per year. Over the years, plastic has entered every aspect of human life, and with that its production rate increased tremendously.

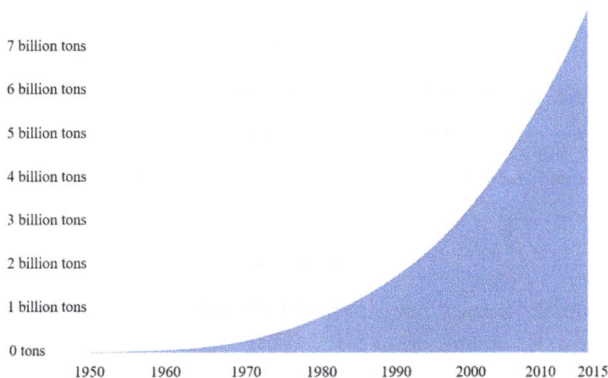

Fig. 9 *Cumulative global plastic production [78].*

Fig. 9 [78] depicts that by 2015 the world has produced nearly 7 billion tons of plastic, to put in context that would be more than one ton of plastic for each person on the planet. From this figure, it is evident that the plastic industry is growing faster than ever.

Degradation of Plastics | Materials Research Forum LLC
Materials Research Foundations **99** (2021) 238-268 | https://doi.org/10.21741/9781644901335-10

Therefore, it can be concluded that there is a fast-expanding market for degradable plastics.

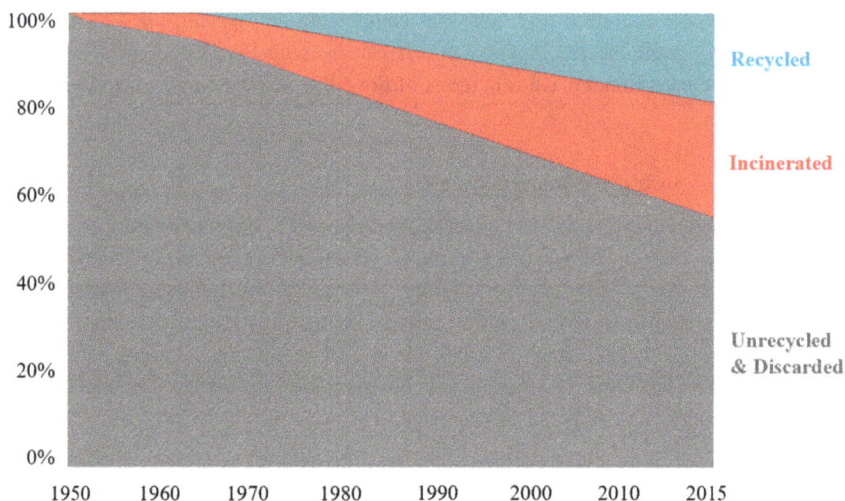

Fig. 10   *Disposal pattern of global plastic waste [78].*

From Fig. 10 [78], a clear pattern can be observed, in 1950, nearly all the produced plastic was dumped, and as newer recycling technologies were introduced the percentage of discarded plastic decreased. However, despite the best effort, nearly 60% of all plastics are still dumped in the environment. This enormous amount of mass, mostly consisting of single-use disposable plastics (e.g., plastic bags, containers) can be produced from degradable plastic. Therefore, it is a clear indication that there is a potential market for degradable plastic to expand. Additionally, the decreasing use of unrecycled plastic suggests that industries and consumers are shifting more towards degradable plastics. It is estimated that with the growth of degradable plastic industries the percentage of discarded plastic would fall below 6% by the year 2050. In order to prevent climate change, the world collectively is transiting towards renewable energy usages [93–99]. Most renewable energy technologies use plastics in production. Therefore, it will create a broader market for biodegradable plastic in the future.

Conclusion

Degradable plastics and biopolymers have been proven as less harmful for the environment than traditional plastic production. Renewable food stocks and agricultural wastes are the main sources of biodegradable films; hence, researchers should devote time and effort on this topic. Unfortunately, the replacement rate of plastics by biodegradable polymers is currently only 1%. The drawbacks using biopolymers are complex downstream processing operations, durability and stability issues and economical scale of production. As future research and development will inevitably improve the quality of degradable plastics, the production of non-degradable will inevitably become obsolete in not too distal future.

Nomenclature

Symbol	Meaning	Symbol	Meaning
PLA	Poly Lactic Acid	FDA	Food and Drug Administration
PCL	Polycaprolactone	MAP	Modified Atmosphere Packaging
PHAs	Poly Hydroxy Alkenoates	PHV	Polyhydroxyvalerate
PBAT	Poly Butylene Adipate-Co-Terephthalate	PHBO	Poly (Hydroxybutyrate-Co-Hydroxyoctanoate)
PBS	Poly Butylene Succinate	PHBD	Poly(3-Hydroxybutyrate-Co-3-Hydroxydeconate)
PE	Polyethylene	PES	Polyethersulfone
PP	Polypropylene	PHB	Polyhydroxybutyrate
PA	Polyamide	PHBV	Poly(3-Hydroxybutyrate-Co-3-Hydroxyvalerate)
PS	Polystyrene	PHBH	Polyhydroxybutylhexanoate
PVC	Polyvinyl Chloride	PTT	Polytrimethylene Terephthalate
PET	Polyethylene Terephthalate	SFC	Seed Film Cultivation
TDPA	Totally Degradable Plastic Additives	HDPE	High-Density Polyethylene

References

[1]   N. Newaj and M. H. Masud, Utilization of waste plastic to save the environment, in International conference on mechanical, industrial and energy engineering, KUET, Bangladesh,(2014), pp. 1–4

[2]   R. Geyer, J. R. Jambeck, and K. L. Law, Production, use, and fate of all plastics ever made, Sci. Adv. (2017), vol. 3, no. 7, p. e1700782, https://doi.org/10.1126/sciadv.1700782

[3]   P. P. Klemchuk, Degradable plastics: A critical review, Polym. Degrad. Stab. (1990), vol. 27, no. 2, pp. 183–202, https://doi.org/10.1016/0141-3910(90)90108-J

[4]   A. G. Buekens and H. Huang, Catalytic plastics cracking for recovery of gasoline-range hydrocarbons from municipal plastic wastes, Resour. Conserv. Recycl. (1998), vol. 23, no. 3, pp. 163–181, https://doi.org/10.1016/S0921-3449(98)00025-1

Materials Research Forum LLC

https://doi.org/10.21741/9781644901335-10

[5]   L. S. Dilkes-Hoffman, S. Pratt, P. A. Lant, and B. Laycock, The role of biodegradable plastic in solving plastic solid waste accumulation, Plastics to Energy (2019), pp. 469–505, https://doi.org/10.1016/B978-0-12-813140-4.00019-4

[6]   M. Mourshed, M. H. Masud, F. Rashid, and M. U. H. Joardder, Towards the effective plastic waste management in Bangladesh: a review, Environ. Sci. Pollut. Res. (2017), vol. 24, no. 35, pp. 27021–27046, https://doi.org/10.1007/s11356-017-0429-9

[7]   Gaelle Gourmel, Global Plastic Production Rises, Recycling Lags | Project AWARE, Published online at projectaware.org(2012), retrieved from: https://www.projectaware.org/update/global-plastic-production-rises-recycling-lags (Accessed: 9 January, 2020)

[8]   EPRO, Plastics – the Facts 2016, An analysis of European plastics production, demand and waste data, Published online at plasticseurope.org (2016),retrieved from: https://www.plasticseurope.org/application/files/4315/1310/4805/plastic-the-fact-2016.pdf (Accessed: 10 January, 2020)

[9]   Masud, M.H., Akram, W., Ahmed, A., Ananno, A.A., Mourshed, M., Hasan, M., Joardder, M.U.H., "Towards the effective E-waste management in Bangladesh: A review", Environmental Science and pollution research, Springer, Vol. 26, no. (2), 2019,https://doi.org/10.1007/s11356-018-3626-2

[10]   D. Hoornweg and P. Bhada-Tata, What a waste: a global review of solid waste management, vol. 15. World Bank, Washington, DC (2012), retrieved from:https://openknowledge.worldbank.org/handle/10986/17388 (Accessed: 11 January, 2020)

[11]   D. Hoornweg, P. Bhada-Tata, and C. Kennedy, Environment: Waste production must peak this century, Nat. News (2013), vol. 502, no. 7473, p. 615, https://doi.org/10.1038/502615a

[12]   J. R. Jambeck et al., Plastic waste inputs from land into the ocean, Science (2015), vol. 347, no. 6223, pp. 768–771, https://doi.org/10.1126/science.1260352

[13]   D. E. MacArthur, D. Waughray, and M. R. Stuchtey, The New Plastics Economy, Rethinking the Future of Plastics, in World Economic Forum (2016), retrieved from: http://www3.weforum.org/docs/WEF_The_New_Plastics_Economy.pdf (Accessed: 8 January, 2020)

[14]   A. L. Andrady, Persistence of plastic litter in the oceans, in Marine anthropogenic litter, in: M. Bergmann, L. Gutow, M. Klages (Eds.), Marine Anthropogenic Litter, Springer,2015, pp. 57–72

[15]   S. Mangaraj, T. K. Goswami, S. K. Giri, and M. K. Tripathi, Permselective MA packaging of litchi (cv. Shahi) for preserving quality and extension of shelf-life,

Postharvest Biol. Technol. (2012), vol. 71, pp. 1–12, 2012,
https://doi.org/10.1016/j.postharvbio.2012.04.007

[16] S. Mangaraj, T. K. Goswami, S. K. Giri, and C. G. Joshy, Design and development of modified atmosphere packaging system for guava (cv. Baruipur), J. Food Sci. Technol. (2014), vol. 51, no. 11, pp. 2925–2946, https://doi.org/10.1007/s13197-012-0860-3

[17] S. Mangaraj, T. K. Goswami, and P. V Mahajan, Applications of plastic films for modified atmosphere packaging of fruits and vegetables: a review, Food Eng. Rev. (2009), vol. 1, no. 2, p. 133, https://doi.org/10.1007/s12393-009-9007-3

[18] G. L. Robertson, Food Packaging, Principles and Practice, CRC. Taylor and Francis, Boca Raton, FL, 2006

[19] J.-D. Gu, D. B. Mitton, T. E. Ford, and R. Mitchell, Microbial degradation of polymeric coatings measured by electrochemical impedance spectroscopy, Biodegradation (1998), vol. 9, no. 1, pp. 39–45, https://doi.org/10.1023/A:1008252301377

[20] H. Webb, J. Arnott, R. Crawford, and E. Ivanova, Plastic degradation and its environmental implications with special reference to poly (ethylene terephthalate), Polymers (2013)., vol. 5, no. 1, pp. 1–18, https://doi.org/10.3390/polym5010001

[21] S. Imam, G. Glenn, B.-S. Chiou, J. Shey, R. Narayan, and W. Orts, Types, production and assessment of biobased food packaging materials, in Environment compatible food packaging (2008), pp. 29–62, https://doi.org/10.1533/9781845694784.1.29

[22] A. N. Malathi, K. S. Santhosh, and N. Udaykumar, Recent trends of Biodegradable polymer: Biodegradable films for Food Packaging and application of Nanotechnology in Biodegradable Food Packaging, Curr. Trends Technol. Sci. (2014), vol. 3, no. 2, pp. 73–79, https://doi.org/10.1.1.428.8076

[23] S. Doppalapudi, A. Jain, W. Khan, and A. J. Domb, Biodegradable polymers—an overview, Polym. Adv. Technol. (2014), vol. 25, no. 5, pp. 427–435, https://doi.org/10.1002/pat.3305

[24] C. J. Weber, Biobased packaging materials for the food industry: status and perspectives, a European concerted action,Published online at biodeg.net (2000), retrieved from:
http://www.biodeg.net/fichiers/Book%20on%20biopolymers%20(Eng).pdf
(Accessed:13 January, 2020)

[25]   S. Mangaraj, T. K. Goswami, and D. K. Panda, Modeling of gas transmission properties of polymeric films used for MA packaging of fruits, J. Food Sci. Technol. (2015), vol. 52, no. 9, pp. 5456–5469, https://doi.org/10.1007/s13197-014-1682-2

[26]   S. Guilbert, B. Cuq, and C. Bastioli, Handbook of Biodegradable Polymers, Shropshire, UK Rapra Technol. Ltd., 2005

[27]   S. Mangaraj, A. Yadav, L. M. Bal, S. K. Dash, and N. K. Mahanti, Application of Biodegradable Polymers in Food Packaging Industry: A Comprehensive Review, J. Packag. Technol. Res. (2019), vol. 3, no. 1, pp. 77–96, https://doi.org/10.1007/s41783-018-0049-y

[28]   J. W. Park, S. S. Im, S. H. Kim, and Y. H. Kim, Biodegradable polymer blends of poly (L-lactic acid) and gelatinized starch, Polym. Eng. Sci. (2000), vol. 40, no. 12, pp. 2539–2550, https://doi.org/10.1002/pen.11384

[29]   W. Y. Jang, B. Y. Shin, T. J. Lee, and R. Narayan, Thermal properties and morphology of biodegradable PLA/starch compatibilized blends, J. Ind. Eng. Chem. (2007), vol. 13, no. 3, pp. 457–464, retrieved from: https://pdfs.semanticscholar.org/f2ef/9771eba63175b68e42ba9a37122416a9d088.pdf (Accessed:14 January, 2020)

[30]   R. A. Auras, S. P. Singh, and J. J. Singh, Evaluation of oriented poly (lactide) polymers vs. existing PET and oriented PS for fresh food service containers, Packag. Technol. Sci. An Int. J. (2005), vol. 18, no. 4, pp. 207–216, https://doi.org/10.1002/pts.692

[31]   K. Leja and G. Lewandowicz, Polymer Biodegradation and Biodegradable Polymers-a Review, Polish J. Environ. Stud. (2020), vol. 19, no. 2, retrieved from: http://yunus.hacettepe.edu.tr/~damlacetin/kmu407/index_dosyalar/2.%20makale.pdf (Accessed:15 January, 2020)

[32]   L. Avérous and E. Pollet, Environmental silicate nano-biocomposites. Springer, 2012

[33]   M. G. A. Vieira, M. A. da Silva, L. O. dos Santos, and M. M. Beppu, Natural-based plasticizers and biopolymer films: A review, Eur. Polym. J. (2011), vol. 47, no. 3, pp. 254–263, https://doi.org/10.1016/j.eurpolymj.2010.12.011

[34]   M. A. Trainer and T. C. Charles, The role of PHB metabolism in the symbiosis of rhizobia with legumes, Appl. Microbiol. Biotechnol. (2006), vol. 71, no. 4, pp. 377–386, https://doi.org/10.1007/s00253-006-0354-1

[35]   M. D. Sanchez-Garcia, A. Lopez-Rubio, and J. M. Lagaron, Natural micro and nanobiocomposites with enhanced barrier properties and novel functionalities for food

Materials Research Forum LLC
https://doi.org/10.21741/9781644901335-10

biopackaging applications, Trends in Food Sci. Technol. (2010), vol. 21, no. 11, pp. 528–536, https://doi.org/10.1016/j.tifs.2010.07.008

[36]   A. Steinbüchel and B. Füchtenbusch, Bacterial and other biological systems for polyester production, Trends Biotechnol. (1998), vol. 16, no. 10, pp. 419–427, https://doi.org/10.1016/S0167-7799(98)01194-9

[37]   N. Galego, C. Rozsa, R. Sánchez, J. Fung, A. Vázquez, and J. Santo Tomas, Characterization and application of poly (β-hydroxyalkanoates) family as composite biomaterials, Polym. Test. (2000), vol. 19, no. 5, pp. 485–492, https://doi.org/10.1016/S0142-9418(99)00011-2

[38]   Y. Tokiwa and T. Raku, "Biodegradable polylactide resin composition." Google Patents, Jan-2006

[39]   E. C. Tweed, H. M. Stephens, and T. E. Riegert, Polylactic acid blown film and method of manufacturing same. Google Patents, Mar-2012

[40]   Q. Xu, Biodegradable packaging materials with enhanced oxygen barrier performance. Google Patents, Jun-2011

[41]   C. A. Lauzier, C. J. Monasterios, I. Saracovan, R. H. Marchessault, and B. A. Ramsay, Film formation and paper coating with poly ([beta]-hydroxyalkanoate), a biodegradable latex, Tappi Journal(1993), ISSN 0734-1415, vol. 76, no. 5, OSTI ID: 6428520

[42]   W. Babel, V. Riis, and E. Hainich, Mikrobielle Thermplolaste: Biosyntheses, Eigenschaften und Anwendung, Plaste und Kautschuk (1990), vol. 37, no. 4, pp. 109–115, retrieved from: http://pascal-francis.inist.fr/vibad/index.php?action=getRecordDetail&idt=19252951 (Accessed:20 January, 2020)

[43]   R. Smith, Biodegradable polymers for industrial applications. CRC Press, Florida, United States, 2005

[44]   G. Chen, G. Zhang, S. Park, and S. Lee, Industrial scale production of poly (3-hydroxybutyrate-co-3-hydroxyhexanoate), Appl. Microbiol. Biotechnol. (2001), vol. 57, no. 1–2, pp. 50–55, https://doi.org/10.1007/s002530100755

[45]   S. Pilla, Handbook of bioplastics and biocomposites engineering applications, vol. 81. John Wiley & Sons, New Jersey, United States, 2011

[46]   Y. Ikada, Interfacial biocompatibility, ACS Publications, Washington, D.C., United States, 1994

[47]   S. Manandhar, Bioresorbable polymer blend scaffold for tissue engineering, Published online at unt.edu, 2011, vol. 50, no. 3, retrieved from:

https://digital.library.unt.edu/ark%3A/67531/metadc68008/m2/1/high_res_d/thesis.pdf
(Accessed:14 January, 2020)

[48]   C.-C. Chu, J. A. Von Fraunhofer, and H. P. Greisler, Wound closure biomaterials
       and devices. CRC Press, Florida, United States, 1996

[49]   R. Langer and M. Chasin, Biodegradable polymers as drug delivery systems,
       Informa Health Care, London, United Kingdom, 1990

[50]   M. Asano et al., Application of poly DL-lactic acids of varying molecular weight
       in drug delivery systems, Drug Des. Deliv. (1990), vol. 5, pp. 301–320

[51]   Y. Tabata, Tissue regeneration based on tissue engineering technology, Congenital
       Anomolies (2004), vol. 44, no. 3, pp. 111-124, https://doi.org/10.1111/j.1741-
       4520.2004.00024.x

[52]   I. Kyrikou and D. Briassoulis, Biodegradation of agricultural plastic films: a
       critical review, J. Polym. Environ. (2007), vol. 15, no. 2, pp. 125–150,
       https://doi.org/10.1007/s10924-007-0053-8

[53]   Y. Ohtake, T. Kobayashi, H. Asabe, and N. Murakami, Studies on biodegradation
       of LDPE—observation of LDPE films scattered in agricultural fields or in garden soil,
       Polym. Degrad. Stab. (1998), vol. 60, no. 1, pp. 79–84, https://doi.org/10.1016/S0141-
       3910(97)00032-3

[54]   E. Chiellini, A. Corti, S. D'Antone, and R. Baciu, Oxo-biodegradable carbon
       backbone polymers–Oxidative degradation of polyethylene under accelerated test
       conditions, Polym. Degrad. Stab. (2006), vol. 91, no. 11, pp. 2739–2747,
       https://doi.org/10.1016/j.polymdegradstab.2006.03.022

[55]   J. Walendziewski and M. Steininger, Thermal and catalytic conversion of waste
       polyolefines, Catal. Today (2001), vol. 65, no. 2–4, pp. 323–330,
       https://doi.org/10.1016/S0920-5861(00)00568-X

[56]   P. T. Benavides, P. Sun, J. Han, J. B. Dunn, and M. Wang, Life-cycle analysis of
       fuels from post-use non-recycled plastics, Fuel (2017), vol. 203, pp. 11–22,
       https://doi.org/10.1016/j.fuel.2017.04.070

[57]   M. U. H. Joardder, P. K. Halder, M. A. Rahim, and M. H. Masud, Solar Pyrolysis:
       Converting Waste Into Asset Using Solar Energy, in Clean Energy for Sustainable
       Development (2017), pp. 213–235, https://doi.org/10.1016/B978-0-12-805423-
       9.00008-9

[58]   A. R. Nabi, M. H. Masud, and Q. M. I. Alam, Purification of TPO (Tire Pyrolytic
       Oil) and its use in diesel engine, IOSR J. Eng. (2014), vol. 4, no. 3, p. 1,
       https://doi.org/10.9790/3021-04320108

[59]   V. Belgiorno, G. De Feo, C. Della Rocca, and R. M. A. Napoli, Energy from gasification of solid wastes, Waste Manag. (2003), vol. 23, no. 1, pp. 1–15, https://doi.org/10.1016/S0956-053X(02)00149-6

[60]   D. S. Scott, S. R. Czernik, J. Piskorz, and D. S. A. G. Radlein, Fast pyrolysis of plastic wastes, Energy & Fuels (1990), vol. 4, no. 4, pp. 407–411, https://doi.org/10.1021/ef00022a013

[61]   M. Sarker, A. Kabir, M. M. Rashid, M. Molla, and A. S. M. D. Mohammad, Waste Polyethylene Terephthalate (PETE-1) Conversioninto Liquid Fuel, J. Fundam. Renew. Energy Appl. (2011), vol. 1, https://doi.org/10.4303/jfrea/R101202

[62]   G. Elordi et al., Catalytic pyrolysis of HDPE in continuous mode over zeolite catalysts in a conical spouted bed reactor, J. Anal. Appl. Pyrolysis (2009), vol. 85, no. 1, pp. 345–351, https://doi.org/10.1016/j.jaap.2008.10.015

[63]   S. Kumar, R. Prakash, S. Murugan, and R. K. Singh, Performance and emission analysis of blends of waste plastic oil obtained by catalytic pyrolysis of waste HDPE with diesel in a CI engine, Energy Convers. Manag. (2013), vol. 74, pp. 323–331,https://doi.org/10.1016/j.enconman.2013.05.028

[64]   R. P. Singh, V. V Tyagi, T. Allen, M. H. Ibrahim, and R. Kothari,"An overview for exploring the possibilities of energy generation from municipal solid waste (MSW) in Indian scenario, Renew. Sustain. Energy Rev. (2011), vol. 15, no. 9, pp. 4797–4808, https://doi.org/10.1016/j.rser.2011.07.071

[65]   L. Tang, H. Huang, H. Hao, and K. Zhao, Development of plasma pyrolysis/gasification systems for energy efficient and environmentally sound waste disposal, J. Electrostat. (2013), vol. 71, no. 5, pp. 839–847, https://doi.org/10.1016/j.elstat.2013.06.007

[66]   Anh Panh, How we can turn plastic waste into green energy, Published online at theconversation.com, (2016),retrieved from: http://theconversation.com/how-we-can-turn-plastic-waste-into-green-energy-104072 (Accessed: 21 January, 2020)

[67]   A. Sanlisoy and M. O. Carpinlioglu, A review on plasma gasification for solid waste disposal, Int. J. Hydrogen Energy (2017), vol. 42, no. 2, pp. 1361–1365, https://doi.org/10.1016/j.ijhydene.2016.06.008

[68]   H. Huang and L. Tang, Treatment of organic waste using thermal plasma pyrolysis technology, Energy Convers. Manag. (2007), vol. 48, no. 4, pp. 1331–1337, https://doi.org/10.1016/j.enconman.2006.08.013

[69]   L. H. Yee and L. J. R. Foster, Polyhydroxyalkanoates as packaging materials: current applications and future prospects, in I. Roy, Visakh P M (Eds.),

Polyhydroxyalkanoate (PHA) based Blends, Composites and Nanocomposites, Royal Society of Chemistry, United Kingdom, 2014, vol. 30, p. 183

[70] M. R. Yates and C. Y. Barlow, Life cycle assessments of biodegradable, commercial biopolymers—A critical review, Resour. Conserv. Recycl. (2013), vol. 78, pp. 54–66, https://doi.org/10.1016/j.resconrec.2013.06.010

[71] D. D. Cornell, Biopolymers in the existing postconsumer plastics recycling stream, J. Polym. Environ. (2007), vol. 15, no. 4, pp. 295–299, https://doi.org/10.1007/s10924-007-0077-0

[72] A. Soroudi and I. Jakubowicz, Recycling of bioplastics, their blends and biocomposites: A review, Eur. Polym. J. (2013), vol. 49, no. 10, pp. 2839–2858, https://doi.org/10.1016/j.eurpolymj.2013.07.025

[73] X. Yang, K. Odelius, M. Hakkarainen, Microwave-assisted reaction in green solvents recycles PHB to functional chemicals, ACS Sustain. Chem. Eng. (2014), vol. 2, no. 9, pp. 2198–2203, https://doi.org/10.1021/sc500397h

[74] J. Myung et al., Disassembly and reassembly of polyhydroxyalkanoates: Recycling through abiotic depolymerization and biotic repolymerization, Bioresour. Technol. (2014), vol. 170, pp. 167–174, https://doi.org/10.1016/j.biortech.2014.07.105

[75] H. Ariffin, H. Nishida, M. A. Hassan, Y. Shirai, Chemical recycling of polyhydroxyalkanoates as a method towards sustainable development, Biotechnol. J. (2010), vol. 5, no. 5, pp. 484–492, https://doi.org/10.1002/biot.200900293

[76] European Bioplastics, What are Bioplastics?, Published online at european-bioplastics.org(2018), retrieved from: https://docs.european-bioplastics.org/publications/fs/EuBP_FS_What_are_bioplastics.pdf (Accessed: 23 January, 2020)

[77] S. Tracy, Humans Have Produced a Whopping 9 Billion Tons of Plastic, Published online at livescience.com(2017), retrieved from: https://www.livescience.com/59862-humans-have-produced-9-billion-tons-of-plastic.html (Accessed: 15 January, 2020)

[78] H. Ritchie and M. Roser, Plastic Pollution, Published online at OurWorldInData.org (2020), retrieved from: https://ourworldindata.org/plastic-pollution (Accessed: 6 January,2020)

[79] M. Weiland, A. Daro, and C. David, Biodegradation of thermally oxidized polyethylene, Polym. Degrad. Stab. (1995), vol. 48, no. 2, pp. 275–289, https://doi.org/10.1016/0141-3910(95)00040-S

[80] J. V Gulmine, P. R. Janissek, H. M. Heise, and L. Akcelrud, Degradation profile of polyethylene after artificial accelerated weathering, Polym. Degrad. Stab. (2003), vol. 79, no. 3, pp. 385–397, https://doi.org/10.1016/S0141-3910(02)00338-5

[81]　J. K. Pandey and R. P. Singh, UV-Irradiated Biodegradability of Ethylene–Propylene Copolymers, LDPE, and I-PP in Composting and Culture Environments, Biomacromolecules (2001), vol. 2, no. 3, pp. 880–885, https://doi.org/10.1021/bm010047s

[82]　E. M. Nakamura, L. Cordi, G. S. G. Almeida, N. Duran, and L. H. I. Mei, Study and development of LDPE/starch partially biodegradable compounds, J. Mater. Process. Technol. (2005), vol. 162, pp. 236–241, https://doi.org/10.1016/j.jmatprotec.2005.02.007

[83]　E. Chiellini, A. Corti, and G. Swift, Biodegradation of thermally-oxidized, fragmented low-density polyethylenes, Polym. Degrad. Stab. (2003), vol. 81, no. 2, pp. 341–351, https://doi.org/10.1016/S0141-3910(03)00105-8

[84]　J. M. Morancho et al., Calorimetric and thermogravimetric studies of UV-irradiated polypropylene/starch-based materials aged in soil, Polym. Degrad. Stab. (2006), vol. 91, no. 1, pp. 44–51, https://doi.org/10.1016/j.polymdegradstab.2005.04.029

[85]　R. D. Stapleton, D. C. Savage, G. S. Sayler, and G. Stacey, Biodegradation of aromatic hydrocarbons in an extremely acidic environment, Appl. Environ. Microbiol. (1998), vol. 64, no. 11, pp. 4180–4184, https://doi.org/10.1128/AEM.64.11.4180-4184.1998

[86]　B. K. Sharma, B. R. Moser, K. E. Vermillion, K. M. Doll, and N. Rajagopalan, Production, characterization and fuel properties of alternative diesel fuel from pyrolysis of waste plastic grocery bags, Fuel Process. Technol. (2014), vol. 122, pp. 79–90, https://doi.org/10.1016/j.fuproc.2014.01.019

[87]　S. M. Alston, A. D. Clark, J. C. Arnold, and B. K. Stein, Environmental Impact of Pyrolysis of Mixed WEEE Plastics Part 1: Experimental Pyrolysis Data, Environ. Sci. Technol. (2011), vol. 45, no. 21, pp. 9380–9385, https://doi.org/10.1021/es201664h

[88]　M. Sarker, M. M. Rashid, and M. Molla, Waste Plastic Conversion into Hydrocarbon Fuel Materials, J. Environ. Sci. Eng. (2011), vol. 5, no. 5, https://doi.org/10.1080/01998595.2011.10389018

[89]　P. T. Williams and E. Slaney, Analysis of products from the pyrolysis and liquefaction of single plastics and waste plastic mixtures, Resour. Conserv. Recycl. (2007), vol. 51, no. 4, pp. 754–769, https://doi.org/10.1016/j.resconrec.2006.12.002

[90]　Y.-H. Lin, K.-K. Chen, and J.-H. Chiu, Coprescription of Chinese Herbal Medicine and Western Medications among Prostate Cancer Patients: A Population-Based Study in Taiwan, Evidence-Based Complement. Altern. Med. (2012), vol. 2012, p. 147015, https://doi.org/10.1155/2012/147015

[91] Y. Uemichi, J. Nakamura, T. Itoh, M. Sugioka, A. A. Garforth, and J. Dwyer, Conversion of Polyethylene into Gasoline-Range Fuels by Two-Stage Catalytic Degradation Using Silica– Alumina and HZSM-5 Zeolite, Ind. Eng. Chem. Res. (1999), vol. 38, no. 2, pp. 385–390, https://doi.org/10.1021/ie980341+

[92] N. Miskolczi, A. Angyal, L. Bartha, and I. Valkai, Fuels by pyrolysis of waste plastics from agricultural and packaging sectors in a pilot scale reactor, Fuel Process. Technol. (2009), vol. 90, no. 7, pp. 1032–1040, https://doi.org/10.1016/j.fuproc.2009.04.019

[93] Ananno, A. A., Masud, M.H., Dabnichki, P., Ahmed, A., Design assnd Numerical Analysis of a Hybrid Geothermal PCM Flat Plate Solar Collector Dryer for Developing Countries, Solar Energy (2019), vol. 196, pp. 270-286, https://doi.org/10.1016/j.solener.2019.11.069

[94] Masud, M.H., Ananno, A. A., Ahmed, N.U., Dabnichki, P., Salehin K.N., Experimental investigation of a novel Waste Heat based Food Drying System, J. of Food Eng. (2020), vol. 281,https://doi.org/10.1016/j.jfoodeng.2020.110002

[95] Masud, M.H., Ananno, A. A., Arefin, A M E, Ahamed, R., Das, P., Joardder, M.U.H., Perspective of Biomass Energy Conversion in Bangladesh, Clean Tech. and Env. Policy (2019), vol. 21, pp. 719-931,https://doi.org/10.1007/s10098-019-01668-2

[96] Masud, M.H., Nuruzzaman, M., Ahamen, R., Ananno, A. A., Tomal, A. A., Renewable Energy in Bangladesh: Current Situation and Future Prospect, International Journal of Sustainable Energy (2019), vol. 39, n. 2, pp. 132-175, https://doi.org/10.1080/14786451.2019.1659270

[97] Masud, M.H., Islam, T., Joardder, M.U.H., Ananno, A. A., Dabnichki, P., CFD analysis of a tube-in-tube heat exchanger to recover waste heat for food drying, International Journal of Energy and Water Resources (2019), vol. 3, pp. 169-186, https://doi.org/10.1007/s42108-019-00032-w

[98] Ananno, A. A., Masud, M.H., Chowdhury, S. A., Dabnichki, P., Ahmed, N., Arefin, A M E, Sustainable food waste management model for Bangladesh, Sustainable Production and Consumption (2021), vol. 27, pp. 35-51, https://doi.org/10.1016/j.spc.2020.10.022

[99] Ananno, A. A., Masud, M.H., Dabnichki, P., Mahjabeen, M., Chowdhury, S. A., Survey and analysis of consumers' behaviour for electronic waste management in Bangladesh, J. of Env. Management (2021), vol. 282, https://doi.org/10.1016/j.jenvman.2021.111943

Degradation of Plastics   Materials Research Forum LLC
Materials Research Foundations **99** (2021) 269-289   https://doi.org/10.21741/9781644901335-11

Chapter 11

Biodegradable Plastics from Cyanobacteria

R.B. Sartori[1], I.A. Severo[1], A.M. Santos[1], L.Q. Zepka[1], E. Jacob-Lopes[1]*

[1]Bioprocess Intensification Group, Federal University of Santa Maria (UFSM), 97105-900, Santa Maria, RS, Brazil

* ejacoblopes@gmail.com

Abstract

High accumulation of generated plastic waste is a growing concern. Alternatively, scientific efforts are underway to drive industrial demand to replace these products with superior quality, adequate biodegradable resources. Cyanobacteria are a versatile group of phototrophic prokaryotes capable of producing polyhydroxyalkanoates (PHAs) using sunlight and carbon dioxide (CO_2) as a form of carbon and energy reserve. PHAs are appealing alternatives for traditional chemical plastics because these have to resemble characteristics, biocompatibility, and complete biodegradability. In this sense, this chapter aims to address the potential of cyanobacteria as a biodegradable alternative solution.

Keywords

Bioplastics, Biopolymers, Sustainability, Polyhydroxyalkanoates, Polyhydroxybutyrate

Contents

1.   **Introduction**

Industrial production of petroleum-based plastics increased significantly after their introduction to the market over 70 years ago. Plastics have gained tremendous importance precisely because they have highly resistant, inert, versatile, and durable properties [1,2]. However, growing concern about the fossil reserves depletion, price volatility, and the environmental impact of conventional plastics is becoming an increasingly severe problem [3].

Currently about 99% of plastics come from petrochemical sources with a global estimate of accumulated plastic waste in the environment at 6.3 gigatons (Gt) by 2015 and may exceed to 12 Gt by 2050 [4,5]. Thus, due to their ecological nature, biodegradable plastics appear as one of the best alternatives to the severe environmental issues caused by synthetic polymers [6].

Biodegradable plastics are biobased polymers used to produce CO_2 as neutral product that is completely degraded in order to minimize their impact on the environment [7]. Bioplastics are suitable substitutes for petroleum-based plastics as they can produce some of the polymers with similar properties, e.g. polyhydroxyalkanoates (PHA), polypropylene substitutes in many applications [8].

Particularly PHA, a group of polyesters of hydroxyalkanoic acids, is considered one of the most promising bioplastics biosynthesized mainly by bacteria such as *Cupriavidusnecator* and *Escherichia coli* [9,10]. However, these methods generate a total production cost of around 50%, with only organic carbon substrates. An alternative way to reduce these costs would be to identify some microorganisms that use cheaper and more efficient carbon sources to produce these biopolymers. In this context, cyanobacteria have gained considerable attention and appear as an ideal platform for bioplastic production because they can capture energy through sunlight and CO_2 via natural fixation, thus eliminating the cost of carbon source [11, 8, 6].

To date, many studies have proven the potential of these cyanobacteria as a sustainable way of producing biopolymers like polyhydroxybutyrate (PHB) (primary representative of polyhydroxyalkanoates). Still, the cost of manufacturing for cyanobacteria bioplastics can be minimized when an integral study of the entire production process is done. Given

Degradation of Plastics   Materials Research Forum LLC
Materials Research Foundations **99** (2021) 269-289   https://doi.org/10.21741/9781644901335-11

that scenario, the objective of this chapter is to provide an overview of biodegradable plastics, biosynthesis pathways, and production processes from cyanobacteria, the main factors affecting them, the economic aspects, and their main commercial applications.

2.   Biodegradable plastics overview

The high consumption of plastic products over the years has raised significant environmental concerns due to their rapid disposal, degradation resistance, and the large waste material that accumulates in landfills or even oceans [12]. This has stimulated the development of new technologies based on alternatives to reduce the degradability problem of plastics, such as the incorporation of additives or obtaining new biodegradable thermoplastic materials [13].

Biodegradable plastics are biologically synthesized and are readily available for use as environmentally friendly polymers. Some of the major biodegradable polymers are produced from renewable source raw materials, thus known to have a shorter life cycle compared to traditional fossil sources [14].

The biopolymers are natural polymers designed to undergo significant changes in their chemical structure, by the action of microorganisms present in nature under various environmental conditions. They can come from renewable sources (such as starch and cellulose), synthesized by microorganisms (as reserve substances), chemical synthesis, or obtained from fossil sources such as petroleum [15,2]. Fig. 1 [16] presents a flowchart with the main biopolymers and their sources. Biopolymers obtained, especially from living organisms, are reported as the only 100% biodegradable polymers.

Among the biodegradable polymers, thermoplastic starch (TPS), polylactic acid (PLA), or the main described in this chapter, polyhydroxyalkanoate (PHA) polyesters, produced from microorganisms are the current highlights in research companies and development [10]. Currently, heterotrophic bacteria are the most widely used microorganisms in industrial biopolymer production. However, the selection of this type of microorganism and the substrates used have a significant influence on the final cost of manufacturing these bioproducts. Besides, biopolymers derived from agricultural raw materials are not yet ideally aligned with sustainable resources due to competition for arable land, freshwater, and food production [17,7].

Given this scenario, cyanobacteria are strong candidates for biopolymers due to their high productivity, fast growth, and adaptation to totally different environments. Cyanobacteria is one of the most abundant groups of gram-negative prokaryotes with a wide diversity in morph and physiology, cell differentiation, besides efficiently mitigate carbon dioxide produced from flue gases [18,19].

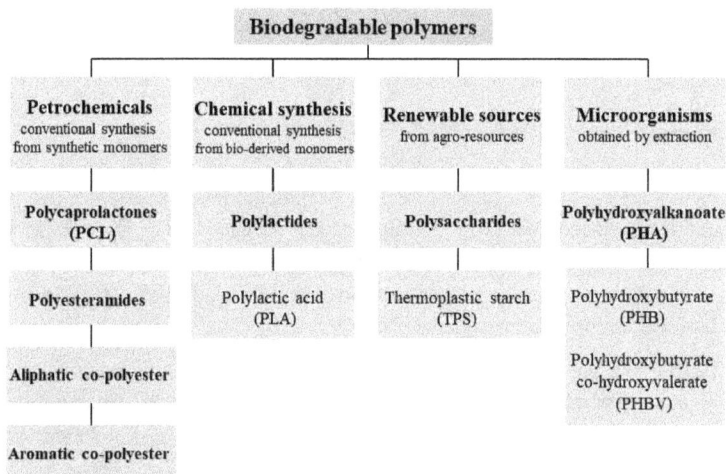

Fig. 1 Classification of some biodegradable polymers according to their source of origin. Adapted from Avérous [16].

Unlike heterotrophic microorganisms, cyanobacteria require only CO_2, sunlight, and minimal nutrients for their growth. As they are naturally alterable, they can be controlled easily to obtain some desirable product. For example, the production of PHB polymers by cyanobacteria is carried out when some basic nutrients are lacking for their development, such as nitrogen and phosphorus [20]. Today, closed systems or photo-bioreactors have the benefit of tailoring these systems to their best performance. Furthermore, the use of wastewater as substrates has been widely used to produce these biopolymers as a sustainable and most economically viable form.

3.   Biodegradable plastics production from cyanobacteria

Cyanobacteria synthesize and accumulate PHA, usually upon entering the lag phase, reaching up to 80% of their cell weight [21]. PHAs are polyesters of 3-, 4-, 5-, and 6-hydroxyalkanoic acids that accumulate like energy storage molecules under excess carbon or when other essential nutrients are limited [11].

Degradation of Plastics
Materials Research Foundations **99** (2021) 269-289

Materials Research Forum LLC
https://doi.org/10.21741/9781644901335-11

PHAs according to the number of carbon atoms in the structure are classified into three major classes. These consist of short-chain length hydroxyalkanoic acids (SCL), ranging from 3 to 5 carbon atoms, medium-chain length (MCL) with 6 to 14 carbon atoms, and long-chain length (LCL) with more than 15 carbon atoms [2].

Among the 150 types of PHA monomers studied up to now, the PHB is the leading representative of PHA and the unique generated under photoautotrophic metabolism referenced so far and therefore, the only one described in this study [8]. Fig. 2 shows the chemical structures of PHB.

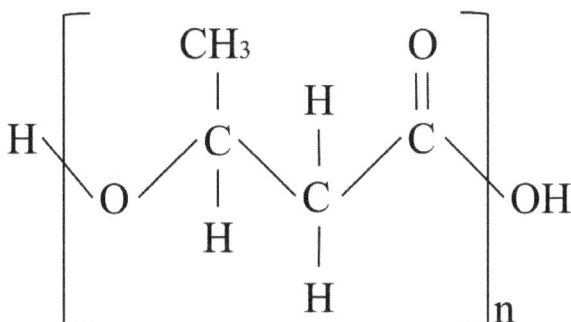

Fig. 2 Chemical structures of PHB.

Among the growth limiting factors, the most important is the nutrient starvation. The reduction of nitrogen concentration in the culture medium is the primary and, therefore, most studied cause for PHB production in cyanobacteria [22,23,8].

A simple growth system to demonstrate the PHB production from cyanobacteria is the growth in three stages in a photobioreactor. In the first stage, the cyanobacteria are in photoautotrophic biomass production in a fresh culture medium, where the Calvin-Benson-Bassham cycle generates an intermediate (glyceraldehyde-3-phosphate) to glycolysis, which is metabolized to pyruvate. This is then converted by pyruvate-dehydrogenase complex to acetyl-CoA, which finally enters the incomplete Krebs cycle, because of the lack of a functional α-ketoglutarate dehydrogenase, generating biosynthesis products like amino acids, nucleotides, etc [24]. These metabolic routes are

Materials Research Forum LLC
https://doi.org/10.21741/9781644901335-11

shown in Fig. 3[24, 8]. This stage, also known as the green stage, usually lasts 5 to 6 days.

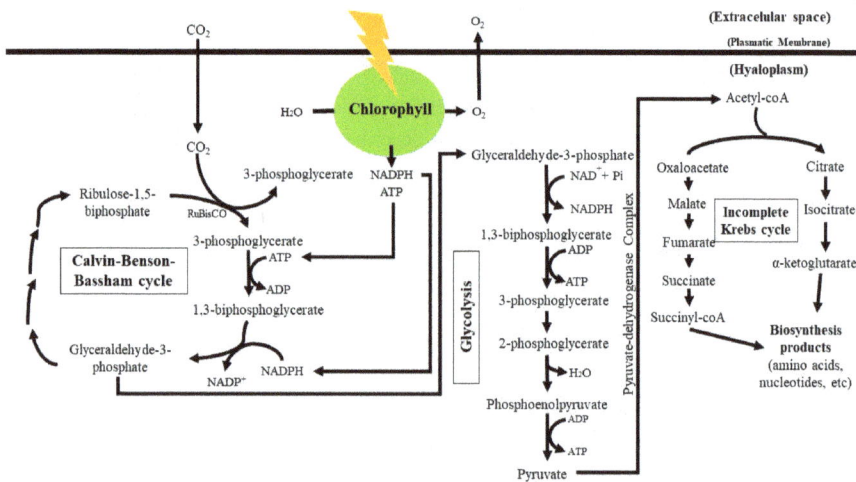

Fig. 3 First stage for the cyanobacterial PHB production in the photobioreactor.

The second stage is characterized by nutrient depletion of culture medium. The cyanobacteria continue in photoautotrophic biomass production, but divert metabolic pathways for the production of reserve compounds, like glycogen (mainly highly branched starch) and PHB [25]. These metabolic routes are shown in Fig. 4 [25, 8]. This stage, also known as the yellow stage, usually lasts 6 to 8 days.

Finally, in the third stage occurs the intracellular conversion of glycogen to PHB in stirred tanks. The glycogen division is realized by the ordered activity of two enzymes, glycogen phosphorylase, which releases glucose-1-phosphate, with the addition of inorganic phosphate (Pi), by sectioning the α-1,4-glycosidic linkages, and glycogen debranching enzyme that ruptures the branch points (α-1,6-glycosidic linkages) and liberates free glucose. Then, via glycolysis is formed the pyruvate, and subsequent conversion to acetyl-CoA by the pyruvate-dehydrogenase complex, and at last, occurs the PHB synthesis [8]. These metabolic routes are shown in Fig. 5 [12, 8]. This stage, also known as the ripening stage, usually lasts 6 to 8 days.

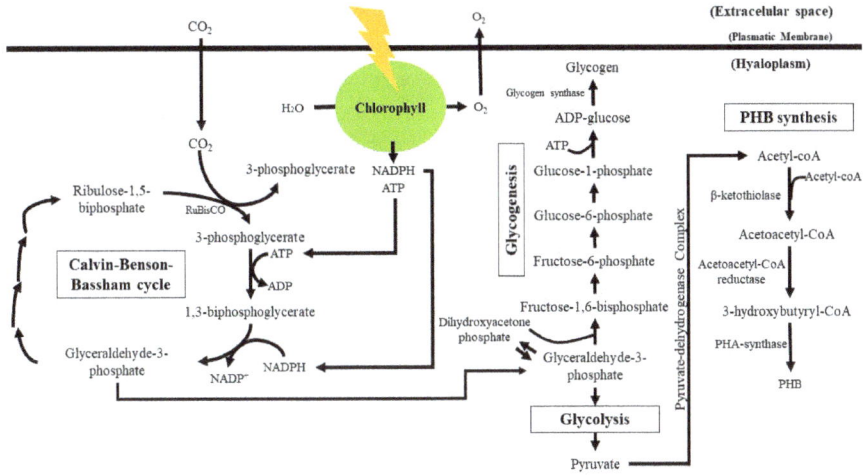

Fig. 4 Second stage for production of cyanobacterial PHB in photobioreactor.

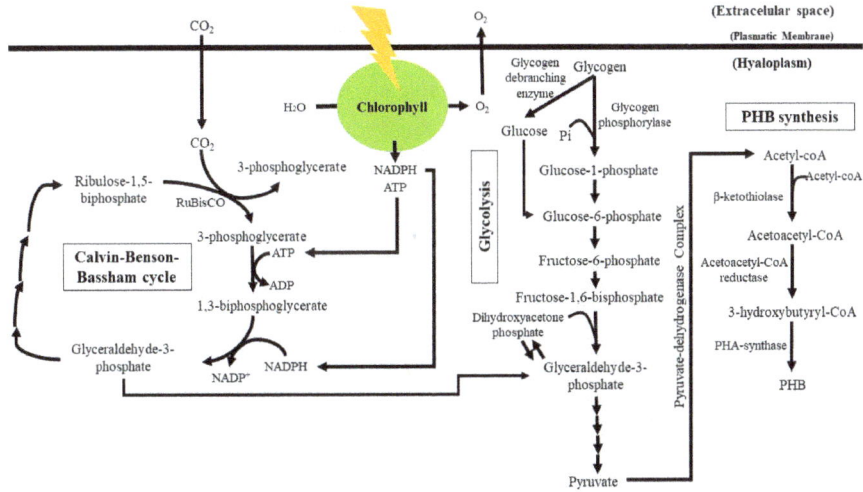

Fig. 5 Third stage for production of cyanobacterial PHB in photobioreactor.

Cyanobacteria demonstrate numerous industrially applicable advantages compared to plants, including a quicker growth rate, increased CO_2 utilization, and better adaptation to genetic engineering [26,25]. Currently, some cyanobacterial species are used for PHA production. *Synechocystis* and *Synechococcus* are extremely small (0.5-2 μm) unicellular cyanobacteria plentiful in just about all illuminated saltwater and freshwater ecosystems. These species in culture medium under nitrogen or phosphorous depletion presented between 4.1 to 11.2% PHB of cell dry weight [8].

Another commonly studied cyanobacterium is the *Arthrospira* (known with *Spirulina*). This species of filamentous cyanobacteria grows naturally in alkaline salt lakes. It has an elevated protein and vitamin content and is mostly grown as a food supplement, and it is the only species made at an industrial scale. The PHB content in non-optimized medium varies 0.5 to 6% PHB of dry cell weight, and in medium, with increased sodium acetate content, the PHB concentration of 2.5% of cell dry weight was achieved. In other research, a strain secluded from the Gujarat coast, India, made 14.7% PHB of dry cell weight under increased salt concentration [8].

Currently, the amount of PHB produced in different strains of photoautotrophic cyanobacteria has been determined. The elevated PHB content was related to the type of strain and not to the genera. From the 137 checked strains, 134 produced PHB [27,8].

4. Factors that influence the biodegradable plastics production from cyanobacteria

Several factors can influence cyanobacteria growth, and changing the growth conditions of these microorganisms may favor the production of compounds of interest, such as PHAs [28,29]. As already reported, nutrient supply limitation, such as nitrogen and phosphorus, is one of the significant factors for the production of biopolymers in cyanobacteria [18]. In a phosphorus-deficient medium, cyanobacteria counteract the scarcity of this nutrient by breaking the up internally stored polyphosphate chains, accumulating intracellular polyhydroxyalkanoate. In the absence of nitrogen, cells are unable to synthesize the proteins needed for reproduction, and therefore, end up saving these compounds in the form of PHB [30,12].

Regarding the supply of nitrogen in the medium, Troschland co-workers[8] demonstrated that a culture containing 0.45 g/L $NaNO_3$ resulted in a 6.6% of PHB dry cell weight (DCW) on *Synechocystissalina*. However, when nitrogen and phosphorus are simultaneously limited, the PHB value reaches 16.4% (DCW) in *Synechocystis*sp. and 22% (DCW) in *Nostocmuscorum* strains [31, 32]. Bhati and Mallick [33] also reported that biosynthesis of this biopolymer by *Nostocmuscorum* resulted in an average yield of

0.098 and 0.101 g L^{-1}d^{-1} per g of biomass under limited phosphorus and nitrogen conditions, respectively. Besides nutrient limitation, other factors such as temperature, light intensity, pH, and salinity influence PHA production in cyanobacteria. What makes PHA attractive is material properties, which are defined during its formation. Therefore, adequate carbon sources, co-substrates (PHA building block precursors), and the selection of highly productive strains are also very relevant factors [34]. In this sense, Table 1 [31,32, 35-47] presents some cyanobacteria accumulating PHAs under specific growth conditions.

Table 1 Different culture conditions of cyanobacteria in the PHAs production. DCW, dry cell weight; N, nitrogen, P, phosphorus;nr, not reported. Adapted from Troschl et al. [8].

Cyanobacteria	Culture conditions	PHA Composition	Content % (DCW)	Ref.
*Anabaena*sp.	Nr	PHB	2.3	[35]
Arthrospiraplatensis	Nr	PHB	6	[36]
*Arthrospira*sp.	Nr	PHB	1	[37]
Arthrospirasubsalsa	N limitation	PHB	14.7	[38]
Aulosirafertilissima	P limitation	PHB	10	[39]
Nostocmuscorum	N and P limitation	PHB	8.7	[40]
Nostocmuscorum	N and P limitation	PHB	22	[31]
Nostocmuscorum	10% CO$_2$	PHB	22	[41]
*Phormidium*sp.	N limitation	PHB	14.8	[42]
Spirulinasubsalsa	Increasedsalinity	PHB	7.45	[38]
Synechococcuselongates	N limitation	PHB	17.5	[43]
Synechococcuselongates	P limitation	PHB	7.02	[43]
*Synechocystis*sp.	N limitation	PHB	8	[44]
*Synechocystis*sp.	N limitation	PHB	4.1	[45]
*Synechocystis*sp.	N limitation	PHB	9.5	[46]
*Synechocystis*sp.	P limitation	PHB	11.2	[46]
*Synechocystis*sp.	N and P limitation	PHB	16.4	[32]

Synechocystis sp.	N and P limitation	PHB	11	[47]

Generally, each cyanobacterial species is defined by its optimum growth temperature. The temperature has visible effects and is a limiting factor sensitive to the metabolic activities and growth of these microorganisms [48]. In a growing medium where thermal stress was reported, PHA accumulation was significantly increased. These results may be related to stress factors, where, in extreme situations, PHAs are formed as a source of energy reserve to acclimate these factors [49].

Light intensity is also another critical parameter for cyanobacterial growth and PHA production. Intensity level and duration of light exposure plays a predominant role in regulating metabolic levels and impose changes in their intracellular content [50]. For example, the higher the light intensity, the greater the amount of biomass produced. However, the photo-inhibition phenomenon can occur and damage the growth of cyanobacteria [51]. At the same time, the light source can strongly influence the relation of photosynthesis of microalgae and, consequently, the process efficiency [52].

Besides, pH is an essential factor for cyanobacterial growth, and salinity also has influence on its composition. Although there are still few studies related to PHA production concerning changes in the medium pH, it is known that for most cyanobacteria, their ideal range is between pH 7-9[53]. However, the typical amounts for salinity may differ from one species to another, depending on your stress and permeability [54].

As for carbon supply, some studies report that an increase in CO_2 concentration may also provide an increase in carbon stock sources such as PHB. In strains of *Spirulina* sp. an increase in the production of these biopolymers was correlated with the highest carbon concentration sources in the culture [55].

Although the most widely used carbon source in raceway ponds is atmospheric CO_2 uptake, the types of open systems today present a critical problem, since at night the CO_2 use is decreased, and carbon stocks and carbon molecules are reduced [56]. In this sense, the use of closed systems, such as photobioreactors, provides an alternative solution to these problems, such as the carbon dioxide utilization in gas cylinders[8]. In these systems, there is a significant advantage over the control and supply of light and propagation, minimization of evaporative water losses, and contamination.

Although cyanobacteria provide a great solution to the environmental problems caused by fossil plastics, the high cost of producing PHAs compared to conventional types is still one of the significant economic barriers to commercialization [3, 57]. The PHB content

Degradation of Plastics
Materials Research Foundations **99** (2021) 269-289

Materials Research Forum LLC
https://doi.org/10.21741/9781644901335-11

produced by cyanobacteria is also usually low, with a concentration of about 10% dry cell weight (DCW) [58]. Thus, to maintain a high PHB content in a continuous mode of operation and to minimize biomass loss, optimizing crop sources seems convincing. Furthermore, to achieve an optimal scalable process, it is crucial to understand all unit operations as well as bioprocess kinetics.

5. Biodegradable plastics production processes from cyanobacteria

Cyanobacteria have the potential to produce PHBs as a reserve material intracellular energy and carbon. In this sense, in terms of energy efficiency and final product quality, the high accumulation of this biomaterial is promising for the production process [6].

Fig. 6 [6] illustrates a generic process flow diagram (PFD) for producing PHAs from cyanobacteria. The PFD shows two main steps: upstream processing and downstream biomass processing. More specifically, these operating stages include (i) strain selection, (ii) reactor cultivation, (iii) harvesting, (iv) drying, (v) extraction, and (vi) purification of the desired target product.

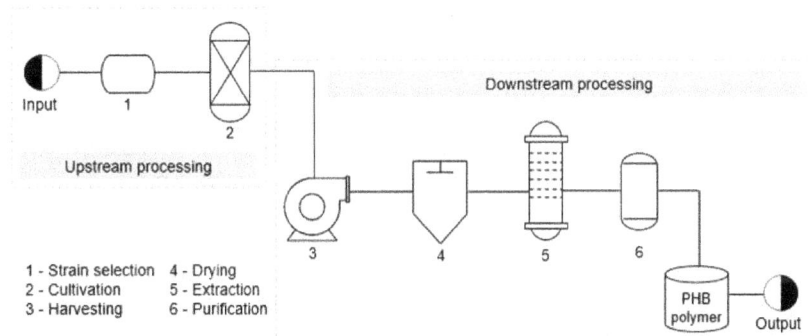

Fig. 6 Generic process flow diagram for the production of PHB from cyanobacteria.

Firstly, one should choose the best strain with maximum productivity, which is directly influenced by the substrate - subsequently, the optimization of the culture medium constituents. The cyanobacteria strain is the prime factor for increasing biopolymer accumulation, whose yields should be in the range of 5 to 70% dry cell weight. Therefore, the production process will only start if the parameters mentioned above are enhanced. Obviously, before this, it is crucial to enlighten the metabolic routes of PHB biosynthesis in cyanobacteria [18].

Cultivation systems are considered the hottest point in the whole process. Today, there are two main ways to cultivate cyanobacteria on a commercial scale, including raceway ponds (open systems) and photobioreactors (closed systems). Raceway ponds are most widely accepted worldwide. Typically, they operate with the aid of a paddle wheel and have a surface area ranging from 300 to 5000 m^2. These systems are easy to build and maintain and are cheaper, but the contamination risk is high enough to affect productivity [59]. According to Samantarayand co-workers[60], the use of raceway ponds is desirable to the wastewater treatment fed fishpond discharge for the cultivation of *Aulosirafertilissima*, whose strain has shown PHB yields of approximately 90 g/m^3. Other open systems for cyanobacteria cultivation of the *Dunaliella* genus are the natural lakes and rivers, thin layer systems, and circular ponds [61]. Already photobioreactors have the most varied configurations with capacities ranging from 100 to 1000 L, are designed with sophisticated operational control parameters (for example, light, pH, temperature, mass transfer, among others) and have higher productivities. However, they are more expensive, and its targeted use only cultivates cyanobacteria species capable of producing fine chemical molecules of high value and the biopolymers. The most commonly used photobioreactors are tubular and flat-plate; however, cyanobacterial PHB production in these systems has rarely been reported.

Upon reaching the end of biochemical reactions in the cultivation system with optimal cell density, upstream processing is over. The next step is the biomass harvest, the first unit operation of downstream processing. The objective of the harvest is to separate the cell broth from concentrating the percentage of solid fraction, i.e., to obtain biomass with about 1 to 15% dry weight. Harvesting is usually done in a centrifuge, and the operation of this device can contribute up to 30% of total production costs. Some other harvesting techniques can also be used, such as sedimentation, electro and chemical flocculation, membrane filtration, and ultrasound-based separation. Therefore, it is essential to intensify research to develop cost-effective cyanobacterial biomass harvesting equipment [62].

Subsequently, the removal of excess water from the wet biomass should be done immediately to prevent degradation and then perform the remaining downstream steps. Drying is an extremely energy-intensive unit operation to increase biomass concentration by 90 to 95% (w/w moisture). Devices used for dehydration include spray drying, drum drying, and lyophilization. Spray drying technology is preferred on a commercial-level production. Nevertheless, it is still a costly bottleneck for the PHBs manufacture from cyanobacteria [61].

The last unit operations of downstream processing are extraction and purification steps. The extraction of intracellular metabolites can be done using chemical and mechanical methods for better cyanobacterial cell disruption. Liquid extraction is the most used operation for the recovery of the metabolites of interest, which is based on the use of high organic solvent volumes (i.e., enzymatic liquid and supercritical fluid extraction treatments). This raises the PHB production costs at the expense of environmental damage. On the other hand, mechanically based technologies such as ultrasound, high-pressure homogenization, and beating can be applied for the extraction of these biopolymers [63]. The application of cell disruption technologies in commercial scale is a testing task since these processes are costly, and morphological characteristics of cyanobacteria influence their efficiency. Once the crude extract is obtained, the chemical molecules are purified [24]. This operation is not economically viable, which can be done by gas chromatography (GC) or high-performance liquid chromatography (HPLC) techniques. These techniques require hydrolysis, methanolysis, or propanolysis of the PHBs in whole cells [6]. Given these circumstances, the PHB purification extraction steps are difficult and time-consuming, so they are offset by the high costs of producing high value-added metabolites.

6.   Commercial applications and economic aspects

The promising advantages of producing bioplastics from cyanobacteria are numerous and already known. Besides, the market for this type of raw material has been expanding quickly. This biologically degradable biopolymer could be targeted for application in many areas, including the food industry, pharmaceuticals, biomedical, and agriculture. In the medical field, for example, bioplastics can be used as suture material, bone plates, surgical meshes, pins, and stents. In the pharmaceutical industry, they are used as drug delivery devices. For food purposes, the use of bioplastics can be directed to packaging [64,65,25].

Concerning the economy, estimates for the current year were that the global PHA market would reach about USD 57 million; while by 2024, the market share should reach USD 98 million at a compound annual growth rate of 11.2% [66]. Despite these potentials, the commercial PHAs production from cyanobacteria is not expected in the short-to-medium term. This is supported by the fact that the price of exploitable bioplastics is up to 17 times higher compared to conventional plastics. Also, they have a very narrow market, despite the potential substitution of 33% of marketable polymeric materials [67]. Genetic engineering approaches and improved fermentation systems capable of increasing productivity could lower the selling price [68].

Still, notably PHAs marketing and industrialization expenses do not decrease considering the issues of low production volume, yield, substrate cost, and recovery technique, despite the parallel exploitation of economic substrates such as waste, glycerol, and cellulose. These factors that impact the production costs of intracellular bioplastics should take into account the adoption of new technologies, such as the co-production of PHAs with other extracellular products, to balance the economic aspects [24]. Finally, market demand for cyanobacterial bulk products, such as bioplastics and trends towards sustainability and circular bio-economy, forces the expansion of the PHAs industry.

7. Final considerations and prospects

The increasing availability of substances with minimal environmental impact, the green footprint, and the implementation of favorable regulatory policies will result in the market expansion of biodegradable polymeric raw materials worldwide.

Cyanobacteria have great potential to produce biopolymers, and the yield of such biopolymers varies with growth conditions. Obtaining cyanobacterial PHAs is related as a way of reducing the environmental problems caused by fossil polymers and provides greater competitiveness over synthetic ones.

Studies on cyanobacterial PHA production are at their beginning and distant from their commercialization. In this sense, intensive research studies are needed to new strategies of the growth and development of better strains and to improve processing steps for PHAs that are more economical.

Some of the newest progress in biopolymer research has focused on genetic modifications of successfully described PHA-accumulating cyanobacteria in *Synechococcus* and *Synechocysti*. Besides, other study topics are addressing the most productive and easy to process types of cyanobacteria.

Finally, it is worth noting that many improvements throughout the upstream and downstream stages still need to be made to boost the cyanobacterial PHBs production. In parallel, genetic improvement issues are paramount for the development of efficient strains with the potential for the accumulation of these biopolymers. In the future, cyanobacterial bioplastics are expected to conquer broader markets.

References

[1]   G.J.M. De Koning, Prospects of bacterial Poly[(R)-3-(Hydroxyalkanoates), Technische Universiteit Eindhoven, The Netherlands (1993). https://doi.org/10.6100/IR403691

[2]   R. Bhati, Biodegradable plastics production by cyanobacteria, in: M. Khoobchandani, A. Saxena (Eds), Biotechnology products in everyday life, EcoProduction, Springer, Cham, 2019, pp. 131-143

[3]   M. Lackner, Bioplastics - Biobased plastics as renewable and/or biodegradable alternatives to petroplastics, in: K. Othmer (Ed), Encyclopedia of Chemical Technology, Wiley, 2015, pp. 1-41

[4]   CIEL (Center for International Environmental Law), Fossils, plastics, & petrochemical feedstocks, in: Fueling Plastics, Center for International Environment Law, 2017, pp. 1-5

[5]   R. Geyer, J.R. Jambeck, K.L. Law, Production, use, and fate of all plastics ever made. Sci. Adv. 3 (2017) e1700782.https://doi.org/ 10.1126/sciadv.1700782

[6]   D. Kamravamanesh, M. Lackner, C. Herwig, Bioprocess engineering aspects of sustainable polyhydroxyalkanoate production in cyanobacter, Bioengineering 5 (2018) 111. https://doi.org/10.3390/bioengineering5040111

[7]   H. Karan, C. Funk, M. Grabert, M. Oey, B. Hankamer, Green bioplastics as part of a circular bioeconomy, Trends Plant Sci. 3 (2019) 237-241. https://doi.org/ 10.1016/j.tplants.2018.11.010

[8]   C. Troschl, K. Meixner, B. Drosg. Cyanobacterial PHA production – review of recent advances and a summary of three years' working experience running a pilot plant, Bioengineering 4 (2017) 1-19.https://doi.org/10.3390/bioengineering4020026

[9]   G.Q. Chen, A microbial polyhydroxyalkanoates (PHA) based bio- and materials industry. Chem. Soc. Rev. 38 (2009) 2434–2446. https://doi.org/10.1039/b812677c.

[10]   M. Koller, L. Maršálek, M.M. de Sousa Dias, G. Braunegg, G, Producing microbial polyhydroxyalkanoate (PHA) biopolyesters in a sustainable manner. New Biotechnol. 37 (2017) 24-38.https://doi.org/10.1016/j.nbt.2016.05.001

[11]   P.M. Halami, Production of polyhydroxyalkanoate from starch by the native isolate Bacillus cereus CFR06, World J.Microbiol. Biotechnol. 24 (2008) 805–812. https://doi.org/10.1007/s11274-007-9543-z

[12]   S. Balaji, K. Gopi, B. Muthuvelan, A review on production of poly βhydroxybutyrates from cyanobacteria for the production of bio plastics, Algal Res. 2 (2013) 278–285.https://doi.org/10.1016/j.algal.2013.03.002

[13]   A.S. Luyt, S.S. Malik, Can biodegradable plastics solve plastic solid waste accumulation, in: S.M. Al-Salem (Ed), Plastics to energy, fuel, chemicals, and sustainability implications, Elsevier, 2019, pp. 403-423

[14]   G.F. Brito, P. Agrawal, E.M. Araújo, T.J.A. Mélo, Biopolímeros, Polímeros
       Biodegradáveis e Polímeros Verdes, Revista Eletrônica de Materiais e Processos 6
       (2011) 127-139.ISSN:1809-8797

[15]   S. Khanna, A.K. Srivastava, Recent advances in microbial polyhydroxyalkanoates,
       Process. Biochem. 40 (2005) 607–619.https://doi.org/ 10.1016/j.procbio.2004.01.053

[16]   L. Avérous, Polylactic Acid: Synthesis, properties and applications. In:
       M.N.Belgacem, A.Gandini (Eds), Monomers, polymers and composites from
       renewable resources, Elsevier, Oxford, 2008

[17]   M.G. Bastos Lima, Toward multipurpose agriculture: food, fuels, flex crops, and
       prospects for a bioeconomy, Global Environ. Politics 18 (2018) 143–150.
       https://doi.org/10.1162/glepa00452

[18]   C. Zhang, P.L. Show, S.H. Ho, Progress and perspective on algal plastics – A
       critical review, Bioresour. Technol. 289 (2019) 121700.
       https://doi.org/10.1016/j.biortech.2019.121700

[19]   D.M. Arias, J. García, E. Uggetti, Production of polymers by cyanobacteria grown
       in wastewater: Current status, challenges and future perspectives, New Biotechnol. 55
       (2020) 46–57. https://doi.org/10.1016/j.nbt.2019.09.001

[20]   B. Drosg, I. Fritz, F. Gattermayer, L. Silvestrini, Photo-autotrophic production of
       poly(hydroxyalkanoates) in cyanobacteria, Chem. Biochem. Eng. Q. 29 (2015) 145–
       156. https://doi.org/10.15255/CABEQ.2014.2254

[21]   V. Mendhulkar, L. Shetye, Synthesis of biodegradable polymer
       polyhydroxyalkanoate (PHA) in cyanobacteria Synechococcus elongates under
       mixotrophic nitrogen- and phosphate-mediated stress conditions, Ind. Biotechnol. 13
       (2017) 85–88. https://doi.org/10.1089/ind.2016.0021

[22]   W. Hauf, M. Schlebusch, J. Hüge, J. Kopka, M. Hagemann, K. Forchhammer,
       Metabolic changes in Synechocystis PCC6803 upon Nitrogen-Starvation: Excess
       NADPH sustains polyhydroxybutyrate accumulation. Metabolites 3 (2013) 101–118.
       https://doi.org/10.3390/metabo3010101

[23]   Y. Nakaya, H. Iijima, J. Takanobu, A. Watanabe, M.Y. Hirai, T. Osanai, One day
       of nitrogen starvation reveals the effect of sigE and rre37 overexpression on the
       expression of genes related to carbon and nitrogen metabolism in *Synechocystis* sp.
       PCC 6803. J. Biosci. Bioeng. (2015) 1–7. https://doi.org/10.1016/j.jbiosc.2014.12.020

[24]   A.K. Singh, N. Mallick, Advances in cyanobacterial polyhydroxyalkanoates production, FEMS Microbiology Letters 364 (2017) 1-13. https://doi.org/10.1093/femsle/fnx189

[25]   E. Markl, H. Grünbichler, M. Lackner, Cyanobacteria for PHB bioplastics production: A review. Y.K. Wong (Ed), Algae, IntechOpen, 2018.

[26]   T. Heidorn, D. Camsund, H.H. Huang, P. Lindberg, P. Oliveira, K. Stensjo, et al., Synthetic biology in cyanobacteria engineering and analyzing novel functions, Methods Enzymol. 497 (2011) 539-579.https://doi.org/10.1016/B978-0-12-385075-1.00024-X

[27]   Kaewbai-ngam, A. Incharoensakdi, T. Monshupanee, Increased accumulation of polyhydroxybutyrate in divergent cyanobacteria under nutrient-deprived photoautotrophy: An efficient conversion of solar energy and carbon dioxide to polyhydroxybutyrate by *Calothrixscytonemicola* TISTR 8095, Bioresour. Technol. 212 (2016) 342–347.https://doi.org/10.1016/j.biortech.2016.04.035

[28]   R.B. Derner, S. Ohse, M. Villela, S.M.F.R. Carvalho, Microalgas, produtos e aplicações, Cienc. Rural 36 (2006) 1959–1967. https://doi.org/10.1590/S0103-84782006000600050

[29]   R.A. Soni, K. Sudhakar, R.S. Rana, Spirulina - from growth to nutritional product: a review, Trends Food Sci. Technol. 69 (2017) 157–171.https://doi.org/10.1016/j.tifs.2017.09.010

[30]   M. Schlebusch, K. Forchhammer, Requirement of the nitrogen starvation-induced protein sl10783 for polyhydroxybutyrate accumulation in *Synechocystis*sp. strain PCC 6803, Appl. Environ. Microbiol 76 (2010) 6101–6107. https://doi.org/10.1128/AEM.00484-10

[31]   B. Panda, L. Sharma, N. Mallick, Poly-hydroxybutyrate accumulation in *Nostocmuscorum* and *Spirulinaplatensis* under phosphate limitation. J. Plant Physiol. 162 (2005) 1376–1379. https://doi.org/10.1016/j.jplph.2005.05.002

[32]   D. Kamravamanesh, S. Pflugl, W. Nischkauer, A. Limbeck, A. Lackner, C. Herwig, Phtosynthetic poly-β-hydroxybutyrate accumulation in unicellular cyanobacterium, *Synechocystis* sp. PCC 6714. AMB Express 7 (2017) 143.https://doi.org/ 10.1186 / s13568-017-0443-9

[33]   R. Bhati, N. Mallick, Production and characterization of poly 3-hydroxybutyrate-*co*-3-hydroxyvalerate co-polymer by a N$_2$-fixing cyanobacterium *Nostocmuscorum* Agardh, J. Chem. Technol. Biot. 87 (2012) 505-512. https://doi.org/10.1002/jctb.2737

[34]   M. Koller, Chemical and biochemical engineering approaches in manufacturing polyhydroxyalkanoate (PHA) biopolyesters of tailored structure with focus on the diversity of building blocks, Chem. Biochem. Eng. Q. 32 (2018) 413-438. https://doi.org/10.15255/CABEQ.2018.1385

[35]   K. Gopi, S. Balaji, B. Muthuvelan, Isolation purification and screening of biodegradable polymer PHB producing cyanobacteria from marine and freshwater resources, Iran. J. Energy Environ. 5 (2014) 94–100.https://doi.org/ 10.5829/idosi.ijee.2014.05.01.14

[36]   J. Campbell, S.E. Stevens, D.L. Balkwill, Accumulation of poly-beta-hydroxybutyrate in *Spirulina platensis*, J. Bacteriol. 149 (1982)361–363

[37]   M. Vincenzini, C. Sili, R. de Philippis, A. Ena, R. Materassi, Occurrence of poly-beta-hydroxybutyrate in Spirulina species, J. Bacteriol. 172 (1990) 2791–2792. https://doi.org/10.1128 / jb.172.5.2791-2792.1990

[38]   A. Shrivastav, S.K. Mishra, S. Mishra, Polyhydroxyalkanoate (PHA) synthesis by *Spirulina subsalsa* from Gujarat coast of India, Int. J. Biol. Macromol. 46 (2010) 255–260. https://doi.org/ 10.1016/j.ijbiomac.2010.01.001

[39]   S. Samantaray, N. Mallick, Production and characterization of poly-hydroxybutyrate (PHB) polymer from *Aulosirafertilissima*, J. Appl. Phycol. 24 (2012) 803–814. https://doi.org/10.1007/s10811-011-9699-7

[40]   L. Sharma, N. Mallick, Accumulation of poly-hydroxybutyrate in *Nostocmuscorum*: Regulation by pH, light-dark cycles, N and P status and carbon sources, Bioresour. Technol. 96 (2005) 1304–1310. https://doi.org/10.1016/j.biortech.2004.10.009

[41]   R. Bhati, N. Mallick, Carbon dioxide and poultry waste utilization for production of polyhydroxyalkanoate biopolymers by Nostoc*muscorum* Agardh: A sustainable approach, J. Appl. Phycol. 28 (2016) 161–168. https://doi.org/ 10.1007/s10811-015-0573-x

[42]   A. Kaewbai-Ngam, A. Incharoensakdi, T. Monshupanee, Increased accumulation of polyhydroxybutyrate in divergent cyanobacteria under nutrient-deprived photoautotrophy: An efficient conversion of solar energy and carbon dioxide to polyhydroxybutyrate by *Calothrixscytonemicola* TISTR 8095, Bioresour. Technol. 212 (2016) 342–347. https://doi.org/10.1016/j.biortech.2016.04.035

[43]   V. Mendhulkar, L. Shetye, Synthesis of biodegradable polymer polyhydroxyalkanoate (PHA) in cyanobacteria *Synechococcus elongates* under

mixotrophic nitrogen and phosphate mediated stress conditions, Ind. Biotechnol.13 (2017) 85–88. https://doi.org/10.1089/ind.2016.0021

[44]   R. Carpine, G. Olivieri, K. Hellingwerf, A. Pollio, G. Pinto, A. Marzocchella, Poly-hydroxybutyrate (PHB) production by cyanobacteria, New Biotechnology (2016) S19–S20.

[45]   G.F. Wu, Q.Y. Wu, Z.Y. Shen, Accumulation of poly-beta-hydroxybutyrate in cyanobacterium Synechocystis sp. PCC6803, Bioresour. Technol. 76 (2001) 85–90. https://doi.org/10.1016/s0960-8524(00)00099-7

[46]   B.B. Panda, P. Jain, L. Sharma, N. Mallick, Optimization of cultural and nutritional conditions for accumulation of poly-b-hydroxybutyrate in *Synechocystis* sp. PCC 6803, Bioresour. Technol. 97 (2006) 1296–1301.https://doi.org/10.1016/j.biortech.2005.05.013

[47]   B. Panda, N. Mallick, N, Enhanced poly-hydroxybutyrate accumulation in a unicellular cyanobacterium, *Synechocystis* sp. PCC 6803. Lett. Appl. Microbiol. 44 (2007) 194–198. https://doi.org/10.1111/j.1472-765X.2006.02048.x

[48]   M.M. El-Sheekh, A.E. El-Gamal, A. Bastawess, A. El-Bokhomy, Production and characterization of biodiesel from the unicellular green alga *Scenedesmus obliquus*, Energ. sources, Part A: Recovery. Utilization and Environmental Effects 39 (2017) 783-793. https://doi.org/10.1080/15567036.2016.1263257

[49]   B. Chen, C. Wan, M.A. Mehmood, J. Chang, F. Bai, X. Zhao, Manipulating environmental stresses and stress tolerance of microalgae for enhanced production of lipids and value-added products – a review, Bioresour. Technol. 244 (2017) 1198–1206.https://doi.org/10.1016/j.biortech.2017.05.170.

[50]   L. Winter, I.T.D. Cabanelas, A.N. Órfão, E. Vaessen, D.E. Martens, R.H. Wijffels, M.J. Barbosa, The influence of day length on circadian rhythms of *Neochloriso leoabundans*, Algal Researc. 22 (2017) 31–38.https://doi.org/10.1016/j.algal.2016.12.001

[51]   X.Y. Wu, T.S. Song, X.J. Zhu, P. Wei, C.C. Zhou,). Construction and operation of microbial fuel cell with *Chlorella vulgari s*biocathode for electricity generation, Appl. Biochem. Biotechnol. 171 (2013) 2082-2092. https://doi.org/10.1007/s12010-013-0476-8

[52]   J.C.W. Lan, K. Raman, C.M. Huang, C. M. Chang, C. M. The impact of monochromatic blue and red LED light upon performance of photo microbial fuel cells

(PMFCs) using *Chlamydomonasreinhardtii* transformation F5 as biocatalyst, Biochem. Eng. J. 78, (2013) 39-43. https://doi.org/10.1016/j.bej.2013.02.007

[53] A.L. Gonçalves, J.C.M. Pires, M. Simões, A review on the use of microalgal consortia for wastewater treatment, Algal Res. 24 (2017) 403–415.https://doi.org/10.1016/j.algal.2016.11.008

[54] L. Xia, J. Rong, H. Yang, Q. He, D. Zhang, C. Hu, NaCl as an effective inducer for lipid accumulation in freshwater microalgae *Desmodesmusabundans*, Bioresour. Technol. 161 (2014) 402–409.https://doi.org/10.1016/j.biortech.2014.03.063.

[55] V.C. Coelho, C.K. da Silva, A.L. Terra, M.G. de Morais, Polyhydroxybutyrate production by *Spirulina* sp. LEB 18 grown under different nutrient concentrations, Afr. J. Microbiol. Res. 9 (2015) 1586–1594. https://doi.org/10.5897/AJMR2015.7530

[56] J.O. Eberly, R.L. Ely, Photosynthetic accumulation of carbon storage compounds under CO_2 enrichment by the thermophilic cyanobacterium *Thermosynechococcus elongates*, J. Ind. Microbiol. Biotechnol. 39 (2012) 843–850. https://doi.org/10.1007/s10295-012-1092-2

[57] D. Kamravamanesh, C. Slouka, A. Limbeck, M. Lackner, C. Herwig, Increased carbohydrate production from carbon dioxide in randomly mutated cells of cyanobacterial strain *Synechocystis* sp. PCC 6714: Bioprocess understanding and evaluation of productivities, Bioresour. Technol. 273 (2019) 277–287. https://doi.org/ 10.1016/j.biortech.2018.11.025

[58] B. Drosg, Photo-autotrophic production of Poly(hydroxyalkanoates) in Cyanobacteria. Chem. Biochem. Eng. Q. 29 (2015) 145–156. https://doi.org/10.15255/CABEQ.2014.2254

[59] L. Christenson, R. Sims, Production and harvesting of microalgae for wastewater treatment, biofuels, and bioproducts, Biotechnol, Adv., 29 (2011) 686-702.https://doi.org/10.1016/j.biotechadv.2011.05.015.

[60] S. Samantaray, J.K. Nayak, N. Mallick, Wastewater Utilization for Poly-β-Hydroxybutyrate production by the cyanobacterium *Aulosirafertilissima* in a recirculatory aquaculture system, Appl. Environ. Microbiol. 77 (2011) 8735–8743. https://doi.org/10.1128/AEM.05275-11

[61] D. Noreña-Caro, M.G. Benton, Cyanobacteria as photoautotrophic biofactories of high-value chemicals, J. CO_2 Util. 28 (2018) 335-366.https://doi.org/10.1016/j.jcou.2018.10.008

[62]   S. Khanra, Downstream processing of microalgae for pigments, protein and carbohydrate in industrial application: A review, Food Bioprod. Process., 110 (2018) 60-84. https://doi.org/10.1016/j.fbp.2018.02.002

[63]   B. Delattre, Production, extraction and characterization of microalgal and cyanobacterial exopolysaccharides. Biotechnol. Adv., 34 (2016) 1159-1179. https://doi.org/10.1016/j.biotechadv.2016.08.001

[64]   W. Vermaas, Production of bioplastics and other biomaterials from the cyanobacterium *Synechocystis*. https://asu.pure.elsevier.com/en/publications/production -of-bioplastics-and-other-biomaterials-from-the-cyanoba, 2019 (accessed 2 December 2019)

[65] K. Meixner, Cyanobacteria for bioplastic production. Conference: Erasmus International Week at Hochschule Darmstadt, DEC 4-8, 2017. At: Hochschule Darmstadt (GER) Affiliation: University of Natural Resources and Life Sciences, Vienna; Institute of Environmental Biotechnology, 2017

[66]   Markets & Markets, Polyhydroxyalkanoate (PHA) Market. https:// www.marketsandmarkets.com /Market-Reports/pha-market-395.html, 2019 (accessed 9 November 2019).

[67]   P. Singh, R. Kumar, Radiation physics and chemistry of polymeric materials, in: V. Kumar, B. Chaudhary, V. Sharma, K. Verma (Eds.) Radiation effects in polymeric materials, Springer, pp 35-68

[68]   B.E. DiGregorio, Biobased performance bioplastic: Mirel, Chem. Biol. Innovat., 16 (2009) 1-2. https://doi.org/10.1016/j.chembiol.2009.01.001

Degradation of Plastics
Materials Research Foundations **99** (2021) 290-324

Materials Research Forum LLC
https://doi.org/10.21741/9781644901335-12

Chapter 12

Plastic Degradation and its Environmental Implications

Manviri Rani[1], Uma Shanker[*2]

[1]Department of Chemistry, Malaviya National Institute of Technology JLN Marg, Rajasthan-India

[2]Department of Chemistry, Dr B R Ambedkar National Institute of Technology Jalandhar, Punjab-India

manviri.chy@mnit.ac.in; shankeru@nitj.ac.in

Abstract

Extensive use of non-biodegradable plastic by anthropogenic activities is posing a severe threat to the global environment in the form of massive waste disposal and problems of closed-landfill sites and rising water and land pollution. Organisms are facing challenge to their lives once they consume plastic in the form of food. In addition, the plastic debris may have additive chemicals having possibility to leach out. Therefore, proper degradation of plastic to lessen negative environmental implications is necessary. Earlier the people were ignorant about this problem however, now-a-days, people are serious about its negative impact. Hence, biodegradable plastics are coming into the trend in the market. Interaction of plastic (adsorbing characteristics) with the environment results in new functional groups on its surface in a dynamic situation. In these circumstances, oxygenated, thermal, bio-based, and photocatalytic degradation of plastics is in high demand. Usually, different-rates along with diverse-pathways have been found in variable polymers like polyethylene, polypropylene, and polyethylene terephthalate. Polyethylene photo-degradation resulted in sharp infrared-peaks of ketones, esters, and acids due to oxidation reaction. Hydrogen peroxide is generated by oxidative action on methylene groups in the backbone of polyethylene terephthalate. The present chapter will provide compiled information about various polymers used in plastic as well as environmental concerns of plastic, finally the chapter concludes with comprehensive details about the degradation of various types of plastics. State-of-the-art on degradable plastics market with an emphasis on principle design for recyclable plastics and biodegradable plastics from renewable raw materials are also included. Factors affecting plastic degradation and metabolic pathways including past and present scenarios have also been discussed in this chapter.

Keywords

Plastic Pollution, Polymer Degradation, Environment Conditions, Biodegradation

Contents

1. Introduction

Since their start in the 1950s, plastics with massive production as well as several applications is transforming human life for convenience and therefore, became an essential part of contemporary culture [1]. Life without it seems almost impossible credited to its versatile nature, molecular structure and additives. The global annual production of synthetic polymers is approaching to 140 metric tons [2]. However, the percentage of recycled or reused plastic is much less, only about 10%. The main reason for such a low percentage is extremely limited disposal options and almost no attention towards solid waste management. Among commodity plastics, polyethylene (PE) with several forms like high density (HDPE), low density (LDPE) and linear low density (LLDPE) are mostly used. Polypropylene (PE), polystyrene (PS), polyvinyl chloride (PVC), polycarbonate (PC), polyethylene terephthalate (PET), and poly(methyl methacrylate) (PMMA) are also in widespread use. These plastic polymers actually contribute nearly 98% share of consumed/produced products worldwide. The plastic consumption demand in Europe followed the order: PE (13,000 kiloton)> PP (8,000 kiloton)>PVC (5000kiloton) and for PET (3,000 kiloton) [3].

The high production along with large consumption of plastic is creating a lot of burden on the environment moreover, due to limited treatment and poor disposal methods also resulting plastic waste. In developing countries with inadequate engineered/planned landfill disposal sites, people used to burn these plastic wastes that leads to evolution of poisonous and destructive gases into the atmosphere. Improper disposal methods such as open-dumping, unrestrained incineration and intuitive composting as well as inappropriate landfill are frequently used in developing countries like India [4]. This practice has further caused concern and called attention for biodegradation of highly stable polyethylene and polystyrene polymers by advanced methods. Furthermore, due to the inert and non-biodegradable nature of plastic, its treatment and disposal is the main problem in urban solid waste management [5]. Plastic polymers release various contaminants and toxic additives once left unprotected in environmental-circumstances [6,7]. It has been reported that due to this 50–80% of debris causes distress in the oceanic environment around the globe [8]. Environment and biodiversity arebeing significantly affected by widespread and unrestrained usage and dumping of such harmful materials [9-11]. Plastics are obstinate in the environment and adsorb contaminants [12-14]. It has been found that microplastics sorb pollutants at a faster rate from environmental matrices (water, air, and soil) [15]. Ioakeimidis and co-workers [16] observed that plastics, with a percentage of 95%, are the major waste at the bottom of the oceans among the total oceanic litter matters. Additionally, plastic debris in the ocean comprises fairly high-levels of organic contaminants e.g., polystyrene having a large number of hexabromocyclododecanes, a flame retardant, and bisphenol A in polycarbonate plastics [17-21].

Polymer degradation is a change in the properties like tensile strength, color and shape of a polymer or polymer-based product under the influence of one or more environmental factors. These factors are heat, light or chemicals such as acids, alkalis and some salts. Such changes are usually undesirable like cracking and chemical disintegration of products or, more rarely, desirable as in biodegradation or deliberately lowering the molecular weight of a polymer for recycling. Degradation of plastic by any methods can be useful in the implication of environment and recycling of plastic waste can be beneficial for further reducing pollution. Chemical structures of various types of plastics are presented in Fig. 1 and 2. Further details about the uses of synthetic and biodegradable plastics are given in Tables 1 and 2).

Table 1 Various types of synthetic plastics and their uses.

Plastic	Use
Polyethylene	Plastic bags, milk and water bottles, food packaging film, toys, irrigation, and drainage pipes, motor oil bottles
Polystyrene	Disposable cups, packaging materials, laboratory ware, certain electronic uses
Polyurethane	Tires, gaskets, bumpers, in refrigerator insulation, sponges, furniture cushioning, and life jackets
Polyvinyl chloride	Automobile seat covers, shower curtains, raincoats, bottles, visors, shoe soles, garden hoses, and electricity pipes
Polypropylene	Bottle caps, drinking straws, medicine bottles, car seats, car batteries, bumpers, disposable syringes, carpet backings
Polyethylene terephthalate (PET)	Carbonated soft drink bottles, processed meat packages peanut butter jars pillow and sleeping bag filling, textile fibers
Nylon	Polyamides or Nylon are used in small bearings, speedometer gears, windshield wipers, water hose nozzle, football helmets, racehorse shoes, inks, clothing parachute fabrics, rainwear, and cellophane
Polycarbonate	Nozzles on paper making machinery, street lighting, safety visors, rear lights of cars, baby bottles and for houseware. Skylights and the roofs of greenhouses, sunrooms and verandahs. Lens in glasses
Polytetraflouroethylene (PTFE)	PTFE is used in various industrial applications such specialized chemical plant, electronics and bearings. It is met with in the home as a coating on non-stick kitchen utensils, such as saucepans and frying pans

Table 2 Various types of biodegradable plastics and their uses.

Plastics	Uses
Polyglycolic acid (PGA)	Specialized applications; controlled drug releases; implantable composites; bone fixation parts
Polylactic acid (PLA)	Packaging and paper coatings; other possible markets include sustained release systems for pesticides and fertilizers, mulch films, and compost bags
Polycaprolactone (PCL)	Long-term items; mulch and other agricultural films; fibers containing herbicides to control aquatic weeds; seedling containers; slow-release systems for drugs
Polyhydroxybutyrate (PHB)	Products like bottles, bags, wrapping film, and disposable nappies, as a material for tissue engineering scaffolds and controlled drug release carriers
Polyhydroxyvalerate (PHBV)	Films and paper coatings; other possible markets include biomedical applications, therapeutic delivery of worm medicine for cattle, and sustained release systems for pharmaceutical drugs and insecticides
Polyvinyl alcohol (PVOH)	Packaging and bagging applications that dissolve in water to release Products such as laundry detergent, pesticides, and hospital washables
Polyvinyl acetate (PVAc)	Adhesives, the packaging applications include boxboard manufacture, paper bags, paper lamination, tube winding and remoistenable labels

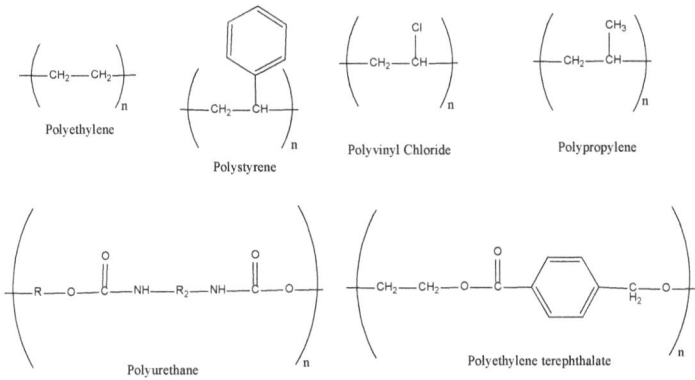

Fig. 1 Structures of conventional plastics polyethylene.

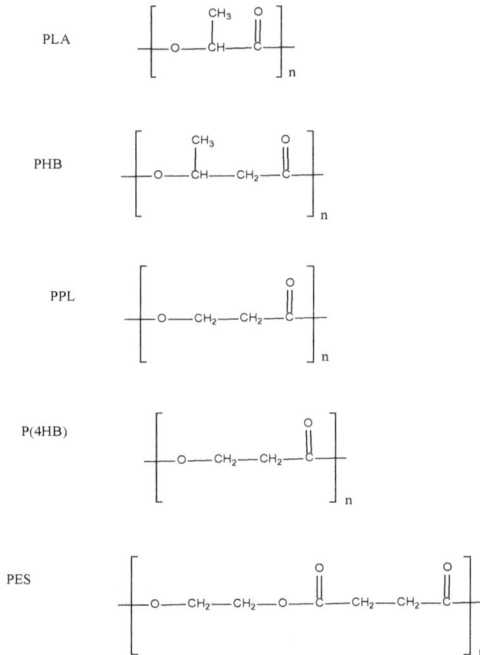

*Fig. 2 Chemical structure of poly(lactide)(PLA), poly(3-hydroxybutyrate)(PHB),
poly(propiolactone)(PPL), poly(ε-caprolactone)(PCL), poly(ethylene succinate) (PES),
poly(butylenes succinate)(PBS), poly(3-hydroxybutyrate-co-3-hydroxyvalerate)(PHBV)
and poly(ester carbonate)(PEC).*

Degradation of Plastics   Materials Research Forum LLC
Materials Research Foundations **99** (2021) 290-324   https://doi.org/10.21741/9781644901335-12

1.1   Commodity polymers

Now a days, polyethylene, polypropylene, polyvinylchloride, polyethylene terephthalate, polystyrene, polycarbonate and poly (methyl)methacrylate represents the seven commodity polymers that are highly used. Such macromolecules have their styles of degradation pathway along with light, chemicals, and temperature-resistant. Some polymers like polyethylene, polypropylene and poly (methyl)methacrylate are highly sensitive to oxidation under UV radiation whereas polyvinylchloride changes its color at high temperature and becomes brittle due to loss of hydrogen chloride gas. Polyethylene terephthalate is affected by the action of water and strong acids, while strong alkalis depolymerize polycarbonates. Polyethylene typically degrades via unsystematic scission. On heating to a temperature above 450 °C, polyethylene is converted into a mixture of molecules similar to gasoline. Poly-α-methylstyrene showed breakage of the polymeric chain, which finally depolymerize into different essential monomers. Fig. 3 represents the worldwide production of various types of polyethylene.

2.   Environmental degradation of polymers

On worldwide scale, limited reports are available on the assessment of degradation of polymers under environmental conditions. Scanning electron microscopic analysis of plastic samples collected from beaches (beached plastic pellets) revealed that relative inertness is generally emphasized but also some degradation and oxidative aging is evident in the plastic pellets. This finally led to complete disintegration of the plastic pellets and dispersal as dust. Further, it was also recommended partial degradation of floating plastic granules while complete degradation of stranded pellet granules of plastic occurs due to exposure of UV radiation [22].

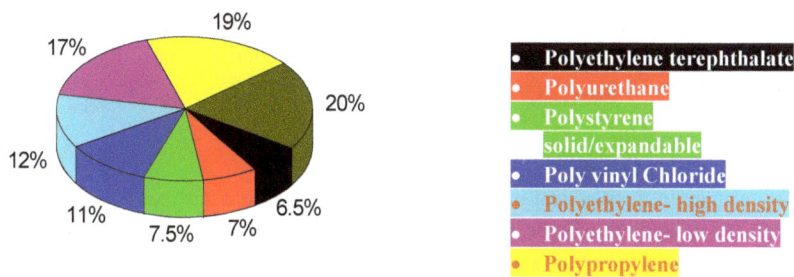

Fig. 3 Production of various types of polyethylene worldwide.

Orhan and Buyukgungor [23] evaluated degradation of low-density polyethylene (that has been blended with starch) baggage in soil and reported that the presence of starch does not affect the degradation of plastic bags even after 90 days. Slight degradation was observed if the soil was injected with fungi. On soil burial for 5 months, such plastic polymers were converted into microplastics. Among the inspected polymeric films, polyethylene blended with starch lost most of the susceptibility followed by low-density polyethylene. Endo et al. [24] studying the sorption properties of plastic pellets suggested that the degree of discoloring can be an index of the residence time of the pellets in the seawater. One of the additives phenolic antioxidants is found responsible for discoloration. These phenolic antioxidants quench free radicals generated by exposure to UV light, high-temperature polymerization, nitrogen oxides in the air, and preventing the polymer from oxidation. Phenolic antioxidant degrades under vigorous conditions into byproducts with quinonoidal structures that impart yellow color. This study has significance while studying the photo-oxidation of polyethylene under UV-light exposure.

Weathering of plastic films of low and linear low-density polyethylene under natural scenario has been investigated by Basfar and Idriss Ali in 2006 [25]. The results showed that added UV-stabilizers and light absorbers protects the plastic films from degradation under UV-exposure. This was evident from analysis of tensile-strength of polymer in amount of 50% within thirty to five hundred ninety days.

Several forms of polyethylene (linear low and high density polyethyene) and polypropylene polymers films with varying film thickness (15 to 80 μm) were degraded under solar light [26]. A significant reduction in molecular weight and almost zero mechanical properties of the polymers was observed in a year. Weathering of plastic was also identified by the increased number of carbonyl groups especially in polypropylene that was having a rapid rate of oxidation than that of polyethylene. Overall, from the above discussion it was concluded that the degradation of plastics is very subjective to the local environmental conditions that are usually a combination of those simulated in laboratory conditions.

3.   Methods for plastic degradation

Any change in the physical or chemical properties due to a change in structure is called degradation and it may be chemical, physicochemical, or biological processes. The polymer degradation mechanisms are mainly due to hydrolysis and oxidation [27]. Photo-degradation is the procedure of decay of the material under sunlight considered as one of the primary sources of damage exerted upon polymeric substrates at ambient conditions. The mechanism of plastic biodegradation under aerobic conditions has been presented as

Fig. 4. Usually, synthetic polymers are vulnerable to degradation started by ultraviolet (UV) and visible light. Generally, wavelength range 400–290 nm of near-UV radiations in the sunlight governs the lifespan of polymers in outdoor uses. The near-UV (400–290 nm) radiations have energies from 3.1 to 4.3 eV that corresponds to 72–97 kcal/mol. required for breaking the chemical bonds in polymers [28]. Generally, pathways induced by photo and thermal degradation are analogous [29]. Most of the polymers degraded under sunlight or UV irradiation yielded lower molecular weight fragments. Chain-growth polymers such as poly(methyl) methacrylate could be degraded at high temperatures ~ 500 °C under reduced pressure ~ 200 bars to yield precursors, like oils, gases, and water. The degradation takes place by hydrogenation and gasification process. Solvolysis was found efficient method for the degradation of step-growth polymers like polyesters, polyamides and polycarbonates to yield lower molecular weight fragments. The hydrolysis was prompted by water comprising an acid or an alkali as catalyst. Polyamides were found sensitive towards the acids. Table 3 summarized the different types of modes by which plastics usually undergo for degradations.

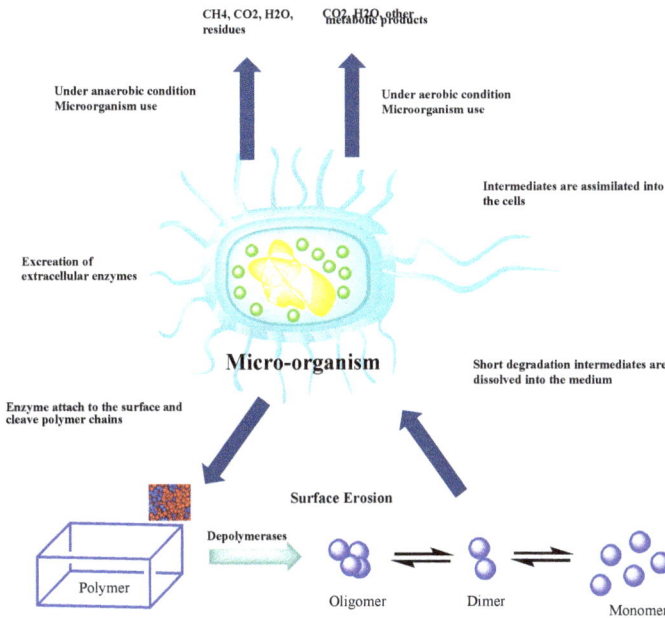

Fig. 4 General mechanism of plastic biodegradation under aerobic conditions.

Table 3 Various polymer degradation routes.

Factors (requirement/ activity)	Photo-degradation	Thermo-oxidative degradation	Biodegradation
Active agent	UV-light or high-energy radiation	Heat and oxygen	Microbial agents
Requirement of heat	Not required	Higher than ambient temperature required	Not required
Rate of degradation	Initiation is slow. But propagation is fast	Fast	Moderate
Other consideration	Environment friendly if high-energy radiation is not used	Environmentally not acceptable	Environment friendly
Overall Acceptance	Acceptable but costly	Not acceptable	Cheap and very much acceptable

(http://www.envis-icpe.com, Plastics recycling-Economic and Ecological

3.1 Biodegradation

Biodegradation is governed by various parameters like nature of pretreatment, polymer features, type of organisms, functional groups, mobility, tacticity, crystallinity, molecular weight, and substituents present in the polymer structure. Addition of plasticizers or additives also plays a significant role during degradation of plastics [30]. Physical and biological forces result in the initial degradation of a polymer [31]. There are various steps of biodegradation, which could be recognized by precise terminology [32]. Deterioration is an apparent degradation that adjusts the mechanical, physical, and chemical properties of a given polymers [33]. Depolymerization that identified by the breakage of polymers into oligomers, dimers, or monomers by catalytic-agents secreted by microorganisms is also a key step. The next step is the integration, and mineralization is the final step of this process. Integration discusses the addition of molecules transported in the cytoplasm in the microbial metabolism to yield energy, new biomass, storage vesicles, and numerous fragments. Mineralization means the secretion of simple and dissimilar salts and complex metabolites which reach the extracellular surroundings. Biodegradable plastics were initially industrialized to resolve precise waste issues of agricultural films or to the assemblage and parting of food waste. Biodegradation constantly follows photo-degradation and chemical degradation [34].

3.2   Polyolefins

These are polymers of simple olefins such as ethylene, propylene, and butenes constituting thermoplastic materials. Polyolefins consist of carbon and hydrogen atoms and they are nonaromatic. The most common polyolefins are polyethylene and polypropylene. The impurities like carbonyls and hydroperoxide groups generated during synthesis are mainly responsible for the photo-degradation of polyolefins [35]. The photodegradation is also dependent on the nature of absorbing impurities and wavelength. The oxidation vulnerability of polyolefins is as follows: (isostatic) polypropylene>low density polyethylene>linear low density polyethylene>high density polyethylene [36,37].

It has been concluded that there is a synergetic interaction between photo-oxidation and biodegradation during the degradation of polyolefins. Photo-oxidation of the polyolefins alters surface area, reduces molecular weight, challenges tensile strength, which helps the bio-degradation [38].

3.3   Degradation of polyethylene

Polyethylene is a thermoplastic polymer consisting of long hydrocarbon chains used in a number of applications including flexible film packaging produced by the blown film process. Substantial dissimilarities in physical properties have been witnessed in linear low-density polyethylene, low-density polyethylene, and high-density polyethylene blown films. The properties of polyethylene's are govern by structural parameters like density/ crystallinity, molecular weight, and its distribution, short-chain branching (SCB)/ long chain branching (LCB) length and amount, and crystalline morphology [39]. The intrinsic factors influencing polymer degradation are the number of branching of the polymer, the molecular weight, the hydrophobicity/hydrophilicity ratio, the crystallinity, and the morphology of the polymer [40]. Numerous reports of degradation of PE and the alteration of plastic surfaces in the laboratory or under controlled conditions in the field [41-47].

The pathway of photo-degradation polyethylene is well understood and can be summarized in Fig. 5 [48]. Light absorption by chromophoric defects consequences in the creation of radicals, that react by different pathways like removal of a hydrogen atom from the polymer chain, adding to an unsaturated group, or joining to oxygen [49]. Primary photoproducts are formed as hydroperoxides. As soon as they are generated, decomposes by the breaking of the weak O–O bond, yields a macro-alkoxy and a hydroxyl radical HO. The alkoxy macroradical can react by numerous pathways like scission with cleavage of the main chain to form aldehydes, abstraction of hydrogen without cleavage of the chain to form hydroxyls, cage reaction between the pair of the radicals formed, i.e., macro-alkoxy radical and hydroxyl radical HO. Ketones reacts photochemically as per Norrish type I or type II reactions [50].

In photo-degradable plastics, the materials are built with a combination of light-sensitive additives or copolymers which weaken the bonds of the macromolecule under ultraviolet radiation. Such plastics are designed to become weak and brittle once exposed to sunlight for a long time. Photosensitizers used include diketones, ferrocene derivatives (aminoalkyferrocene), and carbonyl-containing species. Photodegradable plastics degrade in a two-stage process, with UV light firstly broking few bonds making it more brittle lower molecular weight compounds which further degrade from physical strains [51]. It appears that degradation does expedite the development of microplastics that originally the same polymer but with some of their degraded sides. The most harmful UV wavelength for a specific plastic is subject to which type of bonds present, and the maximum degradation therefore occurs at different wavelengths for various types of plastics, e.g., it is around 300 nm for polyethylene [52].

In most of the studies, photo-degradation results in sharper peaks in the bands which represent ketones, esters, acids, etc. on infrared spectrum. Table 4 [30,31,35, 39-44]., shows the main changes in the surface functional groups of polyethylene under different photo-degradation conditions as were observed using infrared spectrometer.

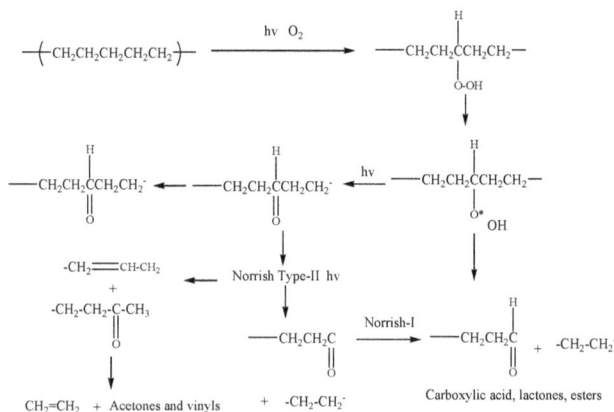

Fig. 5 Photo-degradation of polyethylene.

Table 4 Main changes in the surface functional groups on PE surface after exposure to different conditions [30,31,35, 39-44].

Type of PE	Wavenumber (cm⁻¹)	Functional Group	Degradation mode	Reference
HDPE, LDPE, and LLDPE	1733–1743	Aldehydes or ester	UV-Xenon lamb	[31]
LLDPE		Carbonyl Band	Xenon lamb	[40]
HDPE		Ester	Electron beam and gamma	[30]
HDPE		Ketones and acid	Natural	[41]
LDPE and LLDPE		Aldehydes	Accelerated environmental conditions	[39]
PE with vinyl and t-vinylene groups		Esters and Lactones	Photo-oxidation thermo-oxidation	[35]
Environmentally beached PE pellets		Esters	Environment	[42]
HDPE, LDPE, and LLDPE	1712–1723	Ketones	UV-Xenon lamb	[31]
LLDPE		Acid	Xenon lamb	[40]
HDPE		Ketones	Electron beam and gamma	[30]
HDPE		Ketones and Acid	Natural	[41]
HDPE		Carbonyl	Natural	[43]
LDPE and LLDPE		Ketones	Accelerated environmental conditions	[39]
HDPE		Carbonyl	Natural	[44]
PE with vinyl and t-vinylene groups		Ketones and acid	Photo-oxidation thermo-oxidation	[35]
Environmentally beached PE pellets		Ketones	Environment	[42]

3.4 Biodegradation of polyethylene

Plastic-debris makes available a substrate for aquatic life that has a longer life than natural floating substrates and has been implicated as a vector for transportation of harmful algal species. Such biofilm development changes with the season, substrate, and location [53]. The hhigh hydrophobic level and high molecular weight of polyethylene make it non-biodegradable [54]. Modification in crystalline level, molecular weight, and mechanical properties of polyethylene is needed for its biodegradation [55]. The same can be attained by enhancing the hydrophilic level of PE by shortening polymer chain length via oxidation [56]. Hydro-biodegradation and oxo-biodegradation are the two main mechanisms for the biodegradation of polyethylene [57].

During the last few decades, numerous reports showing that various strains interact with different kinds of polyethylene initiating some kind of degradation. Microbes are capable to colonize the surfaces of polyethylene have diverse effects on its properties. Seven different features were typically examined for alteration to create the degree of biodegradation of the polymer: functional groups on the surface, hydrophobicity/hydrophilicity, crystallinity, surface topography, mechanical properties, molecular weight distribution, and mass balance [58].

Table 5 [25-28,49,51,52] shows the changes in the characteristics of different types of polyethylene. Generally, it is accepted that in the presence of microorganisms, the concentration of the surface functional groups will decrease, which is normally reported as a decrease in the carbonyl groups. Such a phenomenon occurs because microorganisms are normally attached to the carbonyl groups [59-62].

Table 5 Changes found in the literature for polyethylene after exposure to different biodegradation conditions [25-28, 49,51,52].

Type of Polyethylene	Environment	Parameter	Changes	Reference
Low density polyethylene after UV irradiation	LDPE mixed with natural soils for 10 years	Infrared spectrometer	UV irradiation increased carbonyl index, peak at 905–915 cm-1 due to biodegradation	[25]
Polyethylene with Totally Degradable Plastic Additive	Rhodococcus rhodochrous, Cladosporium cladosporioides, and Nocardia asteroides	Scanning electron microscopy Infrared spectrometer Molecular Weight	Surface physically weak, readily disintegrated under mild pressure Band at 1,088 cm-1 due to polysaccharides Increased carbonyl index, decreased molecular weight Reduction in molecular weight	[49]
Low density polyethylene with 60 marine Bacteria from Arabian sea	Monitor environment	Infrared spectrometer Crystallinity	Peak at 905–915 cm-1Reduction in crystallinity	
Low density polyethylene mixed with different natural soils	Monitor environment	Infrared spectrometer Weight loss Tensile strength Elongation brake	Peaks at 1,448–1,470, 2,800–300 cm-1 Weight loss due to Biodegradation Tensile strength decreased elongation brake decreased	[25]
Low density polyethylene with	P. chrysosporium in soils with LDPE with	Infrared spectrometer	1,650–1,860 carbonyl Compounds; 900–1,200 peaks due to biodegradation	[26]

12% starch Low density polyethylene	12% starch Pseudomonas sp. AKS2	Atomic force microscopy Weight loss Tensile strength	Rough surface with cracks and grooves; Time dependent weight loss; Reduction in tensile strength	[51]
Polyethylene food plastic bags	Monitor environment (PE in 2 m depth in the sea)	Hydrophobicity	Decreased hydrophobicity	[27]
High density polyethylene and Low density polyethylene	Monitor environment	Hydrophobicity Atomic force microscopy Infrared spectrometer Weight loss	Decreased hydrophobicity Increased surface roughness; Decreased carbonyl; Index Weight loss with higher rates for LDPE	[28]
High density polyethylene, Low density polyethylene, and Linear low density polyethylene	Rhodococcus rhodochrous ATCC 29672 after UV irradiation	Molecular Weight Infrared spectrometer; Scanning electron microscopy	Reduction in molecular Weight Increased at 1,712 for pre-photo-oxidized samples HPDE film behaves differently than the LDPE and LLDPE and was not so favorable for microbial metabolism	[52]

Reports on microorganism attachment to polyethylene have recognized that the key drawback of the colonization process is the comparatively high hydrophobicity of the polymer contrary to the frequently hydrophilic surfaces of microorganisms [63]. It has been recommended that strains with more hydrophobic surfaces may have an important role in the initial colonization of the polymer. Since polyethylene surface is hydrophobic in nature, hence, the more hydrophobic bacterial cell on the surface, the greater will be the interaction with the polyethylene. Other metabolic version which will be key in polymer colonization is the creation of surfactants that can facilitate the attachment process of microorganisms to the hydrophobic surface. Harshvardhan and Jha [64] quantified that marine bacteria (Kocuria palustris M16, Bacillus pumilus M27, and Bacillus subtilis H1584), the microorganisms, usage the polyethylene surface as a carbon source. This is the first step of polyethylene biodegradation and due to which a decline in molecular weight occurs. Once the size of the molecule is contracted, oxidation is essential to convert the hydrocarbon into a carboxylic acid which finally metabolized by means of β-oxidation and the Krebs cycle.

3.5   Degradation of polypropylene

Polypropylene is a thermoplastic material used in numerous applications that includes packaging, labeling, and textiles. In view of high processability and low cost, polypropylene is one of the most widely manufactured polymers, particularly, for the auto industry. Pristine polypropylene is impervious to photo-oxidation and thermal oxidation at modest temperatures. Furthermore, polypropylene is sensitive to several external stimuli like temperature, light, and irradiation. When polypropylene is exposed to high temperatures or to an irradiation environment, the tertiary hydrogen atoms present in its chains are susceptible to be attacked by oxygen [65]. Oxidation of polypropylene depends on both light irradiation and temperature in outdoor aging circumstances. Polypropylene can also be photo-degraded because several molecular chains are affected in the wavelength range from 310 to 350 nm [66]. Polypropylene is resistant to photo-oxidation at low temperatures while sensitive to many external environments like heat, light, and radiation and hence has a comparatively lower service temperature. When polypropylene is irradiated to a high temperature, the tertiary hydrogen atoms present in its chains are susceptible to be attacked by oxygen [67]. The wavelengths of over 290 nm are sufficient to start the degradation and create discoloration, chalking, and embrittlement of polypropylene. Therefore, assessment of the service life of polypropylene in the natural environment is a well-established practice [68]. Under ultraviolet irradiation cause fast degradation of the polypropylene while the weakening of its strength occurs under sunlight. Numerous efforts have been made to examine the outcomes of photo-oxidation of polypropylene to deliver a good estimation of its continuing service performances. Outdoor weathering test and accelerated weathering test are two methods usually employed for this purpose [69]. Table 6 shows the studies for degradation of polypropylene under different accelerated or environmental conditions.

The photo-oxidation of polypropylene as described is presented in Fig. 3 [70-73]. For degradation of polyolefins in environmental implications, there should be either improvisation of hydrophilic characters or plummeting chain length of polymer, more favorable conditions to microbial degradation [74,75].

Oxidative degradation of polypropylene includes chemical methods, thermal procedures and UV exposure that may be responsible to generation of active functional groups ($>C=O$, $-COOH$ and ester) on surface of polymer. Formation of carbonyl, carboxyl, and functional groups decreases the hydrophobicity of the surface and hence favors biodegradation. Limited studies showing biodegradation of polypropylene under various conditions has been summarized (Table 6 [61-67] and 7 [68, 71-74]).

Degradation of Plastics   Materials Research Forum LLC
Materials Research Foundations **99** (2021) 290-324   https://doi.org/10.21741/9781644901335-12

Table 6 Changes found in the literature for PP after exposure to different biodegradation conditions [61-67].

Environment	Parameters	Changes	Reference
Accelerated and outdoor weathering	Infrared spectrometer Molecular weight Scanning electron microscopy	Different carbonyl products formed during degradation Reduction in molecular weight Increasing cracks in the surface with increasing exposure	[62]
Thermal	Infrared spectrometer chemiluminescence spectroscopy	Initial peak wavelength of chemiluminescence emission at 490 nm remained constant during the early stages of thermal degradation. New emissions developed with time in the red spectral region (i.e., 490, 660, and 740 nm) over an extended oxidation period	[63]
Xisha tropical environment	Macroscopic morphology Chromatic aberration Tensile strength Elongation break	Macroscopic cracks observed after 9 months of exposure; Rapidly increasing chromatic aberration, the sample becomes darker with the exposure; Reduction in tensile strength due to oxidation; Dramatic decrease in elongation; break. The toughness of the polymer is more sensitive than the strength	[64]
Electron irradiated aging in autoclave	Weight and geometry Tensile strength Infrared spectrometer	No significant changes in geometry Reduction of tensile strength proportional to temperature; Ether and carbonyl species due to oxidation	[65]
Accelerated photo-thermal aging	Infrared spectrometer Molecular weight	Different carbonyl species, hydroperoxides, and hydroxyl species due to exposure Reduction in crystallinity, for samples containing nanocomposites increasing crystallinity; Reduction in molecular weight	[66]
Outdoor weathering and accelerated conditions	Tensile strength Intrinsic viscosity Infrared spectrometer	Reduction in tensile strength; Reduction in intrinsic viscosity; The carbonyl contents increased with increasing exposure	[61]
Outdoor weathering and accelerated conditions	Tensile strength Infrared spectrometer	Reduction in tensile strength Carbonyl concentration rose with the irradiation time and the increasing rate depends on UV content	[67]

Table 7 Changes found in the literature for PP after exposure to different biodegradation conditions [68, 71-74].

Environment	Parameter	Changes	Reference
polypropylene and thermally treated PP in soil consortia for 12 months	Infrared spectrometer Surface energy; Mechanical properties	Carbonyl index increased due to thermal treatment; Carbonyl index decreased due to biodegradation; Crystallinity increased, tensile strength decreased	[68]
polypropylene and UV irradiated PP in four different soil consortia	Infrared spectrometer Surface energy; Weight loss	Carbonyl index increased due to thermally treatment Carbonyl index decreased due to biodegradation; Decreased hydrophobicity, highest weight loss for UV irradiated samples	[72]
Thermal and photochemical treatment for PP and then mixed with R. rhodochrous ATCC 29672	Infrared spectrometer Molecular weight; Size exclusion chromatography	Increase in carbonyl index due to abiotic treatment. The loss in molecular weights is lower in the case of additive free polypropylene; R. rhodochrous cells were able to use the oxidized polymer films as carbon source	[73]
PP/starched based materials aged in soil	Thermogravimetry analysis	The kinetic parameters dependent on: (a) the influence of the atmosphere in which the experiment was carried out, (b) the composition of the blend, and (c) the process of degradation in soil	[71]
PP photo-degraded and then Biodegraded	Molecular weight 1H NMR	Decreased molecular weight Biodegradation occurred Anaerobically	[74]

4. Polyethylene terephthalate

(PET, aromatic polyesters) is a significant polymer class that has prevalent commercial uses in several fields in form of fibers, films, food packaging, and beverage containers due to brilliant material characteristics. Polyethylene terephthalate has a very high resistance to atmospheric and biological agents and frequently observed as substantial waste items on land and beaches [76]. Therefore, limited studies of degradation or microbial degradation are reported in literature on polyethylene terephthalate [77,78]. It can be synthesized in laboratory by reaction of ethylene glycol with terephthalic acid (dimethyl terephthalate) followed by 2 or 3-step polymerization, dependent on the desired molecular weight [79,80]. In general, repeated units of PET have a length of about 1.09 nm and a molecular weight of ~200 [81]. Nevertheless, the ester-bond can be simply broken still the presence of an aromatic ring with a short aliphatic chain makes it stiffer in strength if we compare it with other aliphatic polymers (polyolefin and polyamide). Besides, moderately high thermal-stability is credited to the segmental flexibility of polymer chains.

A textile-grade polymer (100 repeating units/ molecule; 100 nm length; molecular weight of ~20,000) needs upper stages of polymerization to yield higher strength fibers. On the other hand, small moisture can affect melt viscosity along with melt-stability and may lead to hydrolytic degradation. The additions of heteroatoms in polymer backbone (e.g., polyesters and polyamines) also improve the susceptibility to microbial or oxidative degradation while inclusion of secondary qualifier in aromatic polymers can make them further tough to degradation [82].

4.1   Photo-thermal degradation of polyethylene terephthalate

On exposure to near-UV-light, polyethylene terephthalate undergoes photo-degradation that involve scission of polymeric chain via Norrish I and II reactions. It may also show further cross-linking if the reaction carried out in presence of linkers and gelling agents. A yellow or brown polyenes are formed when polyethylene terephthalate is thermally degraded that is indicated by discolored final polymer after that increase of more carboxyl terminated-species. Carboxyl groups catalyzed formation of vastly conjugated-species and more discoloration of polymer and hence, quick degradation [81]. Thermo-oxidative stability also decreased in presence of rich carboxyl content [83]. Venkatachalam et al. [81] suggested that the preliminary phase of thermal-degradation is a random scission of the in-chain ester linkage resultant of the generation of a vinyl-ester and carboxyl end groups. Trans-esterification of the vinyl ester yields the vinyl alcohol that converted instantaneously to acetaldehyde. Environmental degradation of polyethylene terephthalate has been scarily reported till now. During the study of degradation of polyethylene terephthalate through irradiation by MeV He$^+$, it was observed that the colorless samples turn brittle and yellow color was witnessed as a function of irradiation [84]. Hydrolytic degradation of polyethylene terephthalate was also reported for its copolymer-containing nitrated units [85,86]. It was observed that the polyethylene terephthalate copolymers having nitrated units degraded with fast rate than semi-crystalline and amorphous PET and the rate of degradation increased with nitrated units. During the study of hydrolytic degradation, significant changes occurred in the v(C–H) bond of polyethylene terephthalate. This might be due to influence of environmental condition on the functional groups, extensive hydrogen-bonding along with aliphatic methyl groups; all factors also responsible for increasing the hydrophilic character as well as degradation of polyethylene terephthalate [85].

4.2   Environmental concern of plastic

There is a continuous enhancement in the production of plastic worldwide as reported in data from 1950 to 2014 (1.7 million tons to 311 million tons) [87]. Plastic has become a widespread and ubiquitous item of marine and land debris credited to extensive production, consumption, and stability [88-90]. Navigational safety and the aesthetic quality of beaches are damaging due to the accumulation of plastic debris and imparting adverse effects on the aquatic system. The estimated financial loss to fisheries, tourism, and time spent cleaning up beaches equal to $13 billion to saline ecosystems caused by plastic pollution every year [91]. Several aquatic species such as marine mammals, birds, sea turtles, crustaceans, and bivalves (approximately 395 species) have been affected by the ingestion/entanglement of marine plastic debris [92].

Marine plastic debris has been acting as vehicle of noxious plastic additives in the environment is an emerging issue. Plastic contains additives and absorbed chemicals that may have possibility to leach into the surrounding environment [93,94]. To enhance the performance of plastics various additives chemicals are added during synthesis processes include antioxidants, plasticizers, etc. Many of them (phthalates and bisphenol A) contain endocrine disrupting properties that may affect the genetics and development of aquatic species [95]. Hydrophobic pollutants (e.g., polychlorinated biphenyls, polycyclic aromatic hydrocarbons, and plastic additives) have the tendency of accumulation on the surface of the seawater. The magnitude of accumulation of persistent organic pollutants of plastics is six times higher in that of ambient seawater [96]. With the floating of plastics, these chemicals or additives may travel from one place to other or from polluted to out-of-the-way clean areas [97,98]. Marine organisms are exposed for those toxic chemicals when they ingested contaminated plastic debris, [99]. These additive chemicals have the ability of assimilation in the stomachs of seabirds via leaching through ingested plastics. In support of that, several controlled studies have been reporting that highlight the transfer of organic chemicals from microplastics and resulted in worsening of physiological functions and health of the targeted organisms [100-105]. Some contradictory studies indicated the negligible bio-accumulation of perilous chemicals by plastic [106,107]. Therefore, to confirm the actual role of plastic as the carrier of persistent organic pollutants, many field-evidence studies are on-going to understand marine microplastic pollution and its aftereffects. Concentration levels/ abundance and chemical profiles were compared in organisms (fish, whales, and birds) with that of microplastics in nearby waters (or gut matters) [108-111]. In spite of this, limited information is available on statements that may say that plastic marine-debris is a vector to transfer dangerous compounds into marine organisms. Therefore, there is much need for plenty of data on this issue to further prove the field evidence.

4.3 Biodegradable plastics

During the biodegradation of recyclable plastics under aerobic conditions, microbes gained carbon and energy. Mechanism occurred during this process can be shown below:

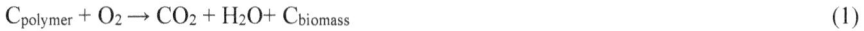

$$C_{polymer} + O_2 \rightarrow CO_2 + H_2O + C_{biomass} \qquad (1)$$

Microorganisms integrated the carbon of the polymer ($C_{polymer}$) into biomass ($C_{biomass}$) followed by fast mineralization into CO_2 and H_2O. This biomass can also be used for development and reproduction (more $C_{biomass}$) which is further mineralized in the long-term because of the succeeding attack of the soil microbial community. Hence, it was observed for the bi-phasic pattern that includes rapid-phase for CO_2 production after that a slower secondary-phase for CO_2 evolution; seen by finally conversion into mineralization of organic-matter [112-114]. The reaction may be expressed as

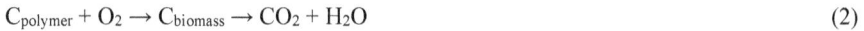

$$C_{polymer} + O_2 \rightarrow C_{biomass} \rightarrow CO_2 + H_2O \qquad (2)$$

Overall, there are two reactions with different kinetics 1) conversion of $C_{polymer}$ into $C_{biomass}$ and 2) conversion of obtained $C_{biomass}$ into CO_2. First one reaction is biodegradation that is indicated by either intake of the reagents or the appearance of the yields and the other one belongs to mineralization. Biodegradation could be monitored and quantified by measuring reagent of reactions (O_2) or the end-product (CO_2) of energy-metabolism. Biodegradation percentage can be defined as ratio of evolved CO_2 over theoretical CO_2 (ThCO_2) i.e. the quantities of CO_2 predictable for total-oxidation of the carbon existent in the plastic sample (Cpolymer) acquaint within the reactor.

$$Biodegradation \% = \frac{C_{evolved\ as\ CO_2}}{Available\ C_{polymer}} X\ 100 \qquad (3)$$

Instead, oxygen intake can be measured with a similar method using the theoretic oxygen demand to designate the maximum-oxygen-uptake for the entire oxidation of the plastics.

Measurement of biomass accurately by a consistent method is still not offered. Accordingly, the assessment of biodegradation has been achieved by mineralization as presented in [Eq. (3)]. As per guidelines, the "ready biodegradable" compounds have to reach 60% (of the $ThCO_2$) in a 10-day window within a 28-day period of the test. The 10-day window begins when the degree of biodegradation has reached 10% and must end before day 28 of the test. This pass level is deliberated to demonstrate practically the complete degradation of the test samples, as the remaining portion of 40% of the test substance is presumed to be adjusted by the biomass or to be present as products of biosynthesis.

A "ready biodegradable" chemical is presumed to undergo quick and final biodegradation in the environment exposure and no supplementary examination of the biodegradability of the polymer, or the probable environmental effects of conversion products, is usually essential. Biodegradable plastics cannot be regarded as "readily biodegradable" as they do not show such fast rates of degradation. The heterogeneous reaction is believed to involve in the biodegradation of solid substances since it occurs on solid/liquid interface, where the microbial enzymes present in the liquid phase interact with the macromolecules accessible at the surface of the solid plastic samples subjected to biodegradation [115]. The polymers in the internal portions of the plastic sample are not involved in the reaction, as they are not available.
Basically:

$$C_{polymer} = \text{Available C} + \text{Unavailable C} \tag{4}$$

Where, "available C" stands for carbon present in the exterior part of plastics, effectively available to the enzymes in the liquid phase and to biodegradation, while the rest is buried in the core of the plastic. Therefore, Eq. (3) when applied to macromolecules rather than to chemicals should be rewritten as:

$$Biodegradation\ \% = \frac{C_{evolved\ as\ CO_2}}{Available\ C_{polymer}} X\ 100 \tag{5}$$

The effect of the surface area of different plastics at laboratory scale under controlled composting conditions on their biodegradation behavior has been examined [116]. Comparison of the biodegradation of polycaprolactone, poly (butylene succinate-co-adipate), poly(L-lactic acid) and poly(butylene succinate) in film and powder form. It was concluded that the shape of the polymer affected the biodegradation [117-119]. The shape of the simply degradable plastics was reported to distress biodegradation only at the commencement of the procedure, while the slowly degrading plastics during the biodegradation. Initial stages of the biodegradation, plastics with a larger surface area were biodegraded faster than the same plastic in film form. Though, at the later stages of biodegradation, the rate of easily biodegradable plastics became closely independent of the form of the samples, while the effect was constant for slowly degrading plastics [120].

Conclusions

From this chapter it was concluded that the most important apprehensions about the natural and synthetic plastics polymers, their kinds, applications, dumping methods, degradability and standards employed in evaluating polymer degradation and interest of biodegradation of new and old plastics. Enhancement in concentration of plastic in land, coastal shores and oceans are triggering their environmental unease. Polyethylene, polypropylene, polystyrene and polyethylene terephthalate are the chief constituents of plastic debris and are regularly employed in several applications, which include clothes and food packaging. Polyethylene terephthalate is extremely unaffected by environmental biodegradation and, therefore, causes several environmental concerns related to its accretion. Degradation of plastics is rarely reported, particularly under natural conditions. Till now, three plastic disposal methods (landfill, incineration, and recycling) are usually employed on an industrial scale with several drawbacks. Both the methods landfill and incineration discharges harmful secondary contaminants into the environment while the landfill method required large portions of land space. The recycling method rectifies such environmental concerns, however, this method is found incompetent and degrading the quality of the polymers. This process was found costly and consequently, less incentive for investment in recycling facilities. Biodegradation was reported as environmentally safe and effective for the disposal of plastic waste. Until now, there is no protocol developed to effectively dispose of polyethylene terephthalate by biodegradation on an industrial level. Though, considerable research & development is still being conducted in the area of biodegradation of polymers. Quantity of CO_2 released on exposure to microorganisms, compost or soil is determined with respirometry technique. Infrared spectroscopy, differential scanning calorimeter, nuclear magnetic resonance, scanning electron microscopy, atomic force microscopy and X-ray diffraction were several techniques used for characterization of plastic surface. Moreover, international standard methods are required to estimate the amount of biodegradation for comparison studies. In light of these, there is a requirement to regulate the parameters in standards to be used worldwide. It has also been observed that many of recalcitrant plastics may be destroyed to significant extent in the proper scenario at the optimum concentration. Although, the prevention measures of the degradation of plastics have already been initiated still more efforts are needed to solve this issue.

Acknowledgments

One of the authors Dr Manviri Rani is grateful for the financial assistance from DST-SERB, New Delhi (Sanction order no. SRG/2019/000114) and TEQIP-III, MNIT Jaipur, Rajasthan, India. Authors are also thankful toTEQIP-phase -III NIT Jalandhar for financial support.

References

[1] A.C. Albertsson, C. Barenstedt, S. Karlsson, Abiotic degradation products from enhanced environmentally degradable polyethylene, Acta Polym., 45 (1994) 97-103. https://doi.org/10.1002/actp.1994.010450207

[2] A.-C. Albertsson, S.O. Andersson, S. Karlsson, The mechanism of biodegradation of polyethylene, Polym. Degradation Stab., 18 (1987) 73-87. https://doi.org/10.1016/0141-3910(87)90084-X

[3] A. Arkatkar, J. Arutchelvi, S. Bhaduri, P.V. Uppara, M. Doble, Degradation of unpretreated and thermally pretreated polypropylene by soil consortia, Int. Biodeterior. Biodegrad., 63 (2009) 106-111. https://doi.org/10.1016/j.ibiod.2008.06.005

[4] A. Arkatkar, J. Arutchelvi, M. Sudhakar, S. Bhaduri, P.V. Uppara, M. Doble, Approaches to enhance the biodegradation of polyolefins, The Open Environmental Engineering Journal, 2 (2009). https://doi.org/10.2174/1874829500902010068

[5] A. Arkatkar, A.A. Juwarkar, S. Bhaduri, P.V. Uppara, M. Doble, Growth of Pseudomonas and Bacillus biofilms on pretreated polypropylene surface, Int. Biodeterior. Biodegrad., 64 (2010) 530-536. https://doi.org/10.1016/j.ibiod.2010.06.002

[6] T. Artham, M. Sudhakar, R. Venkatesan, C.M. Nair, K. Murty, M. Doble, Biofouling and stability of synthetic polymers in sea water, Int. Biodeterior. Biodegrad., 63 (2009) 884-890. https://doi.org/10.1016/j.ibiod.2009.03.003

[7] C.G. Avio, S. Gorbi, M. Milan, M. Benedetti, D. Fattorini, G. d'Errico, M. Pauletto, L. Bargelloni, F. Regoli, Pollutants bioavailability and toxicological risk from microplastics to marine mussels, Environ. Pollut., 198 (2015) 211-222. https://doi.org/10.1016/j.envpol.2014.12.021

[8]   E. Besseling, A. Wegner, E.M. Foekema, M.J. Van Den Heuvel-Greve, A.A.
      Koelmans, Effects of microplastic on fitness and PCB bioaccumulation by the
      lugworm Arenicola marina (L.), Environ. Sci. Technol., 47 (2013) 593-600.
      https://doi.org/10.1021/es302763x

[9]   D.N. Bikiaris, G.P. Karayannidis, Effect of carboxylic end groups on
      thermooxidative stability of PET and PBT, Polym. Degradation Stab., 63 (1999) 213-
      218. https://doi.org/10.1016/S0141-3910(98)00094-9

[10] S. Bonhomme, A. Cuer, A. Delort, J. Lemaire, M. Sancelme, G. Scott,
      Environmental biodegradation of polyethylene, Polym. Degradation Stab., 81 (2003)
      441-452. https://doi.org/10.1016/S0141-3910(03)00129-0

[11] G. Botelho, A. Queirós, S. Liberal, P. Gijsman, Studies on thermal and thermo-
      oxidative degradation of poly (ethylene terephthalate) and poly (butylene
      terephthalate), Polym. Degradation Stab., 74 (2001) 39-48.
      https://doi.org/10.1016/S0141-3910(01)00088-X

[12] M.A. Browne, S.J. Niven, T.S. Galloway, S.J. Rowland, R.C. Thompson,
      Microplastic moves pollutants and additives to worms, reducing functions linked to
      health and biodiversity, Curr. Biol., 23 (2013) 2388-2392.
      https://doi.org/10.1016/j.cub.2013.10.012

[13] E.M. Chua, J. Shimeta, D. Nugegoda, P.D. Morrison, B.O. Clarke, Assimilation
      of polybrominated diphenyl ethers from microplastics by the marine amphipod,
      Allorchestes compressa, Environ. Sci. Technol., 48 (2014) 8127-8134.
      https://doi.org/10.1021/es405717z

[14] M. Djebara, J. Stoquert, M. Abdesselam, D. Muller, A. Chami, FTIR analysis of
      polyethylene terephthalate irradiated by MeV He+, Nuclear Instruments and
      Methods in Physics Research Section B: Beam Interactions with Materials and
      Atoms, 274 (2012) 70-77. https://doi.org/10.1016/j.nimb.2011.11.022

[15] R.M. Donlan, Biofilms: microbial life on surfaces, Emerging Infect. Dis., 8
      (2002) 881. https://doi.org/10.3201/eid0809.020063

[16] [16] R.E. Engler, The complex interaction between marine debris and toxic
      chemicals in the ocean, Environ. Sci. Technol., 46 (2012) 12302-12315.
      https://doi.org/10.1021/es3027105

[17] J.P. Eubeler, M. Bernhard, T.P. Knepper, Environmental biodegradation of
      synthetic polymers II. Biodegradation of different polymer groups, TrAC, Trends
      Anal. Chem., 29 (2010) 84-100. https://doi.org/10.1016/j.trac.2009.09.005

[18]   P. Europe, Plastics the facts 2014/2015: an analysis of European plastics production, demand and waste data, Plastic Europe, Brussels, (2015).

[19]   M.C. Fossi, C. Panti, C. Guerranti, D. Coppola, M. Giannetti, L. Marsili, R. Minutoli, Are baleen whales exposed to the threat of microplastics? A case study of the Mediterranean fin whale (Balaenoptera physalus), Mar. Pollut. Bull., 64 (2012) 2374-2379. https://doi.org/10.1016/j.marpolbul.2012.08.013

[20]   K.N. Fotopoulou, H.K. Karapanagioti, Surface properties of beached plastics, Environmental Science and Pollution Research, 22 (2015) 11022-11032. https://doi.org/10.1007/s11356-015-4332-y

[21]   M. Funabashi, F. Ninomiya, M. Kunioka, Biodegradation of polycaprolactone powders proposed as reference test materials for international standard of biodegradation evaluation method, J. Polym. Environ., 15 (2007) 7-17. https://doi.org/10.1007/s10924-006-0041-4

[22]   S.C. Gall, R.C. Thompson, The impact of debris on marine life, Mar. Pollut. Bull., 92 (2015) 170-179. https://doi.org/10.1016/j.marpolbul.2014.12.041

[23]   D. Hadad, S. Geresh, A. Sivan, Biodegradation of polyethylene by the thermophilic bacterium Brevibacillus borstelensis, J. Appl. Microbiol., 98 (2005) 1093-1100. https://doi.org/10.1111/j.1365-2672.2005.02553.x

[24]   K. Harshvardhan, B. Jha, Biodegradation of low-density polyethylene by marine bacteria from pelagic waters, Arabian Sea, India, Mar. Pollut. Bull., 77 (2013) 100-106. https://doi.org/10.1016/j.marpolbul.2013.10.025

[25]   P.W. Hill, J.F. Farrar, D.L. Jones, Decoupling of microbial glucose uptake and mineralization in soil, Soil Biol. Biochem., 40 (2008) 616-624. https://doi.org/10.1016/j.soilbio.2007.09.008

[26]   E. Huerta Lwanga, H. Gertsen, H. Gooren, P. Peters, T. Salánki, M. van der Ploeg, E. Besseling, A.A. Koelmans, V. Geissen, Microplastics in the terrestrial ecosystem: implications for Lumbricus terrestris (Oligochaeta, Lumbricidae), Environ. Sci. Technol., 50 (2016) 2685-2691. https://doi.org/10.1021/acs.est.5b05478

[27]   D. Jeyakumar, J. Chirsteen, M. Doble, Synergistic effects of pretreatment and blending on fungi mediated biodegradation of polypropylenes, Bioresour. Technol., 148 (2013) 78-85. https://doi.org/10.1016/j.biortech.2013.08.074

[28] D.P. Kint, A.M.n. de Ilarduya, S. Muñoz-Guerra, Hydrolytic degradation of poly (ethylene terephthalate) copolymers containing nitrated units, Polym. Degradation Stab., 79 (2003) 353-358. https://doi.org/10.1016/S0141-3910(02)00299-9

[29] A.A. Koelmans, E. Besseling, E.M. Foekema, Leaching of plastic additives to marine organisms, Environ. Pollut., 187 (2014) 49-54. https://doi.org/10.1016/j.envpol.2013.12.013

[30] A.A. Koelmans, E. Besseling, A. Wegner, E.M. Foekema, Plastic as a carrier of POPs to aquatic organisms: a model analysis, Environ. Sci. Technol., 47 (2013) 7812-7820. https://doi.org/10.1021/es401169n

[31] L.R. Krupp, W.J. Jewell, Biodegradability of modified plastic films in controlled biological environments, Environ. Sci. Technol., 26 (1992) 193-198. https://doi.org/10.1021/es00025a024

[32] M. Kunioka, F. Ninomiya, M. Funabashi, Biodegradation of poly (lactic acid) powders proposed as the reference test materials for the international standard of biodegradation evaluation methods, Polym. Degradation Stab., 91 (2006) 1919-1928. https://doi.org/10.1016/j.polymdegradstab.2006.03.003

[33] E.H. Lwanga, H. Gertsen, H. Gooren, P. Peters, T. Salánki, M. van der Ploeg, E. Besseling, A.A. Koelmans, V. Geissen, Incorporation of microplastics from litter into burrows of Lumbricus terrestris, Environ. Pollut., 220 (2017) 523-531. https://doi.org/10.1016/j.envpol.2016.09.096

[34] E.H. Lwanga, B. Thapa, X. Yang, H. Gertsen, T. Salánki, V. Geissen, P. Garbeva, Decay of low-density polyethylene by bacteria extracted from earthworm's guts: A potential for soil restoration, ScTEn, 624 (2018) 753-757. https://doi.org/10.1016/j.scitotenv.2017.12.144

[35] S. Maaß, D. Daphi, A. Lehmann, M.C. Rillig, Transport of microplastics by two collembolan species, Environ Poll., 225 (2017) 456-459. https://doi.org/10.1016/j.envpol.2017.03.009

[36] J.D. Meeker, S. Sathyanarayana, S.H. Swan, Phthalates and other additives in plastics: human exposure and associated health outcomes, Philosophical Transactions of the Royal Society B: Biological Sciences, 364 (2009) 2097-2113. https://doi.org/10.1098/rstb.2008.0268

[37] T.J. Mincer, E.R. Zettler, L.A. Amaral-Zettler, Biofilms on plastic debris and their influence on marine nutrient cycling, productivity, and hazardous chemical mobility,

in: Hazardous Chemicals Associated with Plastics in the Marine Environment, Springer, 2016, pp. 221-233. https://doi.org/10.1007/698_2016_12

[38] R.-J. Mueller, Biological degradation of synthetic polyesters—enzymes as potential catalysts for polyester recycling, Process Biochem., 41 (2006) 2124-2128. https://doi.org/10.1016/j.procbio.2006.05.018

[39] R.J. Müller, Biodegradability of polymers: regulations and methods for testing, Biopolymers Online: Biology• Chemistry• Biotechnology• Applications, 10 (2005).

[40] L. Nizzetto, S. Langaas, M. Futter, Pollution: Do microplastics spill on to farm soils?, Nature, 537 (2016) 488-488. https://doi.org/10.1038/537488b

[41] E. Oburger, D. Jones, Substrate mineralization studies in the laboratory show different microbial C partitioning dynamics than in the field, Soil Biol. Biochem., 41 (2009) 1951-1956. https://doi.org/10.1016/j.soilbio.2009.06.020

[42] I.G. Orr, Y. Hadar, A. Sivan, Colonization, biofilm formation and biodegradation of polyethylene by a strain of Rhodococcus ruber, Appl. Microbiol. Biotechnol., 65 (2004) 97-104. https://doi.org/10.1007/s00253-004-1584-8

[43] X. Ramis, A. Cadenato, J. Salla, J. Morancho, A. Valles, L. Contat, A. Ribes, Thermal degradation of polypropylene/starch-based materials with enhanced biodegradability, Polym. Degradation Stab., 86 (2004) 483-491. https://doi.org/10.1016/j.polymdegradstab.2004.05.021

[44] M. Rani, S. Hong, W. Shim, G. Han, M. Jang, Leaching characteristics of hexabromocyclododecanes (HBCDs) from expanded polystyrene buoy in water, Organohalogen Compd., 75 (2013) 691-694.

[45] M. Rani, W.J. Shim, G.M. Han, M. Jang, N.A. Al-Odaini, Y.K. Song, S.H. Hong, Qualitative analysis of additives in plastic marine debris and its new products, Arch. Environ. Contam. Toxicol., 69 (2015) 352-366. https://doi.org/10.1007/s00244-015-0224-x

[46] M. Rani, W.J. Shim, G.M. Han, M. Jang, Y.K. Song, S.H. Hong, Benzotriazole-type ultraviolet stabilizers and antioxidants in plastic marine debris and their new products, ScTEn, 579 (2017) 745-754. https://doi.org/10.1016/j.scitotenv.2016.11.033

[47] J.-M. Restrepo-Flórez, A. Bassi, M.R. Thompson, Microbial degradation and deterioration of polyethylene–A review, Int. Biodeterior. Biodegrad., 88 (2014) 83-90. https://doi.org/10.1016/j.ibiod.2013.12.014

[48] M.C. Rillig, Microplastic in terrestrial ecosystems and the soil?, in, ACS Publications, 2012. https://doi.org/10.1021/es302011r

[49] M.C. Rillig, L. Ziersch, S. Hempel, Microplastic transport in soil by earthworms, Sci. Rep., 7 (2017).

[50] C.M. Rochman, M.A. Browne, B.S. Halpern, B.T. Hentschel, E. Hoh, H.K. Karapanagioti, L.M. Rios-Mendoza, H. Takada, S. Teh, R.C. Thompson, Classify plastic waste as hazardous, Nature, 494 (2013) 169-171. https://doi.org/10.1038/494169a

[51] C.M. Rochman, C. Manzano, B.T. Hentschel, S.L.M. Simonich, E. Hoh, Polystyrene plastic: a source and sink for polycyclic aromatic hydrocarbons in the marine environment, Environ. Sci. Technol., 47 (2013) 13976-13984. https://doi.org/10.1021/es403605f

[52] P. Ryan, A. Connell, B. Gardner, Plastic ingestion and PCBs in seabirds: is there a relationship?, Mar. Pollut. Bull., 19 (1988) 174-176.

[53] C. Sammon, J. Yarwood, N. Everall, An FT–IR study of the effect of hydrolytic degradation on the structure of thin PET films, Polym. Degradation Stab., 67 (2000) 149-158. https://doi.org/10.1016/S0141-3910(99)00104-4

[54] R.R. Singhania, G. Christophe, G. Perchet, J. Troquet, C. Larroche, Immersed membrane bioreactors: an overview with special emphasis on anaerobic bioprocesses, Bioresour. Technol., 122 (2012) 171-180. https://doi.org/10.1016/j.biortech.2012.01.132

[55] Y.K. Song, S.H. Hong, M. Jang, J.-H. Kang, O.Y. Kwon, G.M. Han, W.J. Shim, Large accumulation of micro-sized synthetic polymer particles in the sea surface microlayer, Environ. Sci. Technol., 48 (2014) 9014-9021. https://doi.org/10.1021/es501757s

[56] A. Steiner, Emerging issues in our global environment, UNEP Year Book; UNEP: Nairobi, Kenya, (2014).

[57] Z. Steinmetz, C. Wollmann, M. Schaefer, C. Buchmann, J. David, J. Tröger, K. Muñoz, O. Frör, G.E. Schaumann, Plastic mulching in agriculture. Trading short-term agronomic benefits for long-term soil degradation?, ScTEn, 550 (2016) 690-705. https://doi.org/10.1016/j.scitotenv.2016.01.153

[58] M. Sudhakar, M. Doble, P.S. Murthy, R. Venkatesan, Marine microbe-mediated biodegradation of low-and high-density polyethylenes, Int. Biodeterior. Biodegrad., 61 (2008) 203-213. https://doi.org/10.1016/j.ibiod.2007.07.011

[59] K. Tanaka, H. Takada, R. Yamashita, K. Mizukawa, M.-a. Fukuwaka, Y. Watanuki, Accumulation of plastic-derived chemicals in tissues of seabirds ingesting marine plastics, Mar. Pollut. Bull., 69 (2013) 219-222. https://doi.org/10.1016/j.marpolbul.2012.12.010

[60] R.C. Thompson, S.H. Swan, C.J. Moore, F.S. Vom Saal, Our plastic age, in, The Royal Society Publishing, 2009.

[61] P. Tribedi, A.K. Sil, Low-density polyethylene degradation by Pseudomonas sp. AKS2 biofilm, Environmental Science and Pollution Research, 20 (2013) 4146-4153. https://doi.org/10.1007/s11356-012-1378-y

[62] M. van der Zee, Methods for evaluating the biodegradability of environmentally degradable polymers, in: Handbook of Biodegradable Polymers, Smither Rapra, 2014, pp. 1-28.

[63] N. Wanasekara, V. Chalivendra, P. Calvert, Sub-micron scale mechanical properties of polypropylene fibers exposed to ultraviolet and thermal degradation, in: MEMS and Nanotechnology, Volume 2, Springer, 2011, pp. 275-281. https://doi.org/10.1007/978-1-4419-8825-6_40

[64] H.K. Webb, J. Arnott, R.J. Crawford, E.P. Ivanova, Plastic degradation and its environmental implications with special reference to poly (ethylene terephthalate), Polymers, 5 (2013) 1-18. https://doi.org/10.3390/polym5010001

[65] R. Yamashita, H. Takada, M.-a. Fukuwaka, Y. Watanuki, Physical and chemical effects of ingested plastic debris on short-tailed shearwaters, Puffinus tenuirostris, in the North Pacific Ocean, Mar. Pollut. Bull., 62 (2011) 2845-2849. https://doi.org/10.1016/j.marpolbul.2011.10.008

[66] J. Yang, Y. Yang, W.-M. Wu, J. Zhao, L. Jiang, Evidence of polyethylene biodegradation by bacterial strains from the guts of plastic-eating waxworms, Environ. Sci. Technol., 48 (2014) 13776-13784. https://doi.org/10.1021/es504038a

[67] R. Yang, Y. Li, J. Yu, Photo-stabilization of linear low density polyethylene by inorganic nano-particles, Polym. Degradation Stab., 88 (2005) 168-174. https://doi.org/10.1016/j.polymdegradstab.2003.12.005

[68] A. Yano, N. Akai, H. Ishii, C. Satoh, T. Hironiwa, K.R. Millington, M. Nakata, Thermal oxidative degradation of additive-free polypropylene pellets investigated by multichannel Fourier-transform chemiluminescence spectroscopy, Polym. Degradation Stab., 98 (2013) 2680-2686. https://doi.org/10.1016/j.polymdegradstab.2013.09.031

[69] C. Zarfl, M. Matthies, Are marine plastic particles transport vectors for organic pollutants to the Arctic?, Mar. Pollut. Bull., 60 (2010) 1810-1814. https://doi.org/10.1016/j.marpolbul.2010.05.026

[70] H. Zhao, R.K. Li, A study on the photo-degradation of zinc oxide (ZnO) filled polypropylene nanocomposites, Poly, 47 (2006) 3207-3217. https://doi.org/10.1016/j.polymer.2006.02.089

[71] W. Zheng, M. Wen, Z. Zhao, J. Liu, Z. Wang, B. Zhai, Z. Li, Black plastic mulch combined with summer cover crop increases the yield and water use efficiency of apple tree on the rainfed Loess Plateau, PLoS One, 12 (2017). https://doi.org/10.1371/journal.pone.0185705

[72] Y. Zheng, E.K. Yanful, A.S. Bassi, A review of plastic waste biodegradation, Crit. Rev. Biotechnol., 25 (2005) 243-250. https://doi.org/10.1080/07388550500346359

[73] A.-C. Albertsson, S.O. Andersson, S. Karlsson, The mechanism of biodegradation of polyethylene, Polym. Degradation Stab., 18 (1987) 73-87. https://doi.org/10.1016/0141-3910(87)90084-X

[74] A. Ammala, S. Bateman, K. Dean, E. Petinakis, P. Sangwan, S. Wong, Q. Yuan, L. Yu, C. Patrick, K. Leong, An overview of degradable and biodegradable polyolefins, Prog. Polym. Sci., 36 (2011) 1015-1049. https://doi.org/10.1016/j.progpolymsci.2010.12.002

[75] T. Artham, M. Doble, Biodegradation of aliphatic and aromatic polycarbonates, Macromol. Biosci., 8 (2008) 14-24. https://doi.org/10.1002/mabi.200700106

[76] D.K. Barnes, F. Galgani, R.C. Thompson, M. Barlaz, Accumulation and fragmentation of plastic debris in global environments, Philosophical Transactions of the Royal Society B: Biological Sciences, 364 (2009) 1985-1998. https://doi.org/10.1098/rstb.2008.0205

[77] A. Basfar, K.I. Ali, Natural weathering test for films of various formulations of low density polyethylene (LDPE) and linear low density polyethylene (LLDPE),

Polym. Degradation Stab., 91 (2006) 437-443.
https://doi.org/10.1016/j.polymdegradstab.2004.11.027

[78]   M.A. Browne, P. Crump, S.J. Niven, E. Teuten, A. Tonkin, T. Galloway, R.
       Thompson, Accumulation of microplastic on shorelines woldwide: sources and sinks,
       Environ. Sci. Technol., 45 (2011) 9175-9179. https://doi.org/10.1021/es201811s

[79]   I. Carpentieri, V. Brunella, P. Bracco, M.C. Paganini, E.M.B. Del Prever, M.P.
       Luda, S. Bonomi, L. Costa, Post-irradiation oxidation of different polyethylenes,
       Polym. Degradation Stab., 96 (2011) 624-629.
       https://doi.org/10.1016/j.polymdegradstab.2010.12.014

[80]   S. Cheng, F. Dehaye, C. Bailly, J.-J. Biebuyck, R. Legras, L. Parks, Studies on
       polyethylene pellets modified by low dose radiation prior to part formation, Nuclear
       Instruments and Methods in Physics Research Section B: Beam Interactions with
       Materials and Atoms, 236 (2005) 130-136.
       https://doi.org/10.1016/j.nimb.2005.03.272

[81]   E. Chiellini, A. Corti, S. D'Antone, R. Baciu, Oxo-biodegradable carbon
       backbone polymers–Oxidative degradation of polyethylene under accelerated test
       conditions, Polym. Degradation Stab., 91 (2006) 2739-2747.
       https://doi.org/10.1016/j.polymdegradstab.2006.03.022

[82]   M. Edge, M. Hayes, M. Mohammadian, N. Allen, T. Jewitt, K. Brems, K. Jones,
       Aspects of poly (ethylene terephthalate) degradation for archival life and
       environmental degradation, Polym. Degradation Stab., 32 (1991) 131-153.
       https://doi.org/10.1016/0141-3910(91)90047-U

[83]   S. Endo, R. Takizawa, K. Okuda, H. Takada, K. Chiba, H. Kanehiro, H. Ogi, R.
       Yamashita, T. Date, Concentration of polychlorinated biphenyls (PCBs) in beached
       resin pellets: variability among individual particles and regional differences, Mar.
       Pollut. Bull., 50 (2005) 1103-1114. https://doi.org/10.1016/j.marpolbul.2005.04.030

[84]   M. Eriksen, The plastisphere-the making of a plasticized world, Tul. Envtl. LJ, 27
       (2013) 153.

[85]   A. Francois-Heude, E. Richaud, E. Desnoux, X. Colin, Influence of temperature,
       UV-light wavelength and intensity on polypropylene photothermal oxidation, Polym.
       Degradation Stab., 100 (2014) 10-20.
       https://doi.org/10.1016/j.polymdegradstab.2013.12.038

[86]   M. Gardette, A. Perthue, J.-L. Gardette, T. Janecska, E. Földes, B. Pukánszky, S.
       Therias, Photo-and thermal-oxidation of polyethylene: comparison of mechanisms

and influence of unsaturation content, Polym. Degradation Stab., 98 (2013) 2383-2390. https://doi.org/10.1016/j.polymdegradstab.2013.07.017

[87]   B. Graca, M. Bełdowska, P. Wrzesień, A. Zgrundo, Styrofoam debris as a potential carrier of mercury within ecosystems, Environmental Science and Pollution Research, 21 (2014) 2263-2271. https://doi.org/10.1007/s11356-013-2153-4

[88]   M.R. Gregory, Virgin plastic granules on some beaches of eastern Canada and Bermuda, Mar. Environ. Res., 10 (1983) 73-92. https://doi.org/10.1016/0141-1136(83)90011-9

[89]   J. Gulmine, P. Janissek, H. Heise, L. Akcelrud, Degradation profile of polyethylene after artificial accelerated weathering, Polym. Degradation Stab., 79 (2003) 385-397. https://doi.org/10.1016/S0141-3910(02)00338-5

[90]   K. Harshvardhan, B. Jha, Biodegradation of low-density polyethylene by marine bacteria from pelagic waters, Arabian Sea, India, Mar. Pollut. Bull., 77 (2013) 100-106. https://doi.org/10.1016/j.marpolbul.2013.10.025

[91]   H. Hirai, H. Takada, Y. Ogata, R. Yamashita, K. Mizukawa, M. Saha, C. Kwan, C. Moore, H. Gray, D. Laursen, Organic micropollutants in marine plastics debris from the open ocean and remote and urban beaches, Mar. Pollut. Bull., 62 (2011) 1683-1692. https://doi.org/10.1016/j.marpolbul.2011.06.004

[92]   L.A. Holmes, A. Turner, R.C. Thompson, Interactions between trace metals and plastic production pellets under estuarine conditions, Mar. Chem., 167 (2014) 25-32. https://doi.org/10.1016/j.marchem.2014.06.001

[93]   C. Ioakeimidis, C. Zeri, H. Kaberi, M. Galatchi, K. Antoniadis, N. Streftaris, F. Galgani, E. Papathanassiou, G. Papatheodorou, A comparative study of marine litter on the seafloor of coastal areas in the Eastern Mediterranean and Black Seas, Mar. Pollut. Bull., 89 (2014) 296-304. https://doi.org/10.1016/j.marpolbul.2014.09.044

[94]   P. Kandakatla, B. Mahto, S. Goel, Extent and rate of biodegradation of different organic components in municipal solid waste, International Journal of Environment and Waste Management, 11 (2013) 350-364. https://doi.org/10.1504/IJEWM.2013.054262

[95]   H. Karapanagioti, S. Endo, Y. Ogata, H. Takada, Diffuse pollution by persistent organic pollutants as measured in plastic pellets sampled from various beaches in Greece, Mar. Pollut. Bull., 62 (2011) 312-317. https://doi.org/10.1016/j.marpolbul.2010.10.009

[96]   I. Kyrikou, D. Briassoulis, M. Hiskakis, E. Babou, Analysis of photo-chemical degradation behaviour of polyethylene mulching film with pro-oxidants, Polym. Degradation Stab., 96 (2011) 2237-2252. https://doi.org/10.1016/j.polymdegradstab.2011.09.001

[97]   N. Lucas, C. Bienaime, C. Belloy, M. Queneudec, F. Silvestre, J.-E. Nava-Saucedo, Polymer biodegradation: Mechanisms and estimation techniques–A review, Chemosphere, 73 (2008) 429-442. https://doi.org/10.1016/j.chemosphere.2008.06.064

[98]   Y. Mato, T. Isobe, H. Takada, H. Kanehiro, C. Ohtake, T. Kaminuma, Plastic resin pellets as a transport medium for toxic chemicals in the marine environment, Environ. Sci. Technol., 35 (2001) 318-324. https://doi.org/10.1021/es0010498

[99]   T. Muthukumar, A. Aravinthan, K. Lakshmi, R. Venkatesan, L. Vedaprakash, M. Doble, Fouling and stability of polymers and composites in marine environment, Int. Biodeterior. Biodegrad., 65 (2011) 276-284. https://doi.org/10.1016/j.ibiod.2010.11.012

[100] T.-T. Nguyen-Boisse, J. Saulnier, N. Jaffrezic-Renault, F. Lagarde, Miniaturised enzymatic conductometric biosensor with Nafion membrane for the direct determination of formaldehyde in water samples, Anal. Bioanal. Chem., 406 (2014) 1039-1048. https://doi.org/10.1007/s00216-013-7197-2

[101] Y. Ogata, H. Takada, K. Mizukawa, H. Hirai, S. Iwasa, S. Endo, Y. Mato, M. Saha, K. Okuda, A. Nakashima, International pellet watch: global monitoring of persistent organic pollutants (POPs) in coastal waters. 1. Initial phase data on PCBs, DDTs, and HCHs, Mar. Pollut. Bull., 58 (2009) 1437-1446. https://doi.org/10.1016/j.marpolbul.2009.06.014

[102] T. Ojeda, A. Freitas, K. Birck, E. Dalmolin, R. Jacques, F. Bento, F. Camargo, Degradability of linear polyolefins under natural weathering, Polym. Degradation Stab., 96 (2011) 703-707. https://doi.org/10.1016/j.polymdegradstab.2010.12.004

[103] Y. Orhan, H. Büyükgüngör, Enhancement of biodegradability of disposable polyethylene in controlled biological soil, Int. Biodeterior. Biodegrad., 45 (2000) 49-55. https://doi.org/10.1016/S0964-8305(00)00048-2

[104] I.G. Orr, Y. Hadar, A. Sivan, Colonization, biofilm formation and biodegradation of polyethylene by a strain of Rhodococcus ruber, Appl. Microbiol. Biotechnol., 65 (2004) 97-104. https://doi.org/10.1007/s00253-004-1584-8

[105] G. Pastorelli, C. Cucci, O. Garcia, G. Piantanida, A. Elnaggar, M. Cassar, M. Strlič, Environmentally induced colour change during natural degradation of selected polymers, Polym. Degradation Stab., 107 (2014) 198-209. https://doi.org/10.1016/j.polymdegradstab.2013.11.007

[106] J.E. Pegram, A.L. Andrady, Outdoor weathering of selected polymeric materials under marine exposure conditions, Polym. Degradation Stab., 26 (1989) 333-345. https://doi.org/10.1016/0141-3910(89)90112-2

[107] N. Priyanka, T. Archana, Biodegradability of polythene and plastic by the help of microorganism: a way for brighter future, J Environ Anal Toxicol, 1 (2011) 1000111. https://doi.org/10.4172/2161-0525.1000111

[108] B. Rånby, Photodegradation and photo-oxidation of synthetic polymers, J. Anal. Appl. Pyrolysis, 15 (1989) 237-247. https://doi.org/10.1016/0165-2370(89)85037-5

[109] M. Rani, W.J. Shim, G.M. Han, M. Jang, Y.K. Song, S.H. Hong, Hexabromocyclododecane in polystyrene based consumer products: an evidence of unregulated use, Chemosphere, 110 (2014) 111-119. https://doi.org/10.1016/j.chemosphere.2014.02.022

[110] J.-M. Restrepo-Flórez, A. Bassi, M.R. Thompson, Microbial degradation and deterioration of polyethylene–A review, Int. Biodeterior. Biodegrad., 88 (2014) 83-90. https://doi.org/10.1016/j.ibiod.2013.12.014

[111] L.M. Rios, C. Moore, P.R. Jones, Persistent organic pollutants carried by synthetic polymers in the ocean environment, Mar. Pollut. Bull., 54 (2007) 1230-1237. https://doi.org/10.1016/j.marpolbul.2007.03.022

[112] C.M. Rochman, E. Hoh, T. Kurobe, S.J. Teh, Ingested plastic transfers hazardous chemicals to fish and induces hepatic stress, Sci. Rep., 3 (2013) 3263. https://doi.org/10.1038/srep03263

[113] A.A. Shah, F. Hasan, A. Hameed, S. Ahmed, Biological degradation of plastics: a comprehensive review, Biotechnol. Adv., 26 (2008) 246-265. https://doi.org/10.1016/j.biotechadv.2007.12.005

[114] M. Shimao, Biodegradation of plastics, Curr. Opin. Biotechnol., 12 (2001) 242-247. https://doi.org/10.1016/S0958-1669(00)00206-8

[115] B. Singh, N. Sharma, Mechanistic implications of plastic degradation, Polym. Degradation Stab., 93 (2008) 561-584. https://doi.org/10.1016/j.polymdegradstab.2007.11.008

[116] R. Smith, Biodegradable polymers for industrial applications, CRC Press, 2005. https://doi.org/10.1533/9781845690762

[117] G. Swift, Requirements for biodegradable water-soluble polymers, Polym. Degradation Stab., 59 (1998) 19-24. https://doi.org/10.1016/S0141-3910(97)00162-6

[118] S. Taniguchi, F.I. Colabuono, P.S. Dias, R. Oliveira, M. Fisner, A. Turra, G.M. Izar, D.M. Abessa, M. Saha, J. Hosoda, Spatial variability in persistent organic pollutants and polycyclic aromatic hydrocarbons found in beach-stranded pellets along the coast of the state of São Paulo, southeastern Brazil, Mar. Pollut. Bull., 106 (2016) 87-94. https://doi.org/10.1016/j.marpolbul.2016.03.024

[119] E.L. Teuten, J.M. Saquing, D.R. Knappe, M.A. Barlaz, S. Jonsson, A. Björn, S.J. Rowland, R.C. Thompson, T.S. Galloway, R. Yamashita, Transport and release of chemicals from plastics to the environment and to wildlife, Philosophical Transactions of the Royal Society B: Biological Sciences, 364 (2009) 2027-2045. https://doi.org/10.1098/rstb.2008.0284

[120] Y. Zheng, E.K. Yanful, A.S. Bassi, A review of plastic waste biodegradation, Crit. Rev. Biotechnol., 25 (2005) 243-250. https://doi.org/10.1080/07388550500346359

Keyword Index

About the Editors

Dr. Inamuddin is working as Assistant Professor at the Department of Applied Chemistry, Aligarh Muslim University, Aligarh, India. He obtained Master of Science degree in Organic Chemistry from Chaudhary Charan Singh (CCS) University, Meerut, India, in 2002. He received his Master of Philosophy and Doctor of Philosophy degrees in Applied Chemistry from Aligarh Muslim University (AMU), India, in 2004 and 2007, respectively. He has extensive research experience in multidisciplinary fields of Analytical Chemistry, Materials Chemistry, and Electrochemistry and, more specifically, Renewable Energy and Environment. He has worked on different research projects as project fellow and senior research fellow funded by University Grants Commission (UGC), Government of India, and Council of Scientific and Industrial Research (CSIR), Government of India. He has received Fast Track Young Scientist Award from the Department of Science and Technology, India, to work in the area of bending actuators and artificial muscles. He has completed four major research projects sanctioned by University Grant Commission, Department of Science and Technology, Council of Scientific and Industrial Research, and Council of Science and Technology, India. He has published 185 research articles in international journals of repute and nineteen book chapters in knowledge-based book editions published by renowned international publishers. He has published 120 edited books with Springer (U.K.), Elsevier, Nova Science Publishers, Inc. (U.S.A.), CRC Press Taylor & Francis Asia Pacific, Trans Tech Publications Ltd. (Switzerland), IntechOpen Limited (U.K.), Wiley-Scrivener, (U.S.A.) and Materials Research Forum LLC (U.S.A). He is a member of various journals' editorial boards. He is also serving as Associate Editor for journals (Environmental Chemistry Letter, Applied Water Science and Euro-Mediterranean Journal for Environmental Integration, Springer-Nature), Frontiers Section Editor (Current Analytical Chemistry, Bentham Science Publishers), Editorial Board Member (Scientific Reports-Nature), Editor (Eurasian Journal of Analytical Chemistry), and Review Editor (Frontiers in Chemistry, Frontiers, U.K.) He is also guest-editing various special thematic special issues to the journals of Elsevier, Bentham Science Publishers, and John Wiley & Sons, Inc. He has attended as well as chaired sessions in various international and national conferences. He has worked as a Postdoctoral Fellow, leading a research team at the Creative Research Initiative Center for Bio-Artificial Muscle, Hanyang University, South Korea, in the field of renewable energy, especially biofuel cells. He has also worked as a Postdoctoral Fellow at the Center of Research Excellence in Renewable Energy, King Fahd University of Petroleum and Minerals, Saudi Arabia, in the field of polymer electrolyte membrane fuel cells and computational fluid dynamics of polymer electrolyte membrane fuel cells. He is a life member of the Journal of the Indian

Chemical Society. His research interest includes ion exchange materials, a sensor for heavy metal ions, biofuel cells, supercapacitors and bending actuators.

Dr. Rizwana Mobin is working as Assistant Professor in the Department of Industrial Chemistry, Govt. College for Women, Cluster University, Srinagar, India. She received her B.Sc. Hons., Masters and Ph.D (Applied Chemistry) from Aligarh Muslim University, Aligarh, India on the topic "Studies on Thin-Layer Chromatographic Analysis of Surfactants". She has been the recipient of the Gold medal at Masters level. She has published several research articles in international journals of repute and six book chapters in knowledge-based book editions published by renowned international publishers. She has edited books with Materials Research Forum LLC, U.S.A. Her research expertise includes thin-layer chromatography, development of new methodologies involving green solvent system for the analysis of surfactants and food dyes.

Dr. Mohd Imran Ahamed received his Ph.D degree on the topic "Synthesis and characterization of inorganic-organic composite heavy metals selective cation-exchangers and their analytical applications", from Aligarh Muslim University, Aligarh, India in 2019. He has published several research and review articles in the journals of international recognition. Springer (U.K.), Elsevier, CRC Press Taylor & Francis Asia Pacific and Materials Research Forum LLC (U.S.A). He has completed his B.Sc. (Hons) Chemistry from Aligarh Muslim University, Aligarh, India, and M.Sc. (Organic Chemistry) from Dr. Bhimrao Ambedkar University, Agra, India. He has co-edited more than 20 books with Springer (U.K.), Elsevier, CRC Press Taylor & Francis Asia Pacific, Materials Research Forum LLC (U.S.A) and Wiley-Scrivener, (U.S.A.). His research work includes ion-exchange chromatography, wastewater treatment, and analysis, bending actuator and electrospinning.

Dr. Rajender Boddula is currently working with Chinese Academy of Sciences-President's International Fellowship Initiative (CAS-PIFI) at National Center for Nanoscience and Technology (NCNST, Beijing). He obtained Master of Science in Organic Chemistry from Kakatiya University, Warangal, India, in 2008. He received his Doctor of Philosophy in Chemistry with the highest honours in 2014 for the work entitled "Synthesis and Characterization of Polyanilines for Supercapacitor and Catalytic Applications" at the CSIR-Indian Institute of Chemical Technology (CSIR-IICT) and Kakatiya University (India). Before joining National Center for Nanoscience and

Technology (NCNST) as CAS-PIFI research fellow, China, worked as senior research associate and Postdoc at National Tsing-Hua University (NTHU, Taiwan) respectively in the fields of bio-fuel and CO_2 reduction applications. His academic honors include University Grants Commission National Fellowship and many merit scholarships, study-abroad fellowships from Australian Endeavour Research Fellowship, and CAS-PIFI. He has published many scientific articles in international peer-reviewed journals and has authored around twenty book chapters, and he is also serving as an editorial board member and a referee for reputed international peer-reviewed journals. He has published edited books with Springer (UK), Elsevier, Materials Research Forum LLC (USA), Wiley-Scrivener, (U.S.A.) and CRC Press Taylor & Francis group. His specialized areas of research are energy conversion and storage, which include sustainable nanomaterials, graphene, polymer composites, heterogeneous catalysis for organic transformations, environmental remediation technologies, photoelectrochemical water-splitting devices, biofuel cells, batteries and Dr. Rajender Boddula is currently working with Chinese Academy of Sciences-President's International Fellowship Initiative (CAS-PIFI) at National Center for Nanoscience and Technology (NCNST, Beijing). He obtained Master of Science in Organic Chemistry from Kakatiya University, Warangal, India, in 2008. He received his Doctor of Philosophy in Chemistry with the highest honours in 2014 for the work entitled "Synthesis and Characterization of Polyanilines for Supercapacitor and Catalytic Applications" at the CSIR-Indian Institute of Chemical Technology (CSIR-IICT) and Kakatiya University (India). Before joining National Center for Nanoscience and Technology (NCNST) as CAS-PIFI research fellow, China, worked as senior research associate and Postdoc at National Tsing-Hua University (NTHU, Taiwan) respectively in the fields of bio-fuel and CO2 reduction applications. His academic honors include University Grants Commission National Fellowship and many merit scholarships, study-abroad fellowships from Australian Endeavour Research Fellowship, and CAS-PIFI. He has published many scientific articles in international peer-reviewed journals and has authored around twenty book chapters, and he is also serving as an editorial board member and a referee for reputed international peer-reviewed journals. He has published edited books with Springer (UK), Elsevier, Materials Science Forum LLC (USA), Wiley-Scrivener, (U.S.A.) and CRC Press Taylor & Francis group. His specialized areas of research are energy conversion and storage, which include sustainable nanomaterials, graphene, polymer composites, heterogeneous catalysis for organic transformations, environmental remediation technologies, photoelectrochemical water-splitting devices, biofuel cells, batteries and supercapacitors.

www.ingramcontent.com/pod-product-compliance
Lightning Source LLC
Chambersburg PA
CBHW071323210326

41597CB00015B/1331